California Cuisine and Just Food

Food, Health, and the Environment

Series Editor: Robert Gottlieb, Henry R. Luce Professor of Urban and Environmental Policy, Occidental College

Keith Douglass Warner, *Agroecology in Action: Extending Alternative Agriculture through Social Networks*

Christopher M. Bacon, V. Ernesto Méndez, Stephen R. Gliessman, David Goodman, and Jonathan A. Fox, eds., *Confronting the Coffee Crisis: Fair Trade, Sustainable Livelihoods and Ecosystems in Mexico and Central America*

Thomas A. Lyson, G. W. Stevenson, and Rick Welsh, eds., *Food and the Mid-Level Farm: Renewing an Agriculture of the Middle*

Jennifer Clapp and Doris Fuchs, eds., *Corporate Power in Global Agrifood Governance*

Robert Gottlieb and Anupama Joshi, *Food Justice*

Jill Lindsey Harrison, *Pesticide Drift and the Pursuit of Environmental Justice*

Alison Alkon and Julian Agyeman, eds., *Cultivating Food Justice: Race, Class, and Sustainability*

Abby Kinchy, *Seeds, Science, and Struggle: The Global Politics of Transgenic Crops*

Sally K. Fairfax, Louise Nelson Dyble, Greig Tor Guthey, Lauren Gwin, Monica Moore, and Jennifer Sokolove, *California Cuisine and Just Food*

California Cuisine and Just Food

Sally K. Fairfax, Louise Nelson Dyble, Greig Tor Guthey,
Lauren Gwin, Monica Moore, and Jennifer Sokolove

with the Assistance of Matthew Gerhart and Jennifer Kao

The MIT Press
Cambridge, Massachusetts
London, England

MIT Press books may be purchased at special quantity discounts for business or sales promotional use. For information, please email special_sales@mitpress.mit.edu or write to Special Sales Department, The MIT Press, 55 Hayward Street, Cambridge, MA 02142.

This book was set in Sabon by Toppan Best-set Premedia Limited, Hong Kong. Printed on recycled paper and bound in the United States of America.

Library of Congress Cataloging-in-Publication Data

Fairfax, Sally K.
California cuisine and just food / Sally K. Fairfax ... [et.al.].
p. cm.—(Food, health, and the environment)
Includes bibliographical references and index.
ISBN 978-0-262-01811-1 (hardcover : alk. paper)—ISBN 978-0-262-51786-7 (pbk. : alk. paper)
1. Food habits—California—San Francisco Bay Area. 2. Food preferences—California—San Francisco Bay Area. 3. Gastronomy—California—San Francisco Bay Area. 4. Food industry and trade—California—San Francisco Bay Area. 5. Sustainable agriculture—California—San Francisco Bay Area. 6. San Francisco Bay Area (Calif.)—Social life and customs. I. Title.
GT2853.U5F34 2012
394.1'2—dc23
2012008691

10 9 8 7 6 5 4 3 2 1

We dedicate this book to two who are building the next generation in the district, Nikki Henderson and Anya Fernald, and to some of the impressive shoulders on which they stand: Boyd Stewart, Warren Weber, Ellen Straus, Phyllis Faber, Alice Waters, Sibella Kraus, Sue Conley, Peg Smith, Ellie Rilla, and David Evans.

Contents

Series Foreword

I am pleased to present the ninth book in the Food, Health, and the Environment series. This series explores the global and local dimensions of food systems and examines issues of access, justice, and environmental and community well-being. It includes books that focus on the way food is grown, processed, manufactured, distributed, sold, and consumed. Among the matters addressed are what foods are available to communities and individuals, how those foods are obtained, and what health and environmental factors are embedded in food system choices and outcomes. The series focuses not only on food security and well-being but also on regional, state, national, and international policy decisions and economic and cultural forces. Food, Health, and the Environment books provide a window into the public debates, theoretical considerations, and multidisciplinary perspectives that have made food systems and their connections to health and environment important subjects of study.

Robert Gottlieb, Occidental College
Series editor

Foreword

California Cuisine and Just Food takes a deep and comprehensive look at past and new efforts to bring tastier, healthier, locally grown, and ethically produced food to San Francisco Bay Area eaters, poor and rich. The story is inspiring. The authors of this collectively written account, cautious academics as they must be, describe the development of the Bay Area food scene as a "district" rather than as a social movement. But I have no such compunctions. It looks like a social movement to me. This book is about how the Bay Area food movement evolved to what it is today: a vibrant community of highly diverse groups working on highly diverse ways to produce better-quality food and promote a more just, healthful, and sustainable food system—for everyone along the entire system of what it takes to produce, transport, sell, prepare, serve, and consume food.

That much of today's rapidly expanding food movement began in the Bay Area is well recognized, not least because of its rock stars: Alice Waters whose restaurant, Chez Panisse, celebrated its fortieth anniversary in 2010; Eric Schlosser, whose *Fast Food Nation* brought the contradictions of industrial agriculture to public attention; and Michael Pollan, whose *Omnivore's Dilemma* catapulted food system issues into the mainstream. But as this book makes clear, "California cuisine"—generally agreed to encompass local, fresh, seasonal, and utterly delicious food—has never been only about how it tastes. California cuisine also has been and is about politics, and particularly the politics of inequities in how, where, and by whom food is produced and consumed. Some of the political history recounted here will be familiar to anyone engaged in food issues during the past few years, but much of what this book covers is fresh, fascinating, and often surprising. That California is the birthplace of the alternative food movement is ironic in light of that state's preeminence in conventional agriculture. I knew that Upton Sinclair's *The Jungle* launched food safety legislation in the United States in 1906, but had no idea that he lost a campaign for governor of California in 1934 on a platform of ending poverty. And it's good to be reminded of or to discover the century-long efforts of early-twentieth-century hiking clubs and much later environmental groups to rescue critical parts of the Bay Area agricultural landscape from housing developers, and to find out how the Diggers, Black Panthers, and the Berkeley Co-op made food such a central focus of their activities.

My personal experience with the Bay Area food movement began when I was a student at Berkeley in the 1950s. We all knew that something unusually interesting was happening at Hank Rubin's restaurants, the Pot Luck and Cruchon's. When Rubin, arguably the Father of the Berkeley food revolution, died in 2011, his obituaries singled out his accomplishments in both indulgence and innovation. Yes, the food was good—sometimes marvelously so—but his restaurants were also the first in the Bay Area to be fully integrated by race and gender.

More than half a century later, we have come to expect innovations along with indulgence. The Bay Area food scene encompasses everything from the classiest of restaurants to hospitals and schools with organic gardens. It includes urban farms and community-based grocery stores, countless groups working to improve food access and food self-sufficiency for low-income residents, and countless others devoted to improving working conditions and wages for farm and restaurant workers. The activities of these private, nonprofit, and often unsung enterprises, this book maintains, constitute a community whose collective mission is to provide meaningful alternatives to conventional food.

But do such diverse activities constitute a movement? I think so, but a more cautious answer is that it is too soon to tell. The signs are promising, but far more work remains to be done.

I say this because a few years ago, New York University's sociology department offered me an appointment as an affiliated member. Out of respect for that honor, I thought it would be useful to teach a graduate class in food sociology that focused on food as a social movement. At the precise moment I realized I had bitten off far more than I could chew (an appropriate metaphor at this point), a card-carrying sociologist and expert on social movements, Troy Duster, came to the rescue and suggested we coteach the course. And so we did. But this book would have made the experience a lot easier.

The food movement, Duster explained, differs from classic social movements—the civil rights movement, the environmental movement, the women's movement—in having so many goals and such diverse constituencies. And because the purpose of social movements is to change society, it is possible to judge their effectiveness only in retrospect. Did they succeed in changing society? How and to what degree? The need for academic caution became clear.

The fundamental structure of today's society may seem immutable, but it should be obvious to everyone that the food environment in the United States has changed in recent years, and much for the better. In my New York City neighborhood, I have ready access to sources of local food, humanely raised meat, and eggs from free-range chickens. I can buy organic milk and produce not only from the local farmers' market but from almost any nearby grocery. In many city schools, public and private, food is fresher and healthier than in even the recent past. I can easily observe the effects of government programs, limited as they may be, that support organic producers, farm-to-school programs, and young farmers. And it is increasingly evident that public discussion about obesity, formerly focused exclusively on parental and personal responsibility, has now shifted markedly to focusing on analyzing how the existing food system encourages overeating and what to do about it.

Signs of the food movement are evident in the avalanche of books, films, and videos about the failures of our current industrial food system and the work of pioneers in developing alternative ways of producing and distributing foods. Universities all over the country are developing food programs within academic disciplines. My own institution, New York University, not only houses my department's innovative food studies programs (founded in 1996), but also offers closely related programs in environmental studies and animal studies. All teach students to critically analyze problems inherent in industrial food production and to create alternatives that are healthier for people, farm animals, and the planet.

But as this book repeatedly emphasizes, improvements in the food system are most likely to benefit the relatively rich and well educated and trickle down slowly, if at all, to the poor. Many groups described of us got together first in connection with that project, which we pursued with Louise Fortmann, Nancy Peluso, Lynn Huntsinger, Steven Wolf, and Ann Hawkins.

Even thanking those with whom we spoke as part of a formal interview process is complex. Some of our sources were already friends when we started; others have become friends over the years. Many agreed to multiple interviews, often with different coauthors at different times about different topics. Others we talked with when we were working on land acquisition for our previous book, *Buying Nature*. A precious few were willing to read and correct interview notes and even comment on drafts and redrafts of chapters as the project evolved. Nobody actually insisted on remaining anonymous, but we have not intended to connect specific observations to specific individuals unless it made no sense not to. Acknowledging the following individuals is just the tip of the iceberg. But we do so with heartfelt thanks.

Beth Ashley
Lynn Bagley
Marcia Barenaga
Dan Benedetti
Jacquie Berger
Jennifer Bice
Martin Bourque
Bees Butler
Liza Capozzi
Tendai Chitewere
Sue Conley
Jesse Cool
Jeff Creque
Julie Cummins
John Curl
James A. Dahlberg
Mark Dowie
Scott Davidson
Melanie DuPuis
Rekhi Eanni-Rodriguez
David Evans
Dwight D. Evans
Phyllis Faber
Anya Fernald
John Finger
Mary Firestone
Mike and Sally Gale
Kathryn Gilje
Michael Groman
Maisie Greenawalt
Martha Guzman
Dana Harvey
Jamie Harvey
Gail Hayden
Karen Heisler
Nikki Henderson
Eric Holt-Giménez
Lynn Huntsinger
Marcia Ishii-Eiteman
Allen Jacobs
Sibella Kraus
Nancy Langston

expert on social movements, Troy Duster, came to the rescue and suggested we coteach the course. And so we did. But this book would have made the experience a lot easier.

The food movement, Duster explained, differs from classic social movements—the civil rights movement, the environmental movement, the women's movement—in having so many goals and such diverse constituencies. And because the purpose of social movements is to change society, it is possible to judge their effectiveness only in retrospect. Did they succeed in changing society? How and to what degree? The need for academic caution became clear.

The fundamental structure of today's society may seem immutable, but it should be obvious to everyone that the food environment in the United States has changed in recent years, and much for the better. In my New York City neighborhood, I have ready access to sources of local food, humanely raised meat, and eggs from free-range chickens. I can buy organic milk and produce not only from the local farmers' market but from almost any nearby grocery. In many city schools, public and private, food is fresher and healthier than in even the recent past. I can easily observe the effects of government programs, limited as they may be, that support organic producers, farm-to-school programs, and young farmers. And it is increasingly evident that public discussion about obesity, formerly focused exclusively on parental and personal responsibility, has now shifted markedly to focusing on analyzing how the existing food system encourages overeating and what to do about it.

Signs of the food movement are evident in the avalanche of books, films, and videos about the failures of our current industrial food system and the work of pioneers in developing alternative ways of producing and distributing foods. Universities all over the country are developing food programs within academic disciplines. My own institution, New York University, not only houses my department's innovative food studies programs (founded in 1996), but also offers closely related programs in environmental studies and animal studies. All teach students to critically analyze problems inherent in industrial food production and to create alternatives that are healthier for people, farm animals, and the planet.

But as this book repeatedly emphasizes, improvements in the food system are most likely to benefit the relatively rich and well educated and trickle down slowly, if at all, to the poor. Many groups described here are working to redress this imbalance, meaning that they are trying to make fundamental changes in society. No wonder they encounter opposition. The food movement's focus on finding more equitable alternatives to current production systems constitutes an explicit critique of those systems. Food production and food service together are worth more than a trillion dollars a year in sales, just in this country. Vast sums are at stake for everyone producing, marketing, and consuming food.

If this book has one overriding message, it is this: changing the food system is hard work but worth every bit of that effort. Recently a student asked me for advice about how she could become involved in food system work when she felt such despair about her powerlessness to change the system. I told her about the signs of positive change that I see everywhere I look. I'm heartened by how the Occupy Wall Street movement has made so many people aware of income inequities, and how NYU food studies students have come up with their own version, Occupy Against Big Food, and its light-hearted slogan, "Lettuce beet the system."

We won't know the results of current efforts for a long time. In the meantime, there is much work to be done. If young people do not join these efforts, nothing will change. So I say: Get busy. Join the movement. Make change happen. And read this book for inspiration.

Marion Nestle
Paulette Goddard Professor of Nutrition, Food Studies, and Public Health, New York University

Acknowledgments

Even six authors are apparently not enough—we have required a lot of help. In our more than ten years of working on this book, we have had time to impose on a lot of folks' generosity. Hundreds of people have shared their experiences, insights, opinions, and perspectives in many settings. We have been on dozens of farm tours—many sponsored by Marin Agricultural Land Trust, the Center for Urban Education About Sustainable Agriculture, and EcoFarm gatherings, and as part of our own classes. We have attended more food conferences, discussions, and workshops than we can count. Several of us work for and with organizations that are deeply involved in food systems and have participated more or less directly in issues covered in this book. And we have been teaching, talking, and thinking about these issues with our students and colleagues for far longer than we have been working specifically on this book.

We are particularly grateful to Marion Nestle who has conducted a seminar for us on what to read, where to look, and with whom to speak. We are grateful for her generous and insightful answers to our questions over many years. Many students coming through the Fairfax lab on their way to something else have also played an important role. Matt Gerhart and Jennifer Kao never escaped and worked with us long after they graduated. Allison Clark, Michael DeAlessi, and Misa Sugino were a little closer to the door, but we enjoyed their sojourn with us. We have also relied on Agnieszka Fortmann-Roe, who helped us assemble the final draft of this book. And all of us have relied on the support of the Henry J. Vaux Distinguished Professorship. We are also grateful for financial support from a U.S. Department of Agriculture grant (number 98–35401–6123), which enabled us to study "Claiming Rural Places." Some of us got together first in connection with that project, which we pursued with Louise Fortmann, Nancy Peluso, Lynn Huntsinger, Steven Wolf, and Ann Hawkins.

Even thanking those with whom we spoke as part of a formal interview process is complex. Some of our sources were already friends when we started; others have become friends over the years. Many agreed to multiple interviews, often with different coauthors at different times about different topics. Others we talked with when we were working on land acquisition for our previous book, *Buying Nature*. A precious few were willing to read and correct interview notes and even comment on drafts and redrafts of chapters as the project evolved. Nobody actually insisted on remaining anonymous, but we have not intended to connect specific observations to specific individuals unless it made no sense not to. Acknowledging the following individuals is just the tip of the iceberg. But we do so with heartfelt thanks.

Beth Ashley
Marcia Barenaga
Jacquie Berger
Martin Bourque
Liza Capozzi
Sue Conley
Jeff Creque
John Curl
Mark Dowie
Melanie DuPuis
David Evans
Phyllis Faber
John Finger
Mike and Sally Gale
Michael Groman
Martha Guzman
Jamie Harvey
Karen Heisler
Eric Holt-Giménez
Marcia Ishii-Eiteman
Sibella Kraus
Lynn Bagley
Dan Benedetti
Jennifer Bice
Bees Butler
Tendai Chitewere
Jesse Cool
Julie Cummins
James A. Dahlberg
Scott Davidson
Rekhi Eanni-Rodriguez
Dwight D. Evans
Anya Fernald
Mary Firestone
Kathryn Gilje
Maisie Greenawalt
Dana Harvey
Gail Hayden
Nikki Henderson
Lynn Huntsinger
Allen Jacobs
Nancy Langston

Stephanie Larson
Penny Leff
Yael Lehman
Kevin and Nancy Lunny
Philip L. Martin
Andrea Mackenzie
Marion Nestle
Terry Mushovic
Nicolette Hahn Niman
Larry Orman
Paj Patel
Nyles Peterson
Elizabeth Ptak
Ellen Rilla
John Silviera
Paul Starrs
Albert Straus
Pam Tao Lee
Erin Wirpsa-Eisenberg
Hannah Laurison
Dewey Livingston
Kathy Lerza
Jiff Martin
Peter Martinelli
Ellen Moore
Jennifer H. McTiernan
Bill Niman
Bu Nygrens
Martha Page
Duane Perry
Odessa Paper
Margaret Reeves
Daniel Bowman Simon
Peg Smith
Dave Stockdale
Michael Straus
Warren Weber
Heather Wooten

1

Celebrating a Community?

The sun shines through second-story windows at the Ferry Building, illuminating San Francisco's historic transit hub, now transformed into the kind of foodie paradise many people associate with the San Francisco Bay Area (figure 1.1).[1] It is October 2007, and a twenty-fifth anniversary celebration for the nonprofit Pesticide Action Network (PAN), the North American branch of a global federation of sustainable agriculture advocates, is about to begin. Cowgirl Creamery, purveyors of extraordinarily fine cheese, and the Center for Urban Education about Sustainable Agriculture, the nonprofit manager of the sumptuous Ferry Plaza Farmers Market, held here several times a week, have cosponsored the party.

The upper floor of the Ferry Building is frequently leased for weddings and corporate functions; the PAN anniversary gathered an unusual crowd for the site. The first guests to arrive were farmworkers Lidia Sanchez and Victor and Gloria Contreras, who came from the Midwest to join the celebration welcoming PAN's new executive director, Kathryn Gilje.[2] They were soon munching organic appetizers with Karen Heisler, cofounder of San Francisco's community and sustainability-oriented Mission Pie, and colleagues from the Environmental Protection Agency, where Heisler worked for many years before changing gears.

Across the room, Marian Moses, once the personal physician to Cesar Chavez and Dorothy Day, and Consumers Union scientist Michael Hansen join a large contingent of community leaders and high school students from the San Joaquin Valley. The young people are on hand to make a presentation about their work with PAN's Drift Catcher, a tool they use to collect data on pesticides in the air in their community, the epicenter of California's conventional agriculture.[4] Diverse nonprofit and government partners, colleagues from sustainable business enterprises,

Figure 1.1
Main Hall of the Ferry Building. Originally opened in 1898, the transportation hub was central for more than thirty years during which the only way to enter San Francisco, from the north, west, or east, was by boat. The building is still a ferry hub, but it now has a market on the first floor, offices on the second, and a farmers' market several days a week in the surrounding plaza (photograph by Elizabeth Fenwick Photography).

academics, friends, donors, and family members complete the assemblage. When all are fed, Cowgirl founder and owner Peg Smith raises her glass to toast the diverse company as a group that those in the room have dreamed and worked to build. "You are our community," the celebrity chef-turned-cheesemaker exalts, "and we are proud to be here."

Smith's toast frames the central question of this book: Can such a diverse group create a common understanding of alternative food and pursue it as a community? Such an understanding would combine ideas about sustainability and justice with the production of healthy food. Such food would be healthy for the planet and for the people who cultivate, process, and sell it, and it would be accessible to all.[5] But does a maker of expensive, artisanal cheese really share priorities with an impoverished migrant farmworker from the Central Valley? What do they have in common? How do the entrepreneurs who create and sell products available in places like the Ferry Plaza envision and build community with food system workers who want safe and dignified working conditions and fair wages? How do global pesticide campaigners work with urban community organizers who regard food as a focus for self-determination and youth empowerment? It is not easy to establish a small, ethical business that must compete with global firms that prioritize profit, or to maintain a nonprofit that challenges the power of conventional agriculture; nor is it easy to stay centered on a regulatory mission in government agencies that are understaffed, underfunded, and misdirected by politicians in the thrall of agricultural corporations. Mere survival in such difficult environments can be considered a major accomplishment. How could there be energy left over to build connections across diverse missions, resources, and personal priorities?

Our analysis of those and related questions proceeds in nine chapters. The first four form an unusually long introduction. Chapter 2 positions this book in the growing popular and academic conversation about food. Readers who can happily get through the day without ever thinking about economic geographer Michael Storper or social theorist Melanie DuPuis might be tempted to skip what scholars call the "review of the literature." But we recommend that they give it a try, perhaps paired with a dense syrah, because the discussion is useful in general and to the rest of the book. The literature on food has expanded rapidly in the past decade, and sorting out the context for the discussion can be useful. In addition, chapter 2 positions this book within that conversation. Specifically, we adapt Alfred Marshall's concept of an industrial district to explain how relationships between and among actors in close proximity have led over time to the learning and innovation that have made alternative food, and an associated community, both recognizable and possible.

Marshall outlined the basic district concept in the 1890s. His framework has been adapted over the intervening years to describe regions in which a common culture develops around a shared enterprise—such as movies in Hollywood or computers in Silicon Valley—and creates industrial momentum. A belief that "everybody does better when everybody does better" attracts participants to the district and encourages shared learning that fosters innovation. We have found Marshall's framework useful in organizing our story about how people in the Bay Area have challenged, encouraged, and supported one another in demonstrating alternatives to the conventional food system and in framing ever more demanding expectations about food quality.

Chapters 3 and 4 provide a different context for our analysis. In chapter 3, we recount parts of what has become a familiar tale about the evolution of the conventional food system. Our version uses Alfred Marshall's understanding of a shared enterprise to explain how the emerging values and shared enterprise of what we are calling an alternative food district grew in the context of California agribusiness. California is the birthplace of the now-global conventional agribusiness model that provides the context for any work on food in the Bay Area. The brief review also provides justification for presenting conventional food as unsustainable, unhealthy, and unjust.[6] In chapter 4 we look at early phases in the national and global critique of that food system, emphasizing early attention to sustainability, health, and labor issues. Ours is not, the narrative demonstrates, the first generation to oppose conventional food.

Those two chapters set the stage for positioning alternative food specifically in opposition to an identifiable conventional system. They also suggest why we highlight, perhaps unexpectedly, food distribution—the transportation, food handling and processing, wholesaling, and retailing required to move products from the producer to the consumer—as central in the creation of both conventional food and the alternatives to it.

Our introduction ends in the 1970s. If we were talking only about an haute alternative to conventional food commodities, then 1971, when Alice Waters opened Chez Panisse, the Berkeley restaurant that has popularly defined California cuisine, would be an appropriate end of the beginning, that is, the right place to shift from the introduction into the story of our district. If sustainability were the main concern in our discussion, then 1973 would be more appropriate. In that year, the small but expanding community of alternative growers that developed the first standards for organic certification in the nation formally organized as the California Certified Organic Farmers.

Because justice has become central to understanding alternative food, 1969 might be better: that is when the Black Panthers began providing breakfast to schoolchildren in an Oakland church and Glide Memorial Church in San Francisco began feeding poor and homeless people. That was also the year that the nutrition and hunger alleviation fields were reinvigorated by the White House Conference on Food, Nutrition and Health. A decade later would also work: in 1979, the California Agrarian Action Project, based at the University of California (UC), Davis filed a major lawsuit against the university. The litigation dramatized UC's complicity in the conventional agricultural system, charging that its research programs benefited large growers while devastating small farms and farmworkers.

To accommodate the complexity of our understanding of alternative food, we start the story of our district with all of those events. We analyze how expectations about food evolved in the district and became more demanding over the next several decades of innovation.

The analysis in chapters 5 through 8 presents a rough chronology of the district's maturation. Chapter 5 describes innovations that shaped land use in the countryside. The possibility for farming and food production around a major urban area is one legacy of almost a century of civic engagement in land protection. Urban elites advocating for parks and land use planning in the early twentieth century laid the groundwork for an activist civic culture as well as preserving a land base for the district. Following World War II, in a period of rapid suburbanization, the ranching community joined with urban conservationists to develop a new approach to land protection that encouraged agriculture as an element of sprawl prevention. In that supportive political environment, small-scale agriculturalists began a gradual, mutual reskilling process that demonstrated the commercial possibilities of ecologically sustainable production and developed early standards for organic agriculture.

The narrative in chapter 6 is more at the urban core: we use the term *radical regional cuisine* to describe the mixture of politics and food aesthetics that defined alternative food just as the district was becoming recognizable as distinctive. We locate the district's progressive food culture at the intersection of workers' rights and food conspiracies, as well as political radicals and a growing fascination with French produce markets, both of which developed in the Berkeley neighborhood known as the Gourmet Ghetto. Over time, back-to-the-land innovators created alternatives to conventional, increasingly processed food conveniences, and restaurateurs, cooperative markets, and politically and ecologically motivated consumers helped to popularize their products.

California cuisine could not truly take off until alternative producers and restauranteurs learned scale-appropriate distribution: how to get the new produce—distinctive, flavorful varieties harvested at their peak—out of the fields and to urban customers, starting with chefs. While California cuisine began with an emphasis on beautiful food, aesthetics did not stand alone for very long. The district's priorities rapidly broadened to reflect not only questions of distribution, but also the political roots of its food culture: in the process of connecting the back-to-the-land producers to high-end restaurants in the cities, the producers' long-standing concerns about sustainability were amplified by more general public support for environmentalism and moved to the center of conversations about alternative food.

Chapter 7 discusses the maturing of the district, a thickening of institutions and connections around a broader alternative agenda. As conventional agriculture and food processing relied on growth hormones, antibiotics, and a second generation of pesticides, health issues redefined what the district understood as quality food. Although produce and vegetables had initiated the sustainability wave in our district, innovative ranchers and processors soon began to develop alternative approaches to milk and meat. These fledgling operators gained traction as Americans worried about rBGH hormones (synthetic hormones engineered to increase milk production), mad cow disease, and related fallout from concentrated animal feeding operations.[7] At first, addressing health concerns was defined as personal health and relied in part on avoidance: more prosperous consumers could, and did, pay more for "sustainably" grown products, such as those that did not rely on chemicals. However, escalating concerns about *public* health soon required the exploration of a broad range of consumption issues. As the consequences of relying on a highly processed diet with excessive sugars and fats and limited fresh fruits and vegetables became apparent, equitable access to nutritious food became a central element of food quality.[8] The challenges faced by institutional food buyers, particularly schools and hospitals, signal early controversy about scaling up access to alternative food.

Chapter 8 describes the current wave of innovation in the maturing district. We argue that a public health crisis and increasingly diverse and demanding understandings of justice are now driving district innovation. Minority communities have claimed space and voice in the alternative food discussion. Their priorities—universal access to healthy food, personal and community empowerment, economic self-determination, and reskilling around cooking and nutrition—are becoming central to the district's understanding of the essential qualities of alternative food. Notably, even in the face of a sharp economic downturn that could have derailed the district's emerging equity priorities, a new generation of food crafters and entrepreneurs takes the fresh, local, sustainably produced agenda almost for granted and reasserts the importance of affordability goals and local community development. The food crafters have also raised the visibility of a new set of food system workers. While once the celebrity chef was a bit of an oddity, we now see celebrity butchers and pickle makers. Their visibility raises the profile of those who work in food processing, service, and in the fields, helping to reshape the labor debate and returning it to the core of the justice discourse.

Our discussion about food and food quality reflects both a division of labor and periodicity. Particular issues are more compelling in some historic circumstances than in others. As suburbanization spread in the Bay Area, concerned elites advocated first for parks, then for planning and land protection. Around the same time that Rachel Carson's *Silent Spring* appeared in 1962 some organized against the war in Vietnam, some fought in it, some organized to bring down the state, and some turned to making goat cheese or growing carrots and lettuce, often without pesticides or synthetic fertilizers. Farmworkers led grape and lettuce boycotts in California, and in the process, they educated a generation of increasingly sophisticated food consumers in the basics of what has come to be known as ethical consumption. Ethical consumption ripened into "voting with your fork," farmers' markets, and artisanal product as alternative producers responded to a rapid decline in the quality of conventional food and ever clearer consumer demands for fresh, tasty, safe food. Low-income communities and communities of color largely excluded from that process then entered the conversation on their own terms, insisting on access to the same nutritious and safe food that has become available to others.

None of the innovators has seen the range of problems as we now understand them, and none has addressed the issues comprehensively. This is in part because the food industry's growing reliance on chemicals to intensify production and process foods has created a contemporary public health crisis that was not visible in the 1930s or even the 1980s. Although obesity is now generally recognized as a problem, the links joining obesity, diabetes, heart disease, cancer, and other public health issues and our food system are not without controversy, and building the new understandings and coalitions to address that problem has not been an easy process. We should not be surprised that the needs of the well-to-do are served first and best in this struggle, as they have been so often in any effort toward social change.[9] However, the degree to which the priorities of urban communities of color have been embraced by those whose first concerns were food aesthetics and ecological sustainability reflects important learning and fairly consistent openness in a common enterprise in the district.

Tracking the escalating expectations about alternative food is a useful way to understand the change and innovation that have occurred in the district over the past fifty to seventy years. Obviously a more comprehensive conversation about food quality does not create access for all to healthy food, but it is a prerequisite to inclusive collective action.[10] Table 1.1 roughly summarizes the rising, more comprehensive expectations that pull together the narrative that follows.

Our conclusions in chapter 9 are less than what we expected when we started working on this book. It is tempting to assert that somewhere on the path from the food co-ops of the 1930s to the food conspiracies of the 1960s, past heritage varieties of sustainably grown vegetables to

Table 1.1
Escalating food quality expectation in the district

Approximate dates	Theme	Priorities	Examples
1960s	Radical politics	Rejection of corporate commodity foods in favor of bulk, unprocessed, and vegetarian alternatives	Food conspiracies, back-to-the-landers, Diggers, Black Panthers, United Farm Workers
1970s to 1980s	Ingredient quality and environmental sustainability	Sustainably produced, aesthetically pleasing, heritage varieties	California Coalition of Organic Farmers, Star Route Farms, Chez Panisse, Farm-to-Restaurant Project
1990s	Personal health, "not in my backyard" concerns	Health hazards of conventional agricultural processes and products, desire for hormone-free, antibiotic-free, GMO (genetically modified organism) free food	Straus Dairy organic milk, grass-fed beef, artisanal cheese
2000s	Public health and justice	Differential impacts of diet-related health issues, food crafters, food system workers	School lunch activism, food distribution in poor urban areas, micro-enterprises and pop-ups, fairness for food service workers

early-twenty-first-century food policy councils, an alternative food movement has been born. Although we argue that justice issues have moved to the center of food quality in the Bay Area and we see an ever more inclusive conversation in a growing and expanding community, our data do not get us quite that far.[11]

We are guardedly optimistic about fundamental change in the food system, but our narrative does not suggest that it is close at hand. Just the opposite: the narrative reflects the power of the opposition and the barriers to change. That provides some guidance about the possibilities for next steps. We do not believe, for example, that after some decades of "voting with your fork," the time has come to simply get more political and turn to collective action. Clearly where attention to policy has been inadequate, political action is needed. But collective action and government programs have been critical in our alternative food district for a very long time. Simultaneously the conventional food system has prospered with enormous cooperation and support from both federal and state governments. Little taking place in Washington, D.C., and California's state capitol, Sacramento, convinces us that the second decade of this century is a promising period for looking to government for comprehensive or even positive direction.

That does not mean that we do not value collective action. Just that we do believe it is worth noting that few of the late-twentieth-century concerns about food production and consumption are new.[12] Plato and Socrates, it is well known, argued extensively about the virtues of whole grain bread, and more than a century ago Upton Sinclair attempted to galvanize the nation to address health and labor problems in the meatpacking industry that remain unresolved. In spite of intense, intelligent collective action over many decades, the issues seem even less tractable. The progress we report is nevertheless significant: alternatives to the conventional system have been developed and demonstrated, broadly shared expectations about food quality are rising, and inclusive collaborations to achieve alternative food have begun to form. However, even in a place where climate, environment, and tradition combine to support the possibilities for truly alternative food, it seems that the hardest part is still ahead of us.

I

Making A Place for Just Food

Our analysis of the possibilities for a just and sustainable food system focuses on the San Francisco Bay Area over a relatively long period of time. The Bay Area is geographically small, but it has an outsized reputation on issues of food. While it is frequently chided for its precious food obsessions, those preoccupations have never been the most interesting or important part of its story. The discussion in part II devotes considerable attention to exploring what we *do* see as noteworthy—the region's shifting priorities and expectations about what constitutes good food. The conversation that we describe in the Bay Area region has involved an expanding community of people involved in a common enterprise to address food quality. The three chapters in part I provide context for that community and conversation.

It is difficult now to imagine a time when food was not recognized as an important issue. But with very few exceptions until relatively recently, the discussion of food was mostly relegated to cookbooks and specialty magazines. In the academic discussion, primarily technical and economic analysis of agriculture and agricultural production masked larger issues related to producing, distributing, and consuming food. Bay Area writers, activists, and eaters played a significant role in expanding and redirecting the conversation. At first, all but a few notable scholars were on the sidelines; thus, the literature review in chapter 2 is more eclectic than typical, and it turns initially on important popular works—Rachel Carson's *Silent Spring* (1962) and Francis Moore Lappé's *Diet for a Small Planet* almost a decade later (1971).

Agricultural production not only defined the discussion of food for many years, it is also reasonable to argue that the first industrial district in the Bay Area developed around the emergence of what we now regard as industrial agricultural production. Although California agriculture is certainly influenced by the diverse immigrants who came to the state and its favorable climate and wide variation in growing regions, in chapter 3 we examine the web of innovations in finance, institutional organization, transportation, marketing, and labor arrangements that supported the district's first iteration. Much of what we now regard as conventional agriculture has been enabled by a long series of federal and state actions—beginning with federal land grants supporting agricultural colleges and research, but most dramatically during the Great Depression of the 1930s—that have subsidized commodity agriculture and exempted many of its elements from basic operational expectations and regulations that conscribe activities in other sectors of the economy. It also has depended rather consistently on the exploitation of ethnically distinct and readily exploited minorities. Thus, while the Depression era is frequently glossed over as a time of enormous ascendancy for agriculture, power in what was becoming an ever more complex *food system* centralized and consolidated, shifting from farmers to railroads, processors, and finally to retailers in increasingly large-scale and vertically integrated operations.

Dissent and discontent in response to the industrialized food system, which we discuss in chapter 4, arose simultaneously with the conventional system. Consumer advocates, nutritionists, public health practitioners, and surprisingly contentious early soil scientists were only partially successful in addressing growing concerns about the long-term ecological sustainability, personal and public health impacts, and human rights consequences of the standardizing food system. The reasons for their limited effectiveness in slowing the momentum toward industrial, standardized food are complex and numerous; we draw attention to a pattern whereby chemically based commodity production was glossed over as "scientific" agriculture, while traditional practices, although the topic of much research and scientific investigation, were dismissed as women's work, superstition, and organic "gardening."

The three chapters in part I set the stage for the emergence of the San Francisco Bay Area alternative food district.

2

Framing Alternative Food

Both the evolution of the alternative food district and our analysis of it occur in close relationship to a broader discussion that has been taking place, more or less simultaneously, across the nation. Because both the exchange of ideas and the process of continuing dialogue are important to our district, this chapter puts *California Cuisine and Just Food* into the context of what is frequently called "the food literature." That literature did not exist until the 1970s. Prior to that point, most of what we now think of as food and food systems was typically thought about and labeled as "agriculture." Both mainstream and reform discussions of food were dominated by attention to production economics and agricultural technique. It was difficult to discuss food the way we do today for the simple reason that food was hard to envision very much beyond cookbooks.[1] As the more complex idea of food systems became recognizable, discussions expanded to include a fuller picture of the activities involved in feeding a population: cultivation, harvest, processing, packaging, transport, marketing, sale, and consumption, plus disposal of food and food-related items, and, in addition, the labor, financing, and inputs needed and outputs generated in each of those processes.

Until recently equity remained ancillary to the conversation, addressed episodically but only occasionally with sustained intensity. For most of the twentieth century, equity was discussed in the farm-centric context of agricultural labor.[2] Topics now loosely addressed as food security, which turn on access for all to safe, nutritious food, were subsumed into analysis of the problems created by overproduction and agricultural surpluses. The international concept of food sovereignty, which includes an individual right to food and a community right to control its own food system, adds an element of self-determination to questions of food access; but it has only recently begun to shape the U.S. conversation about food.

Perhaps because the early academic discussion of food was so narrowly focused, popular books, newspapers and magazines, and, more recently, blogs, videos gone viral, and feature films, have played an important role in leading the contemporary food conversation. Indeed, the line between academic and trade books has largely eroded in the food literature. Many trade books about food are analytically sophisticated and data rich, and several academics have written best-sellers. That is not unique, but neither is it common, and it may have contributed to the broad array of perspectives on food that is gaining traction as essential to the conversation about the future of the U.S. food system.

Food Systems: An Intersection of Scholarly and Popular Debate

Food systems were not discussed until recently because they were not generally visible. Allen (1993) points out that a focus on farm-related issues largely marginalized the analysis of the social impacts and dynamics of food production, distribution, and consumption. Even when social analysts were involved, a narrow cohort of politically powerful agricultural economists dominated, limiting the range of permissible topics.[3] Academics learned "the dangers of challenging the hegemony of production agriculture and its intellectual articulators, the agricultural economists."[4] For a long time, it was difficult to put together a viable academic career outside a restricted range of technical and economic issues.

It is not surprising, then, that several of the most influential early discussions about food and food systems started outside the academy. Many of the innovators discussed in part II mention two best-selling authors—Rachel Carson and Francis Moore Lappé—as critical to their own values. Carson's *Silent Spring* (1962) compiled data regarding the environmental and health consequences of conventional agriculture's growing reliance on pesticides. Many consider it to have been the opening salvo of the environmental movement. Although the initial response to her work focused on the impact of pesticides on wildlife and ecosystems, Carson's analysis also exposed the disturbing public health consequences of intensifying agricultural production methods.

Lappé's *Diet for a Small Planet* (1971) appeared in the context of a deepening conversation about agricultural intensification, efficiency, and protein. First, the 1967 Report of the President's Science Advisory Committee, *The World Food Problem*, reviewed global crop failures and concluded that "the scale, severity and duration of the world food problem are so great that a massive, long-range effort unprecedented in human history will be required to master it."[5] Promoting the intensive agricultural techniques associated with the "green revolution," the report encouraged more efficient use of protein resources, including a shift toward fish and vegetable sources and away from feed-intensive livestock.[6] The 1973 Arab oil embargo soon added a second concern: conventional agriculture's dependence on oil. Lappé was among the first to suggest a direct and personal response to potential global disasters. She emphasized that although grazing animals can turn low-quality forage into high-quality protein, the reality of conventional beef production was that "*enormous* quantities of *highest*-quality foodstuff were being fed to animals."[7] And she provided a practical guide for strategically combining vegetables and grains to achieve the "complete proteins" that meats provide.

Unlike the presidential commission's report, Lappé's analysis caused change, anchoring two diverse and important elements of the early days of alternative food. First, in tandem with a popular cookbook, *Laurel's Kitchen* (1976), Lappé had a major role in the rapid spread of vegetarianism. Second, Lappé invested her book earnings in a small nonprofit think tank, the Institute for Food and Development Policy (or Food First). Over the years, Food First researchers have critiqued and offered alternatives to U.S. and conventional approaches to food in the globalizing economy. Lappé and Collins's *Food First: Beyond the Myth of Scarcity* (1977) and Lappé, Collins, and Kinley's *Aid as Obstacle: Twenty Questions about Our Foreign Aid and the Hungry* (1981) both focused on impacts in the global South of U.S. agricultural technologies, aid, and consumption, just as the idea of globalization began to shape a general understanding of the world.

A rush of popular books followed Lappé. In *Hard Tomatoes, Hard Time,* Hightower (1973) exposed the close connections between agribusiness and academic researchers at land grant universities.[8] Wier and

Shapiro's *Circle of Poison* (1981) argued that pesticides banned for use in the United States but manufactured for export returned as problems on imported foods. Those early authors helped to shape the more comprehensive food systems story that emerged thereafter.[9]

Academics radicalized by social upheavals in the 1960s gradually changed the focus of scholarly investigation. The global economic crises of the 1970s facilitated this shift, inviting reassessment of economic policies in the United States and Europe and incubating cross-disciplinary discussion of the political economy of agriculture.[10] Two scholarly works published in 1981 proved particularly influential. Friedland and colleagues applied commodity chain analysis (which traces the lifespan of a specific product from production through consumption) to iceberg lettuce in California, examining both who benefited and who was harmed by farm mechanization and consolidation. Sen debunked the dominant agri-centric idea that famine and starvation result from inadequate food production and supply.[11]

Commodity chain analysis encouraged scholars to probe particular products, frequently at a global level. Mintz's masterful *Sweetness and Power* (1985) explored sugar, a product of global trade for centuries before Columbus's voyage, and its role in simultaneously underwriting slavery in the Caribbean and transforming the diet of working-class British. Similar narratives have become both a popular and scholarly genre, enabling attentive readers to see the implications of quotidian food choices. Salt, tomatoes, potatoes, coffee, oranges, sugar, bananas, oysters, milk, and diverse meats are among the products that have been subjected to close inspection.[12]

The commodity chain also anchors a long muckraking tradition in the literature. Sinclair's *The Jungle* (1906) is the archetype: his exposé of the horrors of the Chicago meatpacking industry led to passage of the Pure Food and Drug Act that same year. The equivalent horrors of chicken production have been similarly documented in the early twenty-first century, but to less effect. Striffler's *Chicken: The Dangerous Transformation of America's Favorite Food* (2005), a 2008 *Charlotte* [North Carolina] *Observer* series entitled "The Cruelest Cuts," and the 2009 film *Food, Inc.* illustrate the problematic social and environmental consequences of current industrial practices.[13] Although the chicken industry is changing gradually, this intersection of scholarly and popular work underscores the shift away from the farm-centric, and generally uncritical, themes that dominated twentieth-century agricultural social science.[14]

Sen's pivotal *Poverty and Famines: An Essay on Entitlement and Deprivation* focused on hunger and equity, disproving the standard, durable wisdom that famine and starvation arise from inadequate supplies of food. Sen's review of famine events in Bengal, Ethiopia, and the Sahel demonstrated that access, rather than supply, is most central to famine. Assessing differences in entitlements, he insisted that whether hungry people have food to eat is a function of the power dynamics of resource distribution, not a simple relationship between numbers of people and tons of grain.[15] Subsequent works reinterpreting the Irish potato famine, the late nineteenth-century famines in India, and the "late Victorian holocausts," confirm Sen's essential insights.[16]

If Archer Daniels Midland, Monsanto, and the Gates Foundation are determined to promote supply-side and technological fixes to hunger in the style of the green revolution, it is not for lack of analysis demonstrating the shortcomings of such input-dependent agricultural intensification.[17] Sen's emphasis on food distribution has inspired our own approach to both Berkeley's foodie hub—known as the Gourmet Ghetto—and food justice issues in Oakland: distribution problems that made it difficult to get superior produce to the Gourmet Ghetto are similar to the barriers that make it difficult to provide nutritious foods to hospitals, the elderly, and school children and to what some call "food deserts," that is, neighborhoods without ready access to a healthy diet.[18]

The 1980s launched a broader scholarly inquiry into questions of who is eating what, and why and how foods are produced. Poppendieck's *Breadlines Knee-Deep in Wheat* (1985) demonstrated that U.S. hunger policies begun during the Great Depression have prioritized support for agriculture and dealing with food surplus rather than feeding hungry people. In a slightly different vein, other social scientists have focused on America's changing food habits. Levenstein positioned food more broadly as an appropriate lens for inquiry into social and cultural history: *Paradox of Plenty: A Social History of Eating in America* (1988) explored the shift in the United States from a traditional British menu to a more American diet during the period 1880 to 1930. Belasco explored a similar topic in *Appetite for Change: How the Counterculture Took on the Food*

Industry (1989), documenting Americans' post–World War II embrace of highly processed, standardized food and early efforts to loosen the corporate commodities' grip. By the time Allen's book appeared in 1993, her emphasis on social analysis of the new entity—food systems—was as much evidence of the broadening discussion as a startling call for redirection.

Public events also forced growing attention to the intersection of environmental and food issues and drew both public and scholarly attention to toxins in food and food production. An explosive public debate in 1989 about the growth regulator Alar on apples intensified questions about whether "you are what you eat," and if so, whether we should be concerned about children's tolerance for the chemicals that they consume.[19] The resulting National Academy of Science's *Pesticides in the Diets of Infants and Children* (1993) documented the inadequate oversight in the U.S. food regulatory system. Two 1996 works explored both serious historic abuses in the food system and pending problems. Wargo's *Our Children's Toxic Legacy* demonstrated that "science and the law fail to protect us from pesticides." In *Our Stolen Future*, Colborn raised questions about the hidden, long-term impact of chemicals in food and food production, focusing on human reproduction and bringing the concept of endocrine disrupters—chemicals that interfere with hormone activities in animals, including humans—at least to the outer border of mainstream conversations about food.

Scholars have addressed toxics, access, and health from a variety of directions. Dupuis (2000) identified a "not in my body" politics: advocacy that unites long-standing environmental, public health, and consumer activists with justice advocates' priority on nontoxic, nutritious meals. Scholars in Great Britain first explored the issue of healthy food in what they called "food deserts," neighborhoods that are often served primarily by fast food restaurants and corner stores without a produce section.[20] That idea has defined a major segment of the food justice agenda; for example, Chicago food leader LaDonna Redmond has been quoted time and again observing that in her Chicago neighborhood, it is easier to buy an AK-47 than an organic tomato.[21]

These food systems topics were already on the agenda when three writers—a journalist, a professor, and an academic journalist—produced a wave of work bringing food systems issues into daily conversation.

Atlantic Monthly writer Eric Schlosser's popular volume, *Fast Food Nation: The Dark Side of the All-American Meal* (2001), documented the fast food industry's enormous marketing power and probed the consequences for producers of beef, chicken, and kindred products, from redesigning the potato to redefining school lunch.[22] Schlosser also explored fast food impacts on consumers and on meatpackers and the youth who work in fast food outlets. His analysis inspired Spurlock's 2004 polemical film, *Super Size Me*, in which the producer-director-author-star consumes only McDonald's products for a month, with alarming, albeit not peer-reviewed, health consequences.

Nutrition professor Marion Nestle followed in 2002 with *Food Politics* and a year later with *Safe Food*. Her first book's subtitle, *How the Food Industry Influences Nutrition and Health,* suggests her concerns. Nestle tracked the industry's well-funded marketing and lobbying efforts and concluded that the industry encourages overeating, promotes poor nutritional practices, and confuses basic nutritional guidance. *Safe Food* explored who bears the risks of food safety shortfalls and who benefits from ignoring them. And in 2008, she focused on pet food safety, investigating the problems of monitoring and maintaining basic health standards in extended, globalized food chains.[23]

Completing this trio, journalism professor Michael Pollan has become the ubiquitous face of critical food writing.[24] His *New York Times* articles, best-selling books, including *Omnivore's Dilemma* (2006), and scores of speeches, articles, interviews, and personal appearances have made him a household name. Pollan has even appeared in popular films: *King Korn* (2007), *The Vanishing of the Bees* (2009), and with Schlosser in *Food Inc.* (2008).

These compelling analysts significantly expanded the audience for food discussion at a time when the health consequences of a diet of cheap, processed food were becoming increasingly apparent in the scientific literature and on the street. Halweil's (2004) *Eat Here: Reclaiming Homegrown Pleasures in a Global Supermarket* expresses a vote-with-your-fork response—suggesting that consumers can change the system with careful food choices.[25] But the harder task is addressing access to food and community economic development in historically disenfranchised populations, where folks do not deploy the consumer power in evidence at San Francisco's upscale Ferry Plaza nor do they have a range of food choices easily available. New activist chefs are among those addressing access, development, and deskilling in those communities. For example, Terry's *Vegan Soul Kitchen: Fresh, Healthy, and Creative African-American Cuisine* (2009) weaves together efforts to build community and culture as well as reskill urban workers and urban consumers by growing, cooking, and sharing healthy food and preparing urban populations to find jobs in the growing food sector.

Newly audible voices are adding their own perspectives to the rapidly expanding food systems conversation. The global peasant organization La Via Campesina introduced the notion of food sovereignty at the 1996 World Food Summit in Rome.[26] The concept emphasizes the anticolonial and self-determination elements of food access. In the United States, analysts and community organizers have more commonly pursued what is referred to as food justice. Winne (2008, 2010a) and Gottlieb and Joshi (2010) have probed the rapidly expanding array of organizations, most of them recently formed, that work in the food justice field. On the production side, a new wave of workers' organizations has worked to bring the concerns of low-wage, frequently immigrant food service workers into the food debate. Sen and Mamdouh's *The Accidental American* (2008), the Restaurant Opportunities Centers' *Serving While Sick* (2010) and *Behind the Kitchen Door* (2011) and Liu and Apollon's "The Color of Food" (2011) provide fresh insight into the ever more complex justice arena.

Gottlieb and Joshi's chronicle of new actors and concerns in what will likely become a thick and feisty discourse characterizes food justice as "ensuring that the benefits and risks of where, what and how food is grown and produced, transported and distributed and accessed and eaten are shared fairly."[27] Others who have used and popularized the term *food justice* appear to us, as we discuss in chapter 8, purposefully less comprehensive. They emphasize, as do many of the groups that Gottlieb and Joshi profile, community-level concerns with access to healthy food that weave together self-determination, youth empowerment, economic development, and skill building, frequently with a clear focus on racism and race. Although these concerns are part of the narrative in Gottlieb and Joshi's work, their definition both adds to and subtracts from food justice as we have encountered it.

We therefore take a more inductive approach, using the term *food justice* to reflect as best we can what the advocates in our narrative have told us. When we need a term to convey Gottlieb and Joshi's more comprehensive definition, we use *food democracy:* a participatory decision-making process that addresses the human right to safe, nutritious, justly produced food. That definition emphasizes a comprehensive discussion underlying collective action.[28] It also asserts rights, which, again as we shall discuss in chapter 8, seems particularly important.

The same new perspectives have also confounded the already complex meanings of *local*. At first glance, *local* seems simple and transparently positive. Foods produced more locally will be fresher and the supply chain shorter. That permits a kind of sustainability and accountability that is impossible in a global system that erases the producer and often alters the product. In addition, *local* suggests that consumer dollars will remain in and support the local community. However, just how close to home local must be is a tricky question. For example, under what definition can coffee beans that are grown in Tanzania but roasted in the Bay Area be considered local to the Bay Area? Or under what conditions can milk produced in the Bay Area but packaged in bottles from Canada be considered local?

The confusion arises in part because some understandings of *local* come into the food discussion through global climate change issues, which are every bit as contested in the food systems context as they are more generally. Locavores, who try to eat only food produced within a circumscribed local radius, appear to have borne the brunt of inappropriately dismissive critiques.[29] Similarly, conventional agriculture's reliance on cheap oil for fertilizers and pesticides has become mired in conflicting data about oil used in transport and remains an underaddressed, but significant, issue. The very much related issue of food miles—how far a product travels to get to the point of consumption—is similarly contested, with significant debate about measurement and the environmental impacts of production and processing in one place or another. Sadly, the process of excoriating the locavores has diverted public attention from a more complex and important issue of accountability. Locality provides some mechanisms—perhaps a proxy—for setting and enforcing expectations about food quality that are not available in the long, convoluted supply chains that make conventional processed foods possible.

Early rosy assertions that global warming would benefit agriculture—warmer weather and more carbon dioxide arguably could increase growing seasons and yields in some areas—has given way to less hopeful views.[30] Recent international agricultural assessments anticipate a serious decline in clean and accessible water in developing countries, which would likely be exacerbated by climate change.[31] The United Nations' Intergovernmental Panel on Climate Change won a Nobel Peace Prize in 2007 for reporting, with little visible policy impact, that the worst effects on agriculture may be in the least-food-secure areas: Africa, for example, is expected to experience a 30 percent drop in food productivity over the next several decades.[32] On the positive side, a 2009 global assessment of agriculture found that sustainable production methods that build on site-specific agricultural knowledge could feed the global population and support local economic development around the world. Unfortunately, the report of the International Assessment of Agriculture Knowledge, Science and Technology for Development, *Agriculture at a Crossroads* (2009), has had little impact, although its findings have been amplified in the more recent Food and Agriculture Organization's assessment, *Save and Grow* (2011).[33]

Industrial Districts and Food Systems

We use a simple version of the industrial district concept to focus our narrative. The idea has been a useful tool for explaining innovation and learning in small manufacturing regions for over a century. Districts arise where focused interactions among those involved in a particular trade or activity create a sense of common enterprise. Familiar districts include the actors, writers, financiers, publicists, and related service providers and hangers-on that make Hollywood recognizable as a center of entertainment and the cluster of high-tech folks—the wonks, programmers, engineers, university researchers, funders in the military and venture capital firms, and others—who understand themselves as "working for Silicon Valley" (Saxenian 1994).

Scholars have studied industrial districts in many different places and called them many different things. The idea enjoyed a renaissance in the

1980s as an element of scholarly response to economic globalization. In the conservative political climate that grew out of the economic crises of the 1970s, neoliberals advocated "free market" capitalism without protectionist national trade barriers or subsidies to nationally prioritized industries. Scholars concerned about this round of globalization responded with a wave of analyses exploring the durability and importance of local and regional activity, including industrial districts. Both groups have agreed that industrial districts are important to the businesses involved. Most descriptions of districts have in common an emphasis on the importance of routine face-to-face contact in economic activity, and many of them explore the way regions respond to the globalization process.

The observation that propulsive economic gains stem from close interactions among similar producers has seemed applicable to diverse industrial systems. Brusco (1982) explored the successful industrialization of the Emilia-Romagna region during the 1970s that created the so-called "third Italy."[34] The region became known for its food and machinery industries, comprised of dense networks of small and medium-sized businesses. The Emilian district is very similar to Marshall's original formulation in the1890s. Rebranded again as "flexible specialization," Piore and Sabel used the industrial district frame in 1984 to defend the efficiency of local specificity and creativity against enthusiasm for globalization. Their *The Second Industrial Divide: Possibilities for Prosperity* described an alternative to global mass production: small-scale producers employing craftsmen rather than unskilled, interchangeable labor. In yet another iteration, Saxenian (1994) and Storper (1997) analyzed the formal and informal social practices that produce high-performing industrial regions.

Industrial districts came into discussions on food systems early in the twenty-first century as part of assessing the common assertion that smaller-scale, artisanal production is preferable on its face. Confronted with vibrant regional economies, Goodman and Watts (1994) criticized the presumption that social agency at small scales necessarily withered when confronted with global actors. But scholars did not presume that these more "flexible" or regional forms were superior. Dupuis and Block (2008) underscored, for example, that relocalizing in the milk industry has led to very different outcomes in different places. Winter (2003) argued that efforts to promote embeddedness, local sourcing, and proximate industry can easily obscure or reinforce politics and goals that are neither progressive nor sustainable.[35] Accordingly, most have reasoned that industrial districts suggest possibilities for more creative work and more satisfying workplace practice and politics, but that they do not automatically result in better social conditions or preordain a wide dispersal of economic benefits.[36]

Although most industrial district literature emphasizes producers and manufacturers, Goodman's (2003) description of a "quality turn" away from mass-produced products focused attention on consumers and how their responses to genetically modified organisms, food scares, and increasingly toxic production methods and food products can provide a market for small-scale and artisanal production. In the same vein, numerous scholars have studied the notions of values and trust that appear central to Goodman's idea of quality. Storper (1997) discussed conventions, formal and informal agreements about acceptable social practices, in a district. Reskilling and reshaping practices can underwrite the sale of products based on specific qualities—fresh, organic, artisanal, fair trade—rather than price. Murdoch, Marsden, and Banks (2000) have similarly emphasized that trust, embeddedness, and consumer insistence on food safety reflect local conventions that are critical to understanding shifts in the global agrifood system. Morgan, Marsden, and Murdoch's *Worlds of Food* (2006) expanded on this theme to chart how global, regional, and local cultural, political, and ecological priorities shape the food system in different places. Others have argued that global commodity chains can be part of the quality turn, arguing that they can serve as mechanisms for communicating quality standards across vast distances.[37] Finally, DuPuis and Gillon, summarizing a significant body of recent scholarship, suggest that the values, conventions, and expectations embodied in alternative markets are created and maintained through civic engagement.[38]

Analysts have also pointed to obvious constraints on the possibilities for regional food. For example, the much-discussed idea of more personal relationships between food producers and consumers (as in the slogan, "Know your farmer, know your food") is largely incompatible with the notion of year-round markets with fresh produce, which requires importing many products.[39] Conversely, diverse adaptations of *terroir*—roughly, controlled appellations such as Rocquefort, Champagne, and similar efforts to market regionally distinct products—require integrating industrial districts into global commerce.

Indulgence and Innovation in an Alternative Food District

Our book enters the conversation at this point. We start by admitting that it is more difficult than we anticipated to identify particular acts or items as innovations—leading to alternative food—or as indulgences, and "bads" that we should perhaps oppose, shame, or stamp out. Perspective matters enormously. The local organic peach that may strike farmers' market critics as an indulgence may in fact be one for the shopper, especially when nearby homeless people are looking for food in trash cans. However, the United Farm Workers have regarded eliminating pesticides as a fundamental justice issue for good reason, and farmworkers are unlikely to regard a field without pesticides as an indulgence.

As much as perspective matters, scale is probably more important. It frequently requires time and space to turn an indulgence for the few into a food system innovation. For example, before pasteurized or certified milk was available, the purity of fluid milk was never guaranteed, and many children, rich and poor—but assuredly more of them were poor—died from drinking it. The privileged were the first to enjoy access to consistently safe milk, but it soon became the standard. Moving privileges to scale is a key element of justice.

It is quite common to characterize raising the standards on basic food requirements as elitist, but is it? It is clear whose interests are served when federal regulations permit chicken to be chilled in water baths known as "fecal soup"[40] and a hamburger made with "pink slime"[41] to be the standard. When preventing such outrages signifies a "nanny state," and an air-cooled chicken and a grass-fed beef are characterized as elitist, improving food quality for all is a challenge.[42]

We have adapted the industrial district idea to explain how alternative food is being defined and created in a place where food has been an important focus of social, family, and political activity for generations. Marshall noted long ago that activity in a district is defined by what is in "the air."[43] In our understanding, he meant that in successful districts, routine interactions among people who pursue a common activity generate an air, that is, a shared culture. That culture identifies the district's activity as a common enterprise and invites cooperation that leads to learning. Over time shared learning raises expectations and improves practice. The common enterprise also attracts more and different kinds of related participants, leading to growth and diversification in the district. Finally, the learning and diversification produce innovation, which raises the bar for participants and encourages more learning and innovation.[44] The stories that follow in chapters 5 to 8 track three core elements in the culture of our alternative food district: a common enterprise,[45] institutional thickening,[46] and innovation over time.

A Common Enterprise

An industrial district is not simply an aggregation of similar or related activities in a particular area. Routine, open communications among participants create a recognition that the activity is common and that "everybody does better when everybody does better"—hence our term *common enterprise.* Although competition in a district is real, and an important incentive to innovate, the competition is shaped by the idea that people are in the same field and share and learn together. In our district, much of the earliest important learning involved *re*learning how to produce food without the suite of chemical inputs that underwrite conventional agribusiness.

Elaboration of the culture is encouraged by the fact that boundaries between and among entities and sectors tend to be porous. Individuals and ideas move between firms and sectors, improving practices and creating expectations of mutual support. Thus, while individual firms are recognizable, they are open; for example, someone who works here is married to someone who works over there, and they may work together at a third place soon. Conversely, those in the area who do not regard themselves or act as part of a shared enterprise are recognizable as such, occasionally with repercussions for their activities within the district. The consequences can rise to the level of more or less formal rules and some kind of enforcement, but in our district things have generally been less explicit.

Institutional Thickening

The shared learning and improvement in practice and product encourage an increasing range and number of institutions. This institutional thickening results in part because the common enterprise facilitates economies of regional scale by reducing many kinds of transaction costs within the district. For example, a district can support services that might not be available to isolated operators, and some services may actually become cheaper if most or many people need them. Creameries, slaughterhouses, and accessible commercial kitchens are all important facilities in our district. Similarly, established practitioners get wind of changes and apply them to their own operations, while skilled and talented workers are attracted to districts that provide diverse opportunities. The newcomers contribute their skills and perspectives, pushing the boundaries of the shared enterprise and raising expectations and levels of knowledge.

Districts can encourage thickening by lowering barriers to entry and exit. For example, county planners may delay or even harass the first innovative food incubator or taco truck. But when they begin to regard the activity as beneficial and important, they may develop more sophisticated regulatory programs that can make it easier for others to follow. Similarly, districts can increase the level of public and private investment. Local markets, for example, can help producers accumulate capital within the district. In our case, although innovators developed ways to distribute small lots of product fairly early, it took longer to create ways to accumulate capital and move it within the district, particularly from urban to rural areas. But a new generation of processors has taken advantage of well-established markets in quality products to start small, in mobile and sometimes only occasional food carts (frequently knows as pop-ups) that impose lower expectations and fewer capital requirements.

In our telling, institutional thickening need not be confined to commerce and entrepreneurs. Diverse civic activism that led to both government and private efforts to protect land was central to the beginning of our alternative food district. Entrepreneurs were a critical part of the institutional thickening, but not the initial drivers of it. However, as urban entrepreneurs moved into the countryside, they became important vectors of progressive social values, particularly regarding the treatment of land and labor. This makes our district an excellent place to view and assess the interaction of markets and collective political (not always, but frequently, government) action.

Innovation over Time

The most important dimension of industrial districts is that as they develop, they generate innovations more rapidly and deeply than dispersed actors are able to accomplish. The same local markets, particularly farmers' markets, plus post offices, school yards, coffee shops, and similar gathering places, facilitate organizing, education, and exchange of ideas. Innovation is not an automatic result, but over time, in Marshall's words, "Good work is rightly appreciated; inventions and improvements in machinery, in processes and the general organization of business have their merits promptly discussed. If one man starts a new idea, it is taken up by others and combined with suggestions of their own; and thus it becomes the source of further new ideas."[47]

Risks and Challenges in an Industrial District

An industrial district is not a perpetual innovation machine. Some observers have emphasized the importance of getting from one generation to the next in an industrial district: if the innovations do not gain enough acceptance to get beyond their original creators, or if a district cannot resolve internal conflicts, the district will not endure. Our district was probably inclined from the outset to value longevity. The early planners and park advocates wanted to create a sense of permanence in their community, and the ranchers who joined them in land conservation were experiencing threats to the survival of their family enterprise and way of life. Subsequent generations in those families are almost uniformly determined not to be the ones who lost the farm.

Nonetheless, the sense of shared enterprise can fall victim to either an erosion of consensus or too much of it. Some friction is necessary, of course: without competition within the district and from outside it, it is difficult to create and sustain the risk taking that underlies innovation. But shifting orthodoxies, technical innovation, or changes in parts of the district can erode the common enterprise. Conversely, if participants become guarded against new ideas or stop learning and sharing, the district culture can also disintegrate. In our alternative food district, the embrace of new understandings of justice as defined by the affected communities is not yet accomplished.[48] Farmers' markets are a particular focus of impatient analysis in this vein. Because successful multiracial and multiclass farmers' markets are not common, the genre is frequently dismissed as elitist, racist, or irrelevant to real problems. That may be true in many circumstances, but the outcome is generally not for want of trying: enormous efforts to adapt the farmers' market format to poor, particularly African American, communities, have produced limited success. These outcomes are clear signals of gaps in or threats to the common culture of the district. But they also reflect the enormity of the barriers to altering entrenched food system operations, as well as a host of class and race barriers.

Modifying the Standard Format

We have adapted Marshall's original industrial district format to a more complex setting than is common. Key elements of the district agenda are beyond the reach of even the most diligent innovators. Centuries of racial abuse and economic exploitation and the problems arising from the nation's failure to enact, or even agree on the necessity for, comprehensive health care and social support systems, chemical policies, and cautionary principles that are standard throughout the industrialized world cannot be addressed in the context of a regional analytical frame or, for that matter, even by "fixing" the global or national food system.

The standard industrial district literature also pays little attention to actual people. Although a theoretical cast of characters innovates abstractly, we rarely meet the individuals in the Marshallian work. In our district, we saw a multitude of amazing people every bit as compelling as the innovations they created, so we decided to focus somewhat on a diverse cast of actors: entrepreneurs, academics, governments, activists, farmers and ranchers, distributors, community organizers, and consumers. Perhaps because we have such robust characters, we are unsatisfied with the standard array of roles in networked conceptions of social change: "warriors, builders, and weavers."[49] When distribution is a central element of innovation, for example, it is hard to say what is outreaching and organizing (weavers), what is entrepreneurial (builders), and what is political (warriors). Changing distribution requires people in many different institutions to be all three, frequently simultaneously.

In addition, when institutions are porous and rapidly changing, learning and reskilling are at the heart of a culture that is both cooperative and competitive. Most of the characters have worked those different duty stations at different times.

Instead, our narrative focuses on a group that we call *mavens*.[50] The culture of the district, we have found, did not emerge automatically from frequent interactions or even from a shared interest in accessible food, good pasture management, or fair wages. It was in significant part created by individuals who founded and led institutions—in civil society, business, and government—with the specific intent of building a shared enterprise as an alternative to conventional business as usual. Others have created connections needed for their business plan or shared their knowledge as part of promoting their cause or their product, but mavens have worked actively and self-consciously to nurture the district, make it recognizable to both participants and outsiders, and instill in the district the values and priorities that inspired them to become involved in the first place.

The mavens have been strengthened, we believe, by the fact that profit is less central in our district than in some others. There may be passion for art and glamour in Hollywood and for cutting-edge technology in Silicon Valley, but both are focused, without cavil, on making money, preferably huge amounts of it. Perhaps it is because there is so little money on the table and so little opportunity for a quick buck or a big strike in alternative food, but our district prioritizes different values. Enough profit to stay in business is, of course, mandatory for the entrepreneurs in our narrative, and we do not mean to imply that profit is of no interest. But the mavens in our district specifically nurtured a vision of the district that does not aim at making a large profit. The vision has been about building a different kind of system—one in which healthy food is available to all, one that will protect the land and provide farmers and workers a decent living, and one that builds a community in which they themselves would live and want to remain.

The culture of the district is also supported by the fact that its products tend to compete with each other and with outsiders by raising quality, not lowering price. This could be seen as evidence of a significant profit motive but, in fact, so doing appears to leave room, and may even require, a high respect for both land and skill: quality competition in the district has tended toward nurturing skilled laborers who can create distinctive products and stewarding the land that gives the district its *terroir*, that is, its unique characteristics.

Similarly, although proximity is essential to industrial districts, place and land are not normally necessary. Yet we focus on both. We are interested in the place where the relationships occur—how the people, history, institutions, and physical environment shape the possibilities for innovation and have in turn been shaped by them. We believe that the physical environment and land use practice in a region affect human interactions. Exploring that requires taking a long view, which we have taken. Our district is coming up on five decades depending on where you start, and our background materials add another century. This is unusual, but it permits us to explore changes and constants in the culture of the district.[51]

The land and environment are particularly important components in an industrial district focused on the production, distribution, and consumption of food. To illustrate, for farming purposes, our district is both poorly watered and insufficiently productive to warrant irrigating. Hence our district has not experienced the erosion of community resources that Goldschmidt (1947) and MacCannell (1986) documented. They concluded, four decades apart, that the poverty and class disparity that water projects brought to large-scale corporate farming was not replicated in places without project water where more functional communities featuring smaller, family-owned operations flourished.[52] Much of our district is coastal as well, simultaneously spectacularly beautiful and subject to fogs that enhance its utility for extensive pastures. Those factors give the district a certain stability in the same way that you cannot grow avocados in Vermont or rapidly switch from milk to semiconductors. Significantly, however, the combination of beauty and pasture-enriching fog has led to heated controversy about why and how to protect the land from development.

The focus on land and place has not, however, made us particularly concerned with geographical boundaries, and we do not draw lines to indicate locales that are in or not in the district. When we say "the Bay Area," we mean the nine counties (4.4 million acres) surrounding the San

Francisco Bay, but we are aware that the notion is relatively recently derived and we approach it with some flexibility. We also do not imply that the Bay Area is sealed or isolated. Far from it. The district has long been involved in and dependent on global traffic in ideas, products, money, and people across multiple scales and boundaries. Innovations in the district occur because of and inspired by ideas and events beyond the region. We will, for example, observe the French wine industry's appellation at work in the district and how differently it has played out thus far in neighboring Marin and Sonoma counties. We will also note the inspiration that Oakland's food justice community has taken from Milwaukee and Philadelphia.

Finally, while many districts are in relatively small, isolated, homogeneous areas that focus on a single product, the Bay Area is highly diverse, including both farming communities and major urban areas. This means that justice in our telling must include justice for farmers as well as justice for underserved populations. Although that scope complicates the analysis, it reflects reality. Moreover, bridging the two has been a central task of innovators in the district: adapting urban preservation priorities to agricultural land protection, moving small batches of first funky and then extremely high-quality produce into urban markets, developing markets for new food products, and, more recently, addressing distribution barriers in poor urban neighborhoods.

Having thus modified the concept of the standard industrial district, we then use it to organize a story about the evolution of alternative food in a particular place. District participants have, over more than half a century, worked together in planning processes, teaching, and suasion; leading changes in investment and business strategies; promoting regulatory reform; encouraging individuals to change consumption patterns; and attracting foundation grants and tax dollars. This is one reason why Cowgirl Creamery's Peggy Smith could stand in the Ferry Building in 2007, that temple of "yuppie chow," and welcome organic activists, regulators, academics, health researchers, farmworkers and farmers, and elite chefs as part of the same community. That community has since expanded to include interests and groups previously unconsidered or disarticulated from the alternative food community who have become functional elements of the district, shaping practices and products, environmental quality and markets, and increasingly seeking justice.

We readily grant that seeking justice is not the same as achieving it. It is also very different from ignoring justice and pursuing profit. But while our alternative food district remains a work in progress, it can tell us about what it takes to change the possibilities for healthy, just food in a particular place, about the relationships of markets, nonprofit organizations, and governments—and the people and mavens who inhabit those institutions. It also speaks of the enormous barriers that remain in the path to food democracy.

3

California Agriculture and Conventional Food

Why alternative food? Alternative to what? This chapter describes the conventional side of the food system. The basic story is broadly familiar, and we do not reiterate the abundant literature.[1] Instead we focus on background relevant to our arguments about alternative food. Because we define *alternative* as sustainable, healthy, and just, we need to suggest how conventional food, that is, the food most people eat every day, became unsustainable, unjust, and unhealthy. Because justice has been central in the district over the past decade and is likely to remain so, we emphasize connections between the evolving conventional food system and the struggle for civil and human rights in California and U.S. history.

We begin by arguing, not without precedent, that the way of producing food that is conventional across the United States, and indeed much of the rest of the world today, originated in California. Many of the state's characteristic patterns—specialty crops, large landholdings, and exploited labor—were in clear view during the mission period (late 1760s to mid-1830s). Subsequently, investment following the Gold Rush (1848–1855), construction of the railroads, and significant government subsidies facilitated the creation and distribution of California fruits and vegetables around the world. With such a heavily capitalized and state-supported agriculture, the Jeffersonian vision of hardy, independent yeomen farmers did not apply in California.

We then take a selective look at the complicated period from World War I to the early 1950s. The standard Depression-era narrative reasonably focuses on government action, which we regard as critical in creating an increasingly integrated national food system based on the California model. But in some key areas, Congress acted by not acting or by exempting agriculture from regulation; at a time when it was effectively managing commodity markets, for example, Congress did not extend basic protections to agricultural workers. That reinforced and spread nationwide the exploitative elements of the California approach to labor. Changes in the private sector starting in the 1930s were probably as important as government programs (and inaction) in shaping and nationalizing the conventional model. The growth in grocery stores and supermarkets was a major factor driving consolidation and nationalization of the food system. Californians' beloved automobiles filled suburban parking lots with grocery shoppers who led Americans' rush toward new kinds of ever more processed and convenient food.

Finally, we point to the public health and environmental threats of the "get big or get out" period. Up to the 1950s, it is easy to see the injustices in the food system and many increasingly unsustainable elements, but we have found little evidence that food was reliably bad or bad for you.[2] After World War II, cheap oil and chemical pesticides intensified every aspect of the system, degrading food quality in the process. We present the "green revolution" as a tragic misnomer: there is little about it that was green as we now use that term. The green revolution underwrote the round of federal policies in the 1970s that were *intended* to drive small family farmers off their land. Those hardy yeomen, on whom our core agricultural myths rest, were almost as clearly victims of the system as were the workers in the fields.

From the Missions to World War I

The Emergence of the Conventional Model in California

Analysts of every political stripe and priority note California's remarkable natural resources. Carey McWilliams, for example, wrote in the 1930s that the state's agriculture reflects its "amazing range of environmental factors . . . California has the highest peaks, the lowest valleys, the driest desert, and some of the rainiest sections of the United States."[3] The Sierra mountains, which trace the state's eastern border, and a number of mountain ranges running parallel to the Pacific coast, create hundreds of soil types and microclimates. These mountains encircle the Central Valley, which is the heart of agriculture in California. Extending north nearly to Oregon and south beyond Bakersfield, the valley includes an exceptionally fertile band of farmland and pasture that produces

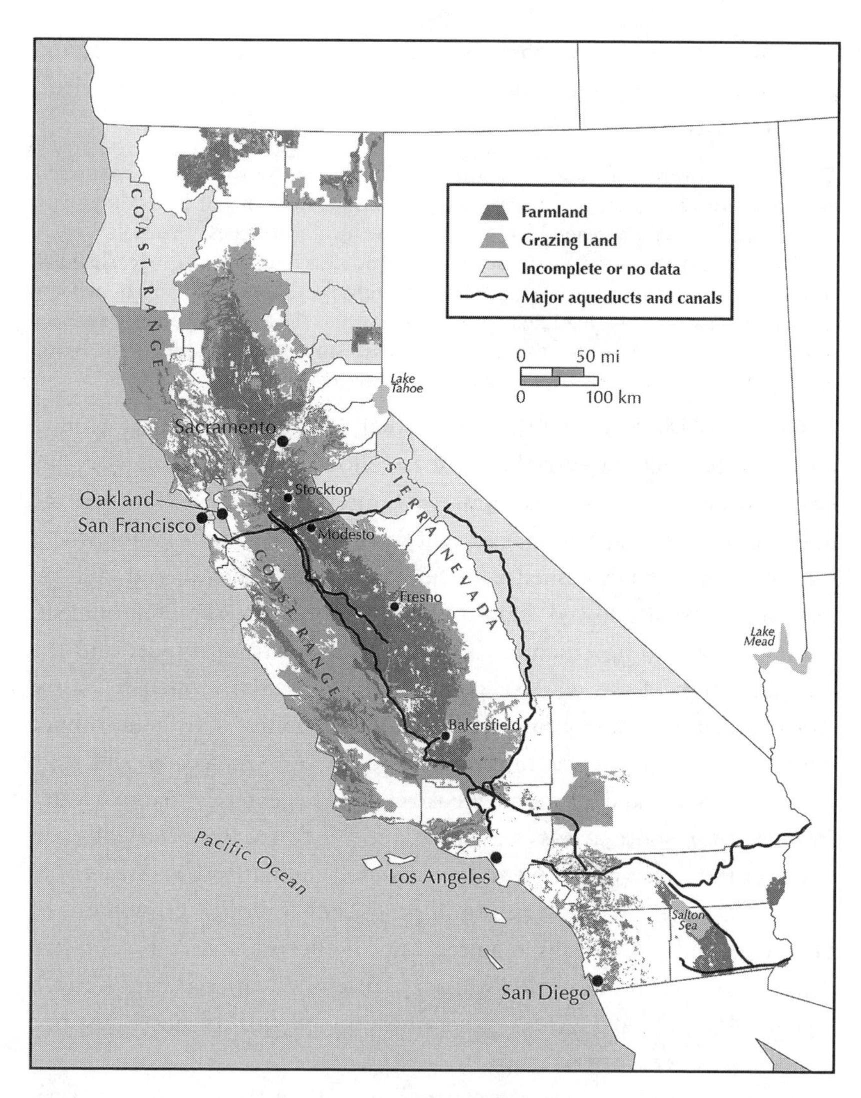

Figure 3.1
California Agricultural Land. California's diverse microclimates have enabled production of an equally diverse array of crops. The state's principal agricultural region, the Central Valley, is formed by the Sierra Nevada and coastal mountains. An extensive irrigation system brings water south and west to the state's agricultural lands. *Source:* California Department of Conservation Farmland Mapping and Monitoring Program.

amazingly huge array of crops. Boosters and land speculators, past and present, have trumpeted California's incredible ecological and agricultural diversity:[4]

> The state accounts for nearly the entire U.S. production of walnuts, almonds, nectarines, olives, dates, figs, pomegranates, and persimmons. It leads the nation in the production of vegetables, including lettuce, tomatoes, broccoli, celery, cauliflower, carrots, lima beans, and spinach, and also of apricots, grapes, lemons, strawberries, plums and prunes, peaches, cantaloupes, avocados, and honeydew melons. It is the nation's leading producer of hay and the second leading producer of cotton. California is also the second ranking state in the production of rice, oranges, tangerines, grapefruit, apples, pears, sweet corn, and asparagus.[5]

However, McWilliams also noted that California's natural bounty only "set the table for everything else to follow," and that human overlay has been decisive. The most impressive intervention has addressed the major flaw in the state's magnificent environment: water is not where the inhabitants have wished it to be. Economic, political, and cultural institutions have allowed Californians to reorganize the environment, building a vast infrastructure to produce and distribute food, and the reorganization of the waterscape has been particularly intense. Native Americans moved water to meet their modest needs.[6] The Spanish built aqueducts to bring water to the missions.[7] However, federal and state governments worked far more extensively and expensively to move water from north to south, across state lines and westward from the Colorado River. Both farms and people followed water, ending the Bay Area's brief role as the state's major agricultural producer.[8] Although growth in San Francisco and its suburbs required one of the more deeply contested water diversions in the nation's history, Yosemite National Park's Hetch Hetchy dam, the Bay Area's generally low soil quality never justified importing water for row crops.[9]

Innovative Californians reorganized more than the waterscape. Agriculture was at the heart of Spanish California, and the mission period inscribed three critical patterns that shape the organization of California's agriculture even today.[10] First, the missions encouraged the accumulation of vast stretches of land by single owners, establishing a pattern that continued under the Spanish, Mexican, and U.S. land grant systems. Although small producers have always played a role in the state's agricultural economy, large holdings have been the political and economic drivers and have accounted for most of the output. Second, Franciscan priests relied on the same exploitation of racially distinct labor that continues today. Franciscans "converted" Indians, who became agricultural laborers, and Mexican and subsequent American landowners enslaved them. Similar race-based practices persisted even as the origins and identities of California growers and workers changed over time. Third, the Franciscans imported livestock and wheat that seeded the ranches of both the Californios and the Americans who succeeded them. They planted many of the European fruits, vegetables, grapes, and olives that became the backbone of the state's diverse horticulture.

Financing "Factories in the Field"

Mission-era landholding, labor, and cropping patterns were in place when the Gold Rush both created demand for food and provided capital for developing the industry. Initially a market in basics like wheat and beef boomed. The expanding population pushed local game species to extinction in the 1850s and the cost of driving meat from Arizona and New Mexico raised prices as much as thirty-fold.[11] Combined with the disposition of Spanish, Indian, and public domain lands this enabled the large-scale, fantastically profitable bonanza ranching that dominated California agriculture after the Gold Rush. The ranches grew to supply not only the miners but also national and international wheat, beef, and mutton markets.

Many institutions and individuals, financiers, railroads, banks, and both successful and disappointed miners, fueled that first rush of capital-intensive agriculture. New mining fortunes as well as East Coast and European banks provided investment capital, abetted by an increasingly sophisticated and more integrated regional financial system.[12] As San Francisco became the first financial center in the western United States, agriculture became one of the most attractive sectors for investment. California farms in 1850 averaged about 4,465 acres compared to a U.S. average of about 202.5 acres.

Although the enormous holdings amassed by California's land barons, including the Henry Miller and Charles Lux empire, did not last and the differential declined, the state's farms remained above the U.S. average until the 1980s. In the 1870s, investment in more intensive and scientific practices began to shift the California industry away from beef and wheat toward labor- and capital-intensive specialty crops such as wine grapes, citrus, vegetables, and nuts.[13] Many large holdings were subdivided and sold, but large holdings did not disappear. Indeed, in many areas and for extensive uses in particular, land consolidation continued. Although the dominance of vast, monopolistic agricultural holdings waned, its large scale remains one of the distinguishing characteristics of conventional California agriculture.[14]

While the banks were critical to the development of large-scale, intensive California agriculture, railroads played an important and unexpectedly varied role. Following the hurried completion of the transcontinental railroad in 1869, improved distribution systems shaped the state's aspirations. California's enormous wheat farms supplied national and international markets for more than thirty years. One-third of the state's grain flowed into Liverpool or Hong Kong at wheat's nineteenth-century peak.[15] Closer to home, rails provided quick access to urban markets around the country for the state's growing bounty of fresh fruits and vegetables. Nonetheless, railroad agents also consciously created their own customers; they actively promoted immigration, lent capital, sold land, and funded irrigation and infrastructure. Railroad promoters invested in large and small operations alike, contributing to the development of communities as well as to massive agricultural enterprises.

The Requirement for Cheap Labor

When land is made expensive by anticipation of profits based on cheap labor, the cost of labor has to remain low if land is to maintain its value or, better yet, appreciate.[16] Certainly both land speculation and the prospects for large-scale production drove California's steadily ascending land prices, but the state's commitment to maintaining cheap agricultural labor was at least as critical.

Prior to statehood, California growers accustomed to the slavery that had sustained mission agriculture flirted with joining the Confederacy. After California was admitted to the Union as a free soil state, the legislature explicitly invited continuing enslavement of Indians. The 1850 Act for the Government and Protection of Indians allowed whites to "obtain control of Indian children" and to purchase at auction Indians who had been "found strolling, loitering where alcohol was sold, begging, or leading a profligate course of life." Clarification of the statute a decade later allowed white owners to "retain the service of Indians until 40 years of age for men and 35 years of age for women." In May 1862, the *Daily Alta California* reported on the common practice of Indian stealing:

> Here, it is well known there are a number of men in this county, who have for years made it their profession to captor [sic] and sell Indians, the price ranging from $30 to $150, according to quality. . . . It is even asserted that there are men engaged in it who do not hesitate, when they find a rancheria well stocked with young Indians, to murder in cold blood all the old ones, in order that they may safely possess themselves of all the offspring. This affords a key to the history of border Indian troubles.[17]

Yet slavery was not optimal for most California growers. Expanding the fruit and vegetable industries depended on cheap *seasonal* labor, and of course it is not profitable to feed or house workers when they are not needed. Hence, California growers have long preferred laborers who are not in a position to demand or require housing or other support during off seasons. When they do need workers, however, growers need a surplus in order to keep wages down.[18] Easily exploited, typically racially distinct immigrant and migrant labor has filled that niche throughout California's history and has enabled specialty crop growers to expand their markets even as land values continued to rise.

Ensuring a sufficient influx of workers to keep wages down while maintaining enough anti-immigrant pressure to prevent them from gaining citizenship or rights has been a difficult balance for California growers, and they have not always succeeded. Aspiring white homesteaders viewed the growers' large landholdings as a barrier to homesteading. Understanding that cheap "coolie labor" was essential to maintaining those holdings, they drove the passage of the Chinese Exclusion Act of 1882, which the growers quite rationally opposed. This first law restricting immigration to the United States prohibited Chinese already in the country from becoming citizens and suspended further immigration from China.[19] Ironically, the Exclusion Act also created a brief period of labor scarcity in which Chinese workers already in the state were able to demand higher wages.[20]

After the Exclusion Act passed, the growers still needed labor. Land reform and sustained upheaval following the restoration of imperial rule in Japan in 1867–1868 encouraged Japanese farmers to migrate and become California's next low-cost labor pool.[21] Unlike the Chinese, the Japanese arrived as families, intending to stay, and until racial prejudice forced otherwise, they did not confine themselves to residential ghettos. That did not sit well with San Franciscans, who established "Oriental Schools" for Japanese children when they rebuilt after the 1906 earthquake.[22]

The Japanese government protested discrimination against its citizens to little effect, but California agricultural labor issues attracted national public attention following the 1913 wheatland riot. When twice as many workers as were needed showed up for jobs picking hops, promised wages were slashed and living conditions at the site became unbearable. Workers organizing to confront the situation were attacked by police and the melee led to four deaths and numerous injuries. Although the riot was blamed on the Industrial Workers of the World, the subsequent outcry prompted California to establish the State Commission on Immigration and Housing to set standards for labor camps and sanitation for fieldworkers. But support for the commission soon withered, and labor issues once again dropped out of sight. Although the plight of agricultural labor in California has caused episodic national scandals, the issue has been easily forgotten or deflected in occasional attacks on immigrants reminiscent of the row over the Chinese Exclusion Act. When McWilliams wrote his classic *Factories in the Field* in the 1930s, farm labor issues in California had already "been lost sight of and rediscovered time and again."[23]

Government Subsidies for "Scientific" Farming

Free land and free soil debates about slavery complicated federal land disposition programs until after the South seceded and left Congress in 1862.[24] With the opposition missing, prodevelopment forces in Congress previously stymied by sectional rivalries quickly adopted three statutes that were critical to the evolution of the conventional model of agriculture. First, the Homestead Act promised 160 acres to any current or prospective citizen, man or woman, who would settle and build a house. A second bill authorized the formation of the Department of Agriculture, whose programs would help the new settlers succeed as farmers. The Morrill Act then granted public domain land to each state to establish "agricultural and mechanical" or "land grant" colleges. In the early twentieth century, the Newlands Act of 1903, opened the way for federal subsidies to irrigation.

Federal support for agriculture was at least nominally intended to encourage smallholders and family farmers, but the programs were easily adapted to meet the needs of California's increasingly powerful large growers. The Homestead Act aimed to support family farming, but the democratic intent generally failed: in California and elsewhere virtually all government land distribution programs were manipulated to facilitate enormous corporate and private holdings.[25] Similarly, the government intended that the land grant colleges and the Department of Agriculture would support smallholders and rural communities. But both acts put the nation on a path to "scientific farming," which was easily turned to the advantage of the largest producers.

George Perkins Marsh's *Man and Nature* (1864) described Americans' love affair with science, but that passion began earlier among agriculturalists, inspired by Justus von Liebig's discoveries about plant nutrition in the 1840s. The German chemist identified nitrogen, potassium, and phosphorus (the NPK in fertilizers) as essential to plant growth. Over the next century, his discoveries underwrote both input-intensive agriculture and a global fertilizer industry. A growing army of land grant college graduates and researchers encouraged not only fertilizer use but also crop specialization and mechanization.[26] The experience and expertise of previous millennia, particularly regarding soil management, was derided as outdated, irrational, or superstitious tradition.[27]

California agribusiness prospered in close relationship to the land grant college system, particularly after Congress added cooperative extension, a state-supported network of specialists who "extended" university research to increase productivity on the nation's farms.[28] Inevitably perhaps, benefits from the enormous federal and state agricultural research and education investments have concentrated on well-organized and powerful agricultural interests, which were best positioned to provide important political support in return for technical and intellectual subsidies.[29]

Similar to the enthusiasm for scientific farming, a commitment to efficiency underwrote efforts to engineer natural resources to suit California agriculture, especially water. The 1903 Newlands Act established the Bureau of Reclamation to relocate, store, and distribute water in the arid lands of the West. As with the Homestead Act, Newlands Act drafters attempted to limit the acreage that could be irrigated with federally "developed" water to family farms no larger than 160 acres. While the water resources did not have a large impact in California until the mid-twentieth century, federal agents had already learned to deflect program benefits to large growers, the bureau's most powerful constituency. Congress finally excised the ineffectual 160-acre limitation from federal statutes in 1982.[30]

Specialty Crop Science and Shifting Scale

Whereas water looms large in the recent history of California agriculture, federal investment in plant breeding was initially more important in shaping patterns of production in the state. During the Victorian era, programs promoted plant breeding and a vision of a "gigantic horticultural garden."[31] An astounding array of fruit, nut, and vegetable cultivars rapidly expanded the mission era suite of specialty crops. As transportation improved, particularly after the first cooled-car shipment of California fruit arrived in New York in 1889, orchards of peaches, pears, and apricots replaced wheat as California's most important agricultural products.[32] That same year, a Southern Pacific agent predicted that California would become "the orchard for the whole world."[33] By 1900, fruit was the state's leading industrial product, and by 1920, California growers produced nearly half of the fruits and nuts consumed in the United States.[34] Although small producers and truck farmers surrounded major cities and remained an important feature of California agriculture at least until the end of World War II,[35] most of California's specialty crop production has taken place on large holdings.[36]

The Cooperatives

Growers' cooperatives solidified growers' political advantages beginning in the early twentieth century. American farmers were "virtually unorganized as a professional group" prior to the Civil War. Of course, they did not need to be: they were the dominant segment of the population and powerful in every state and region.[37] The structure of the national government, a bicameral legislature that apportioned senators by state and not by population, reflected and protected rural interests. However, as the federal government expanded in scope and scale, agriculture organized to support public agencies and programs that would support them.[38]

Changing Expectations about Food Quality: Seasonality and Aesthetics
As refrigerated transport carried California fruits, vegetables, and meats just about anywhere, growers' *marketing* cooperatives—not to be confused with Grange-style organizations of buyers—invested in national advertising and branding of agricultural products that linked California fruit and vegetables to luxury, beauty, and pleasure.[39] Colorful labels distinguished each grower's fruit boxes, even as the fruit itself became an increasingly standardized commodity (figure 3.3).

In the process, the marketing cooperatives redefined the way consumers thought about food: as the refrigerated boxcars improved the appearance and freshness of California products in eastern markets, food quality came to be measured by the way produce looked after shipping.[40] And because California could provide items that could not be grown in winter months in the rest of the country, the notion of seasonality in food consumption began to erode.

As they "took charge of the commodity chain," as Walker styles it, cooperatives became less benign than their name might suggest.[41] Unlike a buyers' cooperative, a cooperative marketing organization must control free riders that might benefit from co-op programs without conforming to quality control and marketing requirements or paying dues. The state soon became intertwined with the marketing cooperatives to such a degree that it blurred distinctions between public and private.[42] For example, in 1917 the state legislature discovered "vital public interests" in marketing the co-ops' products and enacted the growers' own quality requirements into state law. A state-level Enforcement Branch of the U.S. Department of Agriculture soon began backing up the cooperatives' rules, reinforcing their influence.[43] Local farmers no longer served local markets, and co-op rules about shipping and packaging ultimately led to legislation that made direct sales from farmers to consumers illegal.[44]

The marketing cooperatives also needed political help because they were constrained by the 1890 Sherman Anti-Trust Act, which forbid collusion among producers "in restraint of trade." But the peculiar form of cooperative adopted by the growers was specifically designed to restrain trade. Sun-Maid soon ran afoul of federal antitrust law. Federal response established an important new pattern for conventional agribusiness: first the Supreme Court and then Congress, in the 1922 Capper-Volsted Act, exempted agricultural co-ops from antitrust rules.[45] Co-ops

Early Cooperatives: The Rochdale Society and the Grange

Most historians trace cooperatives to a British organization, the Rochdale Society of Equitable Pioneers. The group was founded in 1844 as a buying club intended to lower food prices for members by purchasing in bulk. As the society developed, it promulgated principles for forming a co-op, and the idea spread rapidly.[a]

The cooperative became important in U.S. agriculture when the Grange, formally known as the National Grange of the Order of Patrons of Husbandry, was established in 1867. The organizers sent representatives to England very early in its history to study the Rochdale principles.[b] The Grange was designed to organize farm families to work together toward economic well-being and political power (figure 3.2). Its motto delineates its approach: "In essentials, unity; in nonessentials, liberty; in all things, charity." By the turn of the twentieth century, the Grange had over 1 million members in rural communities across the country. It was a major social and political force in agriculture during the post-1870s build-out of California horticulture, supporting isolated farm families with social events and buying clubs.

During the development of national mail order firms, the Grange made a mutually beneficial buying agreement with the newly established Montgomery Ward: the retailer gave Grange members discounts in return for special access to its members.[c] Although the Grange has lost traction as a political representative for farmers, the cooperative form remains important in the Bay Area food story and some food justice activists have returned to those roots in the past several years, as we shall discuss in chapter 8.

a. Thompson (1994) and Curl (2009).
b. The Grange gained traction during the financial reversals of the 1870s and began lobbying for regulation of railroad and grain storage prices. It also pressed successfully for the 1887 Hatch Act and later for rural mail delivery and, still later, rural electrification. The poster is available through the Library of Congress Web site: http://www.loc.gov/pictures/resource/ppmsca.02956/ (accessed February 2012).
c. Frederick (1997).

Figure 3.2
"Gift for the Grangers," Cincinnati: J. Hale Powers & Co., [1873].

Figure 3.3
As commodities became standardized, growers created distinctive labels that have become a minor art form. Joe Green Estate was located in Cortland, where available packers, wives of German farmers, were white. But the effort to promote the product with reference to that fact reflects the racism that has always characterized California's approach to farm labor. *Source:* From the collection of James A. Dahlberg.

not only adopted political and legal tactics to promote their interests and enforce their rules. They also used violence; the Sun-Maid raisin cooperative, for example, was notorious for its violent efforts to control free riders.[46]

Antitrust exemptions for marketing cooperatives may mark the zenith of growers' power in the conventional system. Key elements of the mission era pattern—large landholdings on which exploited, brown-skinned laborers produced an astounding array of introduced specialty crops—had been institutionalized and capitalized. They had also been backed up by federal support for scientific research and enforced by both federal and state rules regarding cooperatives' packing, shipping, and marketing practices. However, those victories contributed to a crisis of overproduction that devastated farm prices and incomes in the 1920s, and the growers' power within the system declined thereafter.

Depression Era: A California Model for a National Food System

The crises of the Great Depression brought enormous change to the U.S. economy,[47] but the farm sector had been collapsing for a decade before the 1929 stock market crash on Black Friday. Mechanization had put farmers into debt, and resulting overproduction had caused commodity prices and farm income to fall dramatically. Droughts throughout the nation in the early 1930s exacerbated the crisis. The number of farms and farmers had been falling slightly since the 1790s as the nation urbanized; but after a brief uptick following World War I, small farmers were shaken out in droves. (See figure 3.4.)

To stabilize farm prices and incomes, President Herbert Hoover urged passage of the 1929 Agricultural Marketing Act. Rapidly expanding federal participation in the agricultural market added force to a wave of consolidation that shifted the entire nation toward a model of capital-intensive agriculture, based on patterns that had long defined California agriculture. Although Depression-era narratives appropriately emphasized government action, it was government inaction regarding labor that spread that particularly odious part of the model. In addition, the rise and consolidation of a national food distribution sector played a major, arguably decisive, role in the consolidation of a national food system.

Some have suggested that because of California's focus on specialty crops rather than the commodity grains that were the focus of New Deal policy, the state was relatively unaffected by those federal programs.[48] But that is not entirely accurate. We focus on three durable consequences. First, programs to help the hungry were intertwined with and subservient to price supports for agribusiness. Thus, California's poor participate in basically the same food aid and school food programs that distribute surplus foods to the needy elsewhere in the nation. Second, dust bowl migrants flooding the state eroded the position of Mexican laborers and focused grower attention on the threat of unions and Communists. Finally, major changes in retail and distribution turned in part on the car culture and physical infrastructure that developed first in California. A

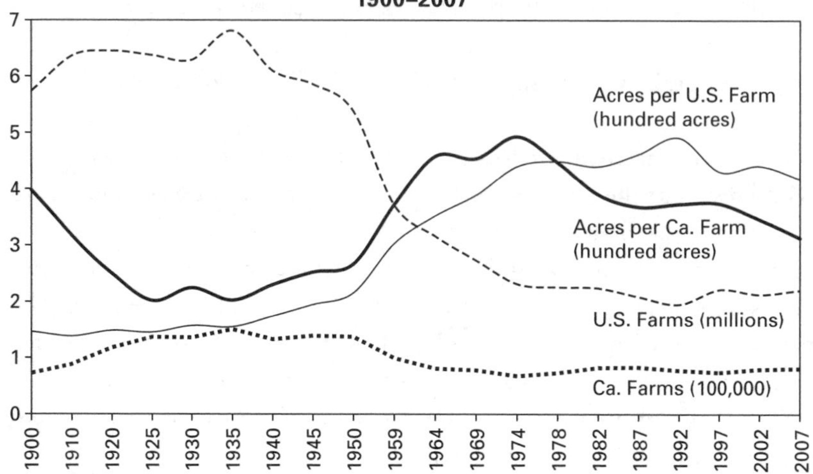

Figure 3.4
Farms and average acres per farm in the United States and California, 1900–2007. The data confirm a dramatic decline in the number of farms in the United States since the mid-1930s—the California model of large landholdings going national—and a corresponding increase in the size of the average U.S. farm at the same time. In California, the swings are less dramatic but a slight uptick in the number of farms since the 1970s and a corresponding but larger decline in the size of farms may signal the impact of the Williamson Act and the back-to-the-landers. Adapted from Hoppe and Banker, 2005, 6; *U.S. Census of Agriculture*, various years. *Note:* Western states include Montana, Idaho, Wyoming, Colorado, New Mexico, Arizona, Utah, Nevada, Washington, Oregon, and California; later, Alaska and Hawaii were added. U.S. Agricultural Census data disagree on the average farm size of western states after 1900. For this figure, 1954 census data are used from 1900 to 1954, and 1969 census data are used from 1959 to 1974. Figures from the 2002 Census are not comparable to data from previous years. See Sumner et al. (2004).

growing consumer movement blew hot and cold about the increasingly dominant retail operations that grew from the margins of agribusiness to take control of the food system.

In spite of the enormous diversity and complexity of the government's market interventions, the basic logic was simple:

> When farmers began to produce too much and prices began to fall, the government would pay farmers to leave some land fallow. When prices threatened to go too high, the payments would end [in theory anyway] and the land would go back into cultivation. . . . The government would also buy excess grain from farmers and store it. In lean years—say, when drought struck—the government would release some of that stored grain, mitigating sudden price hikes.[49]

Food Aid in the Depression

Most of what Americans today encounter as school lunch, food stamps, and federal food and nutrition programs was developed during the financial crises of the 1930s.[50] Those roots have caused continuing problems and controversy in part because the Depression-era programs were specifically not designed or intended to support the needy, but to support agribusiness by distributing surplus commodities that were lowering prices. The poor were allowed to take up the surplus, but no more.[51] The programs reflect Americans' philosophical ambivalence regarding the poor.

Helping the poor has long been regarded as a religious duty.[52] That changed slightly in England following Henry VIII's first divorce and the dismantling of Catholicism. The Elizabethan Poor Laws reallocated the responsibility to local governments. Although who was obligated to do what for whom has shifted over the centuries, the obligations have been fairly consistently structured to be both minimal and socially stigmatizing, and often they benefit those providing the aid. In Elizabethan England, the law distinguished between "sturdy beggars" and "worthy" poor," with the goal of punishing the former while helping the latter, at least a little. Over centuries of war, migration, urbanization, and the rise of industrial capitalism, the need for such aid intensified and altered the details of government programs. But the basic goal of providing minimal support for the worthy poor survived. Utilitarian philosopher Jeremy Bentham summarized justifications for limiting assistance to the barest survival: to discourage laziness and dependence; relief, he said, should be "an object of wholesome horror."[53] British colonists brought those ideas to the United States. Although some jurisdictions provided relatively better care for insane, blind, and deaf people and for children, the fundamental policy goal—preventing destabilizing and unsightly starvation without creating dependence or rewarding the unworthy poor—remained.[54] Where government efforts proved inadequate, families and churches took up some of the shortfall.

When policymakers realized during the Depression that the hungry could absorb agricultural overproduction, the caveat remained: they could take up the excess but no more.[55] The 1933 Commodity Credit Corporation Charter Act authorized the secretary of agriculture to purchase "price-depressing surpluses" and distribute the acquired commodities to the needy. The distribution was explicitly not allowed to compete with "normal market transactions," and it was to be discontinued if it appeared to conflict with the legislation's primary goal of supporting agriculture.[56] The Food Stamp Program, initiated in 1939, was similarly constrained. Only after people on relief had spent "an amount of money representing estimated normal food expenditures" could they use stamps to buy food "determined by the Department to be surplus." The program was not at all designed to ensure that Depression victims gained access to an appropriate mix and amount of food.[57] The relationship between food aid and surplus distribution continues to confound school lunch programs and most related food programs.

Agricultural Labor during the Depression: Government Inaction

But what the government did not do at all has continued as an even more obvious injustice. The dust bowl that decimated the Midwest briefly changed the complexion of California field labor: displaced "Okies" and "Arkies" arrived, providing a new source of cheap labor and tremendous political risk for the growers. With white laborers and families in California's fields and Congress legislating basic rights for workers throughout the industrial economy, the time should have been right to confront the exploitative labor practices at the heart of the state's agriculture.[58]

With growing militancy nationwide, labor leaders recognized the opportunity to organize the state's agricultural sector. Strikes and walkouts by agricultural workers, many of them spontaneous, gained momentum as conditions and wages deteriorated. Organizers came from across the political spectrum, including Communist "agitators" who organized the militant Cannery and Agricultural Workers Industrial Union (CAWIU). In 1933, wages in California fields reached a nadir, and the general atmosphere of union militancy took hold. CAWIU launched a statewide organizing drive that resulted in twenty-four separate strikes and contributed to at least thirteen more throughout the state.

While some of these actions achieved slight wage increases or other concessions, growers' responses were generally brutal: dissenting farmworkers were harassed, beaten, and jailed. The most famous confrontation occurred in Salinas in 1933, when white packing workers temporarily defied growers' traditional, race-based divide-and-conquer strategy. Allying with Filipino field laborers, they demanded union recognition and better conditions. Growers negotiated with the white workers but refused to talk to the Filipinos. Instead, with help from local police, they burned their camps, beat their leaders, and imported scabs. When white workers abandoned their temporary solidarity and looked the other way, they were rewarded with a two-year union contract. Once the growers regained the upper hand, they refused to renew the contract; instead they dismissed protesting workers and brought in more docile replacements. The conflict inspired a new organization, the Associated Farmers of California, to fight unionization and Communist agitation.[59]

In the years that followed, more moderate reformers insisted that if employers offered year-round jobs with decent wages, more white Californians would happily seek employment in the fields. That hypothesis has never been tested.[60] Although Congress was willing to intervene in many aspects of the agricultural economy, it could not reach consensus about labor. The 1930s was a particularly virulent period in U.S. labor history, and California growers echoed leaders in other industries, depicting desperate, largely nonideological workers as a Communist threat. Congress divided, with House members tending to emphasize the Communist elements and favoring restricting workers' rights, while most senators favored including agricultural laborers in the growing suite of rights then being granted to all nonagricultural workers.

Through the Grange, the Farm Bureau, and other organizations, California growers lobbied to exempt agricultural labor from worker protection laws, social security, and similar programs then being legislated. Their argument turned on what is called agricultural exceptionalism:

because of seasonal and weather fluctuations, farmers cannot easily speed up or slow production in response to market factors. Therefore, the growers argued, they required exceptional treatment from governments, including price supports, low interest loans, and exemption from certain taxes and regulations. One of their foremost concerns was labor regulations and immigration law, and they, in alliance with agricultural interests nationwide, demanded special exemptions and policies for agriculture. The growers won many of these concessions, and many of the new protections and rights granted to American workers during the 1930s and 1940s were explicitly denied to those working in the fields.[61]

The Bracero Program

As World War II created labor shortages, California growers were again left with the problem of locating low-wage and easily manipulated seasonal workers. A slight alteration in immigration policy provided the answer. The Roosevelt administration worked with the Mexican government to design the bracero program (from the Spanish *brazo*, or arm), to admit what are now referred to as guest workers. Although Mexican officials tried, as had the Japanese before them, to establish standards to protect the migrants, they fared no better. The program also authorized the U.S. Department of Labor to return the guest workers to their home country after a specified period.[62]

In 1954 the U.S. Immigration and Naturalization Service launched Operation Wetback to return workers without proper documentation (along with numerous U.S.-born Mexican Americans) to Mexico, many of whom returned again and again to labor in California fields. Singer and songwriter Woody Guthrie lamented the nameless victims of a plane crash carrying unneeded workers home. Folk singer and House Un-American Activities Committee target Pete Seeger popularized Guthrie's work, which became a staple in the rising folk and protest repertoire:[63]

Is this the best way we can grow our big orchards?
Is this the best way we can grow our good fruit?
To fall like dry leaves to rot on my topsoil
And be called by no name except "deportees"?[64]

We will encounter alternative approaches to agricultural labor markets in chapters 7 and 8, but the basic pattern remains. California growers have long invested energy and resources not in employees but in politicians who will ensure that national immigration policy meets their exacting requirements. Martin summarizes the strategy with a neat analogy between labor policy and agricultural water management:

> If water is cheap, farmers flood fields with water; if water is expensive, farmers may invest in drip irrigation systems. The analogy to recruitment and retention [of labor] is clear: farmers more often work collectively to flood the labor market with workers, usually by getting border gates opened or left ajar, instead of recruiting and retaining the best farm workers for their operation.[65]

Chain Stores, Grocery Stores, and Self-Service Supermarkets

In spite of the growers' significant and durable Depression-era victory, they were losing traction as a power center in the conventional food system. A revolution in food distribution—transportation, processing, and retailing—that continues today began to develop after the Civil War. Specialty shops and mobile retail—butchers, greengrocers, bakeries, peddlers, and milkmen—were the heart of most U.S. food distribution through the mid-twentieth century (figure 3.5 a, b, c, and d). That changed rapidly after World War II. Railroads and cooperatives catalyzed the initial shift, and the marketing cooperatives soon joined sanitation and public health advocates to create a cleaner, more controlled food distribution system.

San Francisco's arrangement was typical, if accelerated by natural disaster. After the 1906 earthquake, food distribution was limited to the city's wholesale produce market, where it could be effectively inspected and regulated.[66] By the end of World War I, fewer than one hundred wholesalers controlled the supply of agricultural products to stores and restaurants throughout the city. Indeed, the San Francisco district handled the fresh food supply from Santa Clara to Sonoma. Wholesalers distributed California's standardized commodities among small specialty retailers—greengrocers, bakers, and butchers.[67] The chain store groceries that rose to prominence in the 1920s, including the Great American Tea Company, began a revolution that changed all that.

Large and rapidly consolidating grocery store chains took advantage of agricultural overproduction to squeeze producers on price. The introduction of self-service around World War I intensified the shift: mass selling of standardized, prepackaged commodities in retail outlets

a

b

Figure 3.5
Specialists including peddlers, greengrocers, and butchers were central to food distribution through the mid-twentieth century. (a) Chinese vegetable peddler in Oakland, c. 1890. *Source:* Moses Chase Album, Joseph R. Knowland Collection, Oakland Public Library; (b) San Francisco green grocer during World War II (photograph by Dorothea Lange, *War Years North Beach (Italian Sector)*) *Source:* The Dorothea Lange Collection, Oakland Museum of California, City of Oakland, gift of Paul S. Taylor; (c) Richmond, California butcher shop, 1914. *Source:* Richmond Local History Photograph Collection, Richmond Public Library; (d) Farmer selling melons and peaches at a farmers' market in Albany, California, in 1945. *Source:* Courtesy of the Albany Library Historical Collection.

c

d

Figure 3.5
(continued)

From the Great American Tea Company to the A&P

The Great American Tea Company began in New York City just before the Civil War as a purveyor of fine teas in the butcher and baker specialty shop tradition of food distribution. But as the United States developed a national economy, its founder began buying teas directly from China, eliminating diverse middlemen, and organized merchants across the nation into "clubs" to sell the high-quality product at a low price.[d] When the transcontinental railroad opened, the clubs became a string of "grocery stores," and the renamed Great Atlantic and Pacific Tea Company became an early grocery store, selling diverse condiments and household items as well.[e]

Additional outlets made the A&P a chain store that benefited from economies of scale and central management. Significantly, it also offered a limited selection of predictable, uniform products and set prices. Chain stores were also a new sort of buyer, well suited to the large-scale, marketing co-op-based marketing that was emerging in California. The A&P eliminated standard service features of retail selling—credit sales, home deliveries, purchase stamps, and premiums—and offered instead a reliable, pleasant, and anonymous experience for shoppers. Its "cash-and-carry" service was efficient, cheap, and immensely popular.[f] At its peak in 1930, A&P operated at 15,137 sites. At midcentury, it began to lose market share to discount superstores and was purchased by a German retailer in 1979. The company filed for bankruptcy in 2010.

d. W. Walsh (1986).
e. In a parallel universe, similar changes were coming to the meat industry: Gustavus Swift used the railroads to revolutionize meat distribution, bypassing the local butchers who had long dominated the market.
f. W. Walsh (1986).

was designed for high volume and minimal staffing. In 1920, chain stores accounted for about 4 percent of retail food sales in the United States. By 1935, the four top chain operations did 25 percent of all American retail, and consolidation had just begun.[68] Food distribution and retail operations, increasingly national in scope, soon entered the processing and distribution businesses and began offering their own brands of basic commodities, a virtual death knell for milkmen and vegetable peddlers who sold local products along an established route. Butchers were similarly displaced as chain buyers demanded not sides of beef or pork, but rather precut and wrapped portions of meat.[69] (See figure 3.5.)

Supermarkets, Parking Lots, and a New Breed of Consumers

The model that A&P started led to one adopted throughout the country by imitators, including Kroger and Piggly Wiggly. But the center of innovation in food retail soon shifted to California, where automobiles, low-density development, and a rapidly expanding highway system were beginning to reshape the scale, form, and location of food distribution. The immense modern supermarket, occupying huge warehouses surrounded by even bigger parking lots, could attract shoppers from miles around with an immense array of low-cost products. This model was especially suited to the automobile-oriented culture and environment of midcentury California.[70]

Vons of Los Angeles and Safeway of Oakland provide a window into this expanding and ultimately revolutionary breed of retail (see figure 3.6 a and b). Like A&P, Vons started small, with a single warehouse-style operation, where the emphasis was on volume and efficiency, rather than service or décor, in Los Angeles. As it began adding new locations, Vons leased space at the front of each store to independent specialty vendors. But they did not last long as part of the model. And as the industry continued to consolidate, the Oakland-based Safeway chain acquired Vons. Safeway had also started small, when Sam Seelig opened one location in Los Angeles in 1915. As he began adding new sites, Seelig soon lost control of the business and merged with the Skaggs chain of Idaho. A confusing series of acquisitions and mergers designed in the mid-1920s by Merrill Lynch founder Charles Merrill combined numerous chains in the western states under the Safeway name with Skaggs at the helm. Two decades after its founding, Safeway was the second largest chain in the nation, having transformed its acquisitions from grocery stores into a new breed of retail: supermarkets.[71]

The modern self-service supermarket took economies of scale to a new level, emphasizing high volume and low price during the Depression. But by the 1950s, the convenience it offered was equally compelling. Supermarkets were a major driver in the rise of food processing, promoting fast and easily prepared foods for busy shoppers. Supermarkets were also instrumental in promoting frozen foods in the 1950s, which were cheap, durable, and completely impervious to seasonality.[72] Spacious and well lit, supermarkets often featured tall, eye-catching displays of (sanitary) canned goods emphasizing the mantra, "Pile it high, sell it low."[73]

Figure 3.6
California-based Safeway spearheaded the shift to the automobile-oriented supermarket as the basis of food distribution after World War II. (top) A San Francisco storefront c. 1936. (bottom) A supermarket in the Marina district, 1959. *Source:* San Francisco History Center, San Francisco Public Library.

Supermarket chains grew dramatically and gained enormous power in the decades following World War II. By the 1970s, a handful of major regional supermarket chains dominated as both purchasers of food and food retailers.[74] Equally profound, the self-service operations encouraged Americans to regard themselves as consumers. Packaging, branding, and marketing became important components of food choice, and the freedom to consume entered as a critical subtext of a transformed marketplace.[75]

Car Culture and Processed Foods

New Deal efforts to employ the jobless in road-building projects subsidized expansion of motor, as opposed to railroad, transport. The intersection of supermarket and highway was supported by yet another regulatory exemption for agriculture. Farmers had been among the first and most enthusiastic mechanizers. They began to load their own vehicles to transport their own products in the 1920s, and farmers lobbied to protect their own participation in food transport as New Deal legislators considered regulation of the expanding trucking industry.

The 1935 Motor Carrier Act went well beyond that, granting agriculture broad exemptions from trucking regulation and labor law for products in transit.[76] The act helped lift the Teamsters Union into power, but trucking of agricultural products proceeded outside New Deal rules that stabilized wages and costs.[77] Independent food system truckers were not helped by their political success. In spite of a mystique of freedom and free enterprise, their reality was quite grim. Owner-operator truckers took on enormous risk and committed themselves to pursuing unpredictable, low-wage work. The conditions were reinforced by an antiunion ethos embedded in growing consumer advocacy.

Postwar Agricultural Intensification

In the generally prosperous decades following World War II, consolidation, intensification, and integration of the food system proceeded as if on, and then actually on, steroids. Small farms continued to disappear as government programs penalized family operators who did not "get big or get out." Small producers turned to contract farming, and popular demand for processed foods and fast foods turned meat and potatoes into the burger and fries, and then this "all American meal," into a health hazard. Food quality plummeted as producers adopted high-tech farming practices designed to produce larger quantities of commodities, and processors and fast food came to the center of the American diet.

Food and Cars

Yet another burst of road building during the Eisenhower administration intensified Americans' preoccupation with automobiles. Because transportation is the second-largest expense in agricultural production, maintaining low-cost transportation became a core priority in the postwar industrial food system. Agribusiness and rural interests allied with big trucking firms to promote the Highway Trust Fund. Established in 1956, it allocated gas tax receipts to subsidize the National Defense Highway System, a comprehensive interstate highway system.[78] Mobility along highways also allowed firms to relocate operations such as meatpacking and processing away from centers of union power toward less union-friendly operations in the South and West.[79]

The roads also encouraged food retailers to create destinations for motorists and contributed to a new, and ultimately destructive, diet: fast food. Especially popular in California, the auto-oriented food culture underwrote the emergence of the drive-in.[80] Stands selling prepared food had long been ubiquitous but seasonal. Nonetheless, in southern California, where "it felt like summer all year long . . . a whole new industry was born." The idea "combined girls, cars, and late-night food." But Anaheim burger vendors Richard and Maurice McDonald went a step further. They adapted the basic theme of supermarket success to create fast food. They replaced curb service with their Speedee Service System that promised potential franchise owners: "No Carhops—No Waitresses—No Dishwashers—No Bus Boys—The McDonald's System is Self-Service."[81] When Ray Kroc bought out the McDonald brothers, the modern fast food approach to integrated food production, sourcing, processing, and distribution was born.[82] Currently more than half the ground beef consumed in the United States is eaten away from home, mostly in restaurants, and the nation's puny vegetable consumption consists primarily of chips, fries, and catsup.[83]

The Population Bomb and the Green Revolution

Amazingly, Depression-era crop subsidies designed to respond to massive overproduction were soon defended and justified by scarcities. First wartime shortages and then postwar Malthusian horror stories drove continuation of the subsidies. After Pearl Harbor, rationing focused attention on limits: FDR revived World War I price controls and curtailed the sale of meat, butter, and other consumer goods. Victory gardens sprouted across the nation. The scarcities did not last, but their shadow hung over agricultural policy throughout the developed world, driving further intensification and ever more destructive overproduction even while the subsidies continued.

The skewed incentives did not go wholly unnoticed: President Truman's secretary of agriculture Charles F. Brannan proposed a plan that would have ended price supports and allowed the market to set prices for perishable agricultural commodities. Brannan also proposed focusing on the disappearance of small farmers by limiting government aid to producers at or near the average U.S. family farm size.[84] The plan was extensively debated in Congress but soundly defeated. Farm policy rhetoric continued to exploit Americans' longstanding love affair with small family farms, even as government programs ultimately benefited increasingly larger and more integrated food conglomerates.[85]

Paul Ehrlich's 1968 blockbuster, *The Population Bomb,* predicted overpopulation leading to global disaster and provided intellectual and political cover for further intensification and subsidies. The scarcity analysis has been familiar ever since Thomas Malthus argued that unchecked population would rapidly outpace food supplies and lead to mass suffering and starvation. Ehrlich's analysis added the fledgling environmental movement's concerns to the familiar Malthusian mix, emphasizing the fragile earth, and an era of limits, resource degradation, and disease that not even the wealthy could escape. Ehrlich's polemic was broadly discussed and deeply polarizing, attracting criticism for its racist overtones and for stirring up anti-immigration sentiment.[86] However, the dispute provided cover for an avalanche of technological innovation in the food system that soon drowned out basic facts about overproduction.

Without regard for the fact that the farm crises of the previous several decades had been caused by overproduction, the green revolution resulted

Cattle Production after World War II

California's bonanza beef industry was displaced by specialty crops in the nineteenth century, but cattle returned to the region after World War II in massive feeding facilities. One of the nation's first confined animal feeding operations (CAFO) was designed by a former Safeway supermarket executive who sensed opportunity in the demand for backyard steaks and the fast food hamburgers. Located in Bakersfield, California, the operation fed up to 50,000 animals at a time using rigidly controlled grain-based "rations." Beef marbled with streaks of fat soon defined quality for suburbanizing Americans enjoying cookouts. Grain surpluses resulting from a combination of favorable growing conditions, high-yielding new hybrid varieties, and federal subsidies underwrote the shift to CAFOs along with an influx of investment encouraged by federal tax incentives. Tax breaks encouraged thousands of wealthy Americans, including cowboy-actor John Wayne, to join clubs (actually limited partnerships) "that bought livestock, placed them in feedlots, and then sold them for club members."[g]

The CAFOs led to changes in meat processing. Both Armor and Swift abandoned their Chicago stockyards in the 1950s and relocated in rural, usually union-free areas, where they could get tax breaks to build a new generation of packing plants. As transportation fully shifted from rail to truck, huge slaughter plants were built near the new feedlots throughout rural America. By the early 1960s, roughly 40 percent of the cattle slaughtered in the United States were finished in highly automated, western "beef factories."[h] Consolidation continued, and by the twentieth-century's end, four companies controlled more than 80 percent of meat processing.

g. Hamilton (2008).
h. See Greenberg (1970) and Skaggs (1986). Hoy (1978) describes farming to create tax losses as a perennial problem.

from several decades of government and private research into *increasing* crop yields. Scientists developed strains of rice that were particularly responsive to fertilizers.[87] Simultaneously chemical firms discovered ways to produce large amounts of inexpensive fertilizer from fossil fuels. Soon they also figured out that steroids would speed animal growth. And they used antibiotics prophylactically to control the diseases that arose in the crowded concentrated animal feeding operations that were developed for dairy and beef cattle in California and spread rapidly across the nation. Although the environmental and public health consequences of these shifts raised early alarms, the dominant response was initially as euphoric as Americans' embrace of scientific farming a century earlier. The input and capital-intensive approach to agriculture was exported along with still troubling U.S. surpluses to the Third World, where the alleged population bomb was looming.[88]

Input Suppliers

The growing reliance on chemical fertilizers and pesticides added yet another set of politically powerful and increasingly concentrated corporate players to the food system roster: the companies that produced the inputs. Their continuing importance was baked into the technology: as target pests became resistant to one set of chemicals, industry developed another set to maintain yields.[89] Similarly, as pesticides and fertilizers displaced natural soil fertility and killed beneficial insects, soil bacteria, and, increasingly, pollinators, new and improved technological fixes were required.

In the post–World War II United States, important scientific discoveries in low-tech and low-cost biological pest control were overtaken and eventually largely displaced by increasing reliance on chemical pesticides. Technological advances displaced a century and a half of cumulative farmer efforts to adapt Old World crops and practices to a new continent. Farmers' skill sets shifted from understanding natural processes to managing chemical inputs.[90]

Get Big or Get Out: Cheap Food and Food Dumping

In the 1970s, the long-standing federal commitment to overproduction intensified, specifically embracing not small family farmers but increasing scale and intensity, which were glossed as on-farm efficiency.[91] The priorities of Secretary of Agriculture Earl Butz (1971–1976) were shaped by major shifts in the global economy. In 1975, the United States experienced a positive trade balance for the last time. At a point when fundamental, continuing problems of import and export ratios became a major problem for the nation, agriculture attracted attention because it was the only sector in which the United States generated an actual trade surplus. But to remain a dominant player in the world market, Butz insisted that the United States would have to accelerate the trend toward a smaller number of large, least-cost operators.[92] Bad weather in 1972 produced poor harvests all over the world, and Butz seemed briefly like a genius. Policymakers urged farmers to "plant from fencerow to fencerow."[93] Farmers responded enthusiastically, going deep into debt in order to "get big."

The Butz era precipitated enormous environmental destruction that required exempting agribusiness from yet another round of fundamental regulatory reform in the United States: the national movement to limit and internalize the environmental costs of production. During the "environmental decade," Congress made "a nearly unbroken series of decisions to exclude farms and farming from the burdens of federal environmental law," creating "a vast 'anti-law' of farms and the environment." Many states followed suit.[94] For example, although the Clean Water Act (CWA) "prohibits the 'discharge of any pollutant by any person,' . . . this prohibition is riddled with important exemptions for farms." The CWA's definition of *pollutant* includes "agricultural waste discharged into water," but outside the definitions section, the statute puts agricultural discharges "largely beyond regulatory reach."[95] Regulating the pollution generated by CAFOs has, for example, proven near impossible.[96] Because CAFOs are defined legally as farms, not factories, they are exempt from many relevant air and water pollution regulations and actually qualify for government subsidies to deal with agricultural waste.[97]

The economic conditions of the 1970s were particularly difficult for small farmers, many of whom turned to contracting as a strategy for bare survival. They agreed to supply specific manufacturers or retailers, particularly fast food restaurants and still-consolidating grocery stores, with product grown to strict specifications in exchange for a guaranteed price. Most of the risks remained with the farmers, who invested heavily in intensification required by the processors and retailers who increasingly controlled on-farm practices. Integrated buyers would provide the seed, chicks or piglets, strict protocols for managing their growth, and specifications for the finished product. As integrated global corporations offered lower and lower prices to the heavily leveraged suppliers, justice for farmers became a heightened concern for many family farm advocates.[98]

Conclusion

Walker urges us to admire the genius of California and, by extension, American agribusiness, before we evaluate or criticize it.[99] The system is a product of two and a half centuries of enterprise, insight, organizing and political action, consumption, and very hard work. California growers seized the opportunities in railroad construction to create products and marketing mechanisms that allowed them to dominate global food production. They redefined how we grow food, sell food, and indeed what we consider to be food. The growers were soon overpowered, however, by a sequence of equally innovative distributors, marketers, processors, contractors, and input providers who have dominated and redefined the conventional food system.

While admiring the genius, we also note that the same choices have given us increasingly unsafe food, underpaid and undignified jobs for disenfranchised and easily exploited migrants, and environmental devastation. For most of the twentieth century, the image of the independent homesteader and family farmer sustained and concealed political and economic power sufficient to avoid basic regulatory constraints on the exploitation of labor, externalize enormous environmental and public health consequences, devour oil in diverse forms, and create many rather tasteless, nutritionally deficient, foodlike substances.

Clearly there is nothing inevitable about the conventional agricultural system. It is the product of conscious choices, sustained effort, risk taking, and public and private investment. The basic model for this system began to take shape as the earliest Europeans settled in California. It was underwritten by the enormous resources circulating in the state after the Gold Rush and during the capitalist penetration of the U.S. West. When overproduction encouraged by that trajectory brought the system to its knees, the federal government intervened in the market to support prices, effectively enabling another round of consolidation and intensification that made the overproduction problem worse still. The system was further encouraged by the war economy and then by the adaptation of chemical weapons to the production of food.

Although it threatens human and environmental health, most clearly where the two intersect, the conventional food system is not about to collapse under its own weight. It would make things easier for reformers if we could believe that people just began to notice and protest the public health, social, and environmental costs of our food system, but that is not the case. Since late in the nineteenth century, soil scientists, nutritionists, and public health professionals have joined consumers, workers' advocates, and social reformers in demonstrating the system's increasingly negative consequences and offering more promising alternatives. The efforts of those early reformers are treated in the next chapter.

4

The Discontents

For more than half a century before what we describe as an alternative food district began to take shape, people were challenging—in very different ways and venues—the conventional food system as it emerged. Here we address four strands of criticism in roughly chronological order. The first was raised by what we would now call sustainability advocates, although the term was not widely used until the 1980s. An 1890s dust-up in the field of soil science marginalized those interested in soil fertility and inscribed the discussion of sustainable agriculture as less than scientific. Opposition to the reckless, almost random, use of chemicals to control pests, particularly using lead as a pesticide on apples, also became an issue in the early 1900s.[1] Activists in that arena were similarly sidelined and ineffective, their concerns also dismissed as unscientific. When Rachel Carson is celebrated for inspiring Earth Day or sparking the modern environmental movement, or both, it suggests that midcentury sustainability advocates were more successful than their predecessors, but that is not wholly correct. Congress debated a response to pesticides for a dozen years after Carson's book was published, and even then enacted only highly compromised legislation. By that time, innovators in California had developed the nation's first standards for certification of organically grown products—although they never managed to entirely shed their "unscientific" reputation.

Consumer advocates responded to the consolidation of conventional food system on two fronts. Public health issues became an important part of the conventional food system critique almost simultaneously with soil fertility. Consumer interest in health has been pivotal in part because the fields of public health and nutrition have had inconsistent impact on policy, allowing room for engagement by non-professionals. Early advocacy for cleaner, safer food was inspired more by an early-twentieth-century novel as by any scientific analysis. Upton Sinclair's meatpacking best-seller, *The Jungle*, finally crystallized long-standing efforts to regulate food processing. Despite improvements that resulted from the 1906 Pure Food and Drug Act that resulted, the same uphill battle to ensure safe food continues in our own century. Consumers perhaps had more success in addressing the cost of food. Depression-era enthusiasm for alternative retail organizations, including buying clubs and food co-ops across the country, challenged the emerging conventional supermarket system.

Third, episodic labor activism, strikes, and public expressions of concern for farmworkers came to a head in the 1960s. As the bracero program ended, slowing the immigration pipeline, California agriculturalists still insisted on the necessity of low-wage workers.[2] With labor unusually scarce, union organizers gained traction, focusing national attention on the appalling working conditions in American agribusiness. The United Farm Workers' critique of California agriculture achieved a protracted and riveting moment of public concern.

Finally, food security and food miles have a different focus in the early twenty-first century, it is important to note that both were discussed in the1970s. First, Richard Nixon's Department of Defense discussed food security in the context of nuclear war, giving the term a very different spin from today's use of the term. Second, Carter-era oil price shocks precipitated a first brush with food miles and the consequences of oil dependence in the U.S. food system.

The Roots of Sustainable Agriculture

Government Soil Science and Organic Gardening

Sustainability advocacy probably began in the unexpectedly fractious field of soil science. In the late nineteenth century, orthodoxy in soil science asserted that a soil's "parent material" naturally replenished soil. Accordingly, mapping those materials was essential, and soil erosion—wearing away of the parent material—was the critical soil problem. In this view, soil productivity was permanent and easily adjusted with commercial fertilizers. A U.S. Department of Agriculture (USDA) soil fertility expert, F. H. King, was dismissed in 1904 from the Bureau of Soils for asserting the importance of the relationship between the nutrients in soil and crop yields. He used his free time to study agriculture in Japan, Korea, and China and became convinced that "judicious and rational methods of fertilization" were critical to maintaining soil productivity. Commercial fertilizers were generally unavailable where he worked, and farmers used compost and legumes to manage the soil.[3] King's interest in compost fertilization, however, was outside the scientific consensus and ran afoul of the growing fertilizer industry. In the next decades, a British soil scientist, Sir Albert Howard, pursued King's work. Based on research in Indore, India, he also concluded that composting was essential to sustaining agricultural productivity and that chemical fertilizers were not an adequate substitute.[4] But he too was sidelined.

Hugh Hammond Bennett, on the other hand, was a mainstream soil scientist. He worked his way up in the USDA until his 1928 USDA circular, *Soil Erosion, A National Menace,* caught the attention of Congress. Small appropriations supported his research until 1933, when Congress established the Soil Erosion Service with Bennett in charge. He spent the rest of his life fighting soil erosion within the government establishment and is widely regarded as the Father of Soil Conservation. While we do not dispute that soil erosion was and remains a critical issue, soil fertility is also important. Yet, erosion absolutists and chemical fertilizer manufacturers marginalized scientists addressing significant fertility issues.

Soil management and composting became the domain of true believers, who were frequently regarded as crackpots. Some American farmers looked briefly toward compost during World War I, when the nitrogen then used to manufacture chemical fertilizers was diverted to use in weapons. But after the war, farmers refocused on scientific farming and the chemical fertilizers heavily promoted in land grant college and extension programs. Efforts to sustain organic farming were probably not helped when, in 1924, Austrian philosopher Rudolf Steiner introduced the concept of biodynamic farming, arguing that farming involved working with the cosmos and spiritual beings.[5] Biodynamic farming has become an important element of "beyond organic" discussions, but the spiritual beings verbiage did not win converts among many scientists or farmers in the United States.

In 1942, J. I. Rodale started publishing *Organic Farming and Gardening* magazine, with Albert Howard as an associate editor, as another nonscientist, Eve Balfour, conducted experiments that compared the productivity of chemical and organic soil management. Her research, reported in *The Living Soil* (1943), demonstrated unambiguously that traditionally managed soil was more productive than chemically fertilized soil.[6] Nonetheless, the era of the British amateur naturalist had passed, and, as Beatrice Potter could tell you, even at its peak, it rarely empowered female amateurs—Balfour's findings did not deflect the intensification of postwar agriculture.

Rodale's small book, *Pay Dirt: Farming and Gardening with Composts* (1945), became a best-seller in a small circle of believers, and a second generation of Rodales settled in Erasmus, Pennsylvania, to continue experimenting and promoting composting.[7] Much of the early organizing in the sustainable agriculture community that we discuss in chapter 6 involved rediscovering Howard, Balfour, and the rest, sharing, reskilling, and demonstrating successful tools and approaches for specific crops in specific places. Although their experimenting was not regarded as science, their outsider status did give the alternative growers a sense of themselves as a community and a movement that continues to flavor alternative food advocacy.

The Early Organic Movement

Rachel Carson's 1962 *Silent Spring* added pesticides to the sustainability discussion. Although agriculturalists had adapted heavy metals and arsenicals to pest control use in the late 1800s (the most familiar perhaps being Paris Green, a copper acetoarsenite also used as a dye), reliance on chemical pesticides intensified significantly after World War II. Synthetic chemicals developed as weapons were redeployed against agricultural pests. Carson's book aggregated studies that had been accumulating in the scientific literature and brought them to the attention of a far broader audience. Her work, originally serialized in the *New Yorker*, documented the unintended consequences of releasing vast amounts of poisons into the environment.

The risks from relying on chemical pesticides had been sufficiently obvious that state-supported research on alternatives began in the early 1950s. A Division of Biological Control was established at the University of California (UC), Berkeley to develop pest management strategies that did not require synthetic chemicals. It was among the earliest state-supported centers pursuing ecological agricultural practices. Although it was more difficult to dismiss its findings as not real science, the response was not surprising: the industry worked, and still does, to vilify Carson, and growers defended their practices in spite of compelling evidence of major health impacts.[8]

Farther south and about a decade later, in 1967, at the newly opened UC campus in Santa Cruz, British master gardener Alan Chadwick started the Student Garden Project. The Santa Cruz area had enjoyed a reputation as a center of utopian community experimentation in the early 1900s, and Chadwick's project prospered there. In 1974, the UCSC "Farm," a larger and more formal garden, was established. From its inception, the Santa Cruz program has trained leaders, advocates, and practitioners in organic and sustainable farming, and they soon constituted a supportive community for the growing sustainable agriculture movement in California.[9] Again, the Santa Cruz farm was regarded as neither science nor agriculture—those activities took place at UC Davis and UC Berkeley—but it was nevertheless critical in promulgating organic agriculture as an organized, self-conscious movement in the United States.

More broadly, one segment of the disaffected dropouts of the 1960s, the back-to-the-landers, picked up on the earlier work by King, Howard, Rodale, and Balfour on soil productivity and began their own experiments.[10] In 1974, a small number of growers in the Santa Cruz area established a support group for organic farmers to build skills, find each other, and build a community. Populated in significant measure by those involved with Alan Chadwick's UC Santa Cruz garden project, they published the first edition of *The California Certified Organic Farmer* while collectively recovering, refining, and sharing a wealth of information about organic practices. The old soil fertility crowd enjoyed a renaissance as alternative producers struggled to demonstrate the utility of less input-dependent, chemically dominated processes. The alternative growers began distributing their product by driving around to the alternative markets and co-ops that we discuss in the next chapter.

The alternative producers began very early to develop institutions that reflected and shared their experience. On this initial foundation of common practice and reskilling, a serious generation of alternative growers developed one of the first organizations in the nation with a set of conventions for certifying farmers as organic. Long before the state, and later the U.S. Congress, adopted working definitions and practices that would guarantee consumers that food labeled as organic was indeed that, the rapidly expanding California Certified Organic Farmers (CCOF) had developed criteria that defined organic and began educating farmers and consumers about the benefits of organic farming. In 1979, CCOF led a successful campaign for legislation that adopted its general approach into state law. The California Organic Food Act defined an agricultural commodity as organic if it was "produced, harvested, distributed, stored, processed, and packaged without the application of synthetically compounded fertilizers, pesticides, or growth regulators."[11]

Consumer Food Advocacy: Health and Cost

Consumers first organized effectively in an effort to protect the public from toxic food. Public health and nutrition concerns have been translated into action in significant measure by a century of consumers' rights advocates, in part because the relevant scientists—public health and nutrition practitioners—had some of the same problems as King and Howard: they were not always recognized as scientists and have frequently been marginalized by the mainstream field of medicine. Professionals from both fields have played an important role in the critique of conventional food, but the distinctions between them and medical practitioners have given doctors the upper hand.

Public health emphasizes disease prevention for populations. Medicine approaches illness on a case-by-case basis and attends to individuals who are experiencing specific symptoms and can afford to pay for help. The public health profession enjoyed a golden era of its own in the late nineteenth and early twentieth centuries in a stunning series of victories over major diseases: smallpox, diphtheria, cholera, tuberculosis, yellow fever, and malaria. However, the profession lost stature following the flu pandemic that raged after World War I, killing somewhere between 20 and 40 million people across the globe. That fall from grace opened the door to individualized health care, and the medical profession rapidly gained ascendancy.[12]

That power shift, and the medical profession's emphasis on curing individual patients, has encouraged a related idea: good health is an individual responsibility. That notion gained momentum in late-twentieth-century market-driven politics along with an apparent corollary: ill health reflects bad personal choices. Those ideas have focused national debate on care and health insurance for individuals rather than on disease prevention and public health maintenance. They have also constrained collective action against diseases related to environmental degradation, such as airborne and waterborne toxins, and, indeed, the public's ability to see many afflictions as environmentally related.[13]

Sadly, the field of nutrition is no better positioned than the public health profession. Human nutrition began weak: it was initially seen as women's work—an extension of the mother's role in managing the home. As a profession, it developed in home economics departments, which were created primarily as a repository for female scientists unwanted elsewhere.[14] Until recently, nutrition was regarded as irrelevant to medical care and was not part of medical education.[15] The relative weakness of both public health and the nutrition field and the long indifference of the medical profession to diet-related illness has left the easily manipulated USDA in charge of nutritional standards and allowed every stripe of food processor and vendor to discredit public health professional and nutritionists' efforts to advocate on behalf of basic human nutrient requirements.[16]

Consumer Advocacy and Toxins in Food

Passage of the Pure Food and Drug Act of 1906 is discussed, with some justification, as a victory for consumer advocates. Their effort began in response to the most obvious alcohol-based potions promoted at "medicine shows." Although snake oil and its salesmen are now glossed as colorful Americana, numerous fatalities were on the congressional agenda when Sinclair tied that public concern to the safety of the industrializing food system.[17] Consumers' rights groups embrace the 1906 statute as their first success.[18]

Toxins in food nevertheless continue to be a major public concern. Two Depression-era books took up where Sinclair left off. The first, Kallet and Schlink's widely read 1933 book, *100,000 Guinea Pigs: Dangers in Everyday Foods, Drugs, and Cosmetics,* argued that manufacturers were conducting population-level experiments, marketing products without any understanding of their health consequences.[19] The authors founded the Consumers Union and began agitating for major reforms in the 1906 statute. Their effort led to another compromise: the 1938 amendments to the Pure Food and Drug Act established enforceable food standards, set tolerances for some poisonous ingredients, and authorized factory inspections and injunctions.[20] But the law also created wiggle room to distinguish between "adulterants," made illegal in the 1906 act, and "additives," which are controlled but legal. Many issues involving toxic and foodlike substances arise in the crevasse between an adulterant and an additive.[21] Moreover, the 1906 act focuses on the adulterations that poison consumers. Those may be difficult to trace in our global system, but tying the food to the death is relatively straightforward. Dealing with conventional food that becomes a health hazard as it dominates a diet over many years is more challenging.

Diet-Related Illness

Weston Price, a researcher and a dentist, probed those difficult questions about the consequences of consuming a diet of highly processed foods. His 1939 comparisons of primitive traditional and Western diets led him to argue that foods "as Nature makes them have much more nutritional value than after they are processed." His book was positively reviewed when it appeared, although also criticized for belaboring the obvious. It has since been through many editions and continues to influence advocates today of "nutrient-dense" foods.[22] It did not, however, inspire political action similar to the post-*Jungle* era, and it has taken far longer for diet-related illness to be recognized as a public health issue, as important as more transparently toxic food.

The Cost of Food

Consumers were not enthusiastic about New Deal efforts to curb overproduction by raising food prices. Subsets of consumers responded differently, and the results were frequently incompatible. Some supported the chain stores, believing that they offered more affordable food than the surviving local retailers. Others sought to regulate the chains, while still others tried to get outside the rapidly integrating retail behemoths with buying clubs and cooperative markets. A fading cadre of independent retailers presented chains as an "evil, monopolistic conspiracy that was responsible for impoverishing local communities, depressing farm prices, cheating consumers, [and] destroying small businesses."[23] Louisiana populist Huey P. Long agreed, declaring that he "would rather have thieves and gangsters than chain stores in Louisiana."[24]

Although most consumers tended to support the chains, others pledged to "put the High Cost of Living on trial" at congressional hearings.[25] In response, retailers organized themselves into the National Association of Food Chains in 1933.[26] Hearings before the Senate Judiciary Committee in 1936 provide an early scorecard: small grocers' pleas for controls on "food monopolies" were countered by "housewives" who praised the low prices, quality food, and "sanitary and hygienic" practices of the centralized chains.[27] A proposed federal tax on chains failed, giving both the retailers and some consumers groups early victories.

The Early Co-ops

Nevertheless, other consumers sought a lower-cost alternative to the chains. Cooperative buying clubs, the kind that were started by the Grange during the 1880s, were not common in the early twentieth century, but they played an important role in addressing the food needs of pockets of poverty in the United States and Canada.[28] When *Jungle* author Upton Sinclair ran for governor of California in 1934, his EPIC platform (End Poverty in California) was ignored in most news media but circulated widely in hundreds of EPIC clubs, which frequently morphed into buying clubs.[29] Sinclair lost a surprisingly close election, but the time had become ripe for cooperatives.

The basic co-op format has two characteristic elements: member or worker ownership and member education. Members invest some capital and share benefits, and the education can be focused on the products the co-op sells, or provide general survival skills, or include almost anything from literature to current events.[30] The Berkeley Co-op in California soon became the largest and most successful co-op in the nation. The operation wove together numerous buying clubs in and around Oakland and Berkeley, and many of them had roots in Sinclair's EPIC campaign.[31]

The Co-ops and Food Quality

Although co-ops were typically local organizations, as federations they could achieve some of the economies of scale that the chains enjoyed, providing storage facilities, transportation, and connections to small farmers' cooperatives and canneries.[32] Also like the larger retailers, the co-op organizations could develop their own brands and could get directly involved in milk processing, lumber mills, and the sale of electrical appliances.

In spite of the scale efficiencies provided by organizing nationally, the co-ops' priorities were quite different from those of the large chain stores. The co-ops were, for example, deeply involved in testing for adulterated products. As commodity retailers watered down vinegar, bleach, and maple syrup and reduced the count, but not the price, of packages of paper napkins, the co-ops tested and tried to stock full-strength, full-count product.[33] The Berkeley Co-op moved in 1942 to ensure that it sold only lean ground chuck as "hamburger," whereas in conventional stores, "hamburger" might mean "anything that would go through the meat grinder."[34] Similarly, the Berkeley Co-op was a strong participant in consumer advocacy. It issued statements "to regulatory agencies and legislatures" regarding labeling of orange juice and orange juice products, fish protein, and frozen raw breaded shrimp, and invested significantly in policy "efforts to enhance food safety and labeling standards."[35]

Alemany Farmers' Market

By the end of the 1930s, buying co-ops were becoming familiar, but San Francisco was soon distinguished by a farmers' market. During World War II, newspapers headlined high food prices and food rationing in tandem with farmers dumping a bumper pear crop that the grocery chains could not absorb. The San Francisco Board of Supervisors adopted an ordinance in 1943 that created exemptions from the web of federal and state packing and shipping regulations. That shift allowed farmers to sell directly to consumers and allowed a farmers' market to open. The board responded to protests from chain stores and wholesalers with assurances that the market was only a temporary wartime emergency measure.

But the market began with a bang and contined to gather momentum. A large crowd greeted the first trucks of Marin and Sonoma county produce in a mortuary parking lot. In the next three years, the temporary market sold $6 million worth of fruit and vegetables, often at half the price of similar items elsewhere in the city. A public referendum was required to make the market permanent. When it passed in 1945 by a six-to-one margin, the city appropriated money to provide a permanent location. Farmers from San Diego to Siskiyou County on the Oregon border were soon selling at the San Francisco market. The market cultivated goodwill between San Francisco and its hinterland at a time when state politics often pitted urban and rural interests against one another.

For thirty years, the Alemany Market looked like a major exception to the packing and shipping regulations that prevented direct sales from farmer to consumer. In the 1950s and 1960s, industrial representatives worked in Sacramento to tighten restrictions on the sale of agricultural products. Nevertheless, San Francisco's city government continued to skirt state packing and labeling requirements with city exemptions. For many years, Alemany was the only farmers' market venture operating in northern California.

Supermarkets in the 1960s

That exception notwithstanding, by the 1950s, supermarkets were well established in urban and suburban areas alike. They had moved rapidly into automobile-oriented low-density neighborhoods that were the main product of the post–World War II housing and road building booms and federal policies encouraging suburbanization.[36] Their convenience and low prices had driven smaller food retailers out of business in many places.[37] However, as they began to compete among themselves in the 1960s, the chains turned to gimmicks, including an "avalanche of trading stamps."[38] When the price of food rose sharply, shoppers began to regard the ploys as wasteful and aggravating. In October 1966, the Denver-based Housewives for Lower Food Prices charged the chains with collusive price setting and denounced their "frills, stamps and gimmicks."[39]

As the boycotts spread, industry dismissed the growing movement as a "temporary spasm" by misguided housewives. But the protests gained momentum.[40] Consumer advocacy perhaps peaked in the 1960s.[41] In 1963 President Lyndon Johnson created a special assistant for consumer affairs in the White House and appointed labor organizer Esther Peterson to fill the position. For two decades, Peterson fought to inform consumer choices. She advocated for truth in advertising, standardization in packing and pricing, and consumer rights across industries. In the food context, information that we now regard as standard—"sell before" dates on perishable items, nutritional content labels, and unit-pricing information on grocery store labels—all stem from her time in office.[42] Pressure from food retail interests ultimately forced her to resign.[43] Despite industry efforts to blame these innovations for the rising cost of food, congressional hearings produced considerable data supporting collusive pricing allegations against the chain stores.[44]

As industry supporters scrambled to counter these charges, a different crisis was forming: a nationwide exodus of supermarkets from urban areas was gaining momentum. Another round of price increases in 1968 resulted in more pickets and boycotts and raised policy questions about food imports and agricultural price supports, as well as supermarket practices.[45] Just a hint of early twenty-first-century food justice emerged at the time. Volunteer social workers collected evidence of differential pricing for poor areas and demonstrated that supermarkets raised prices the day that welfare checks were issued, and they publicized this data on TV.[46] But the expansive, consolidated chain store that dominated food retail in the 1960s, 1970s, and 1980s never fit comfortably into the physical space of crowded cities. Public policies encouraging suburbanization gave the retail industry no incentives to overcome the obstacles to urban investment.[47] Supermarket redlining coincided with white flight into the suburbs and a dramatic increase in segregation in the United States. The food deserts that were left behind are treated in chapter 8.

The Golden Era for Farmworkers

Workers' critique of the conventional food model was quite different from that of the housewives, and occasionally it was in conflict with a priority on low prices.[48] Although Cesar Chavez has dominated public memory, Filipino activists affiliated with the AFL-CIO Agricultural Workers Organizing Committee (AWOC) began the strike against grape growers in the area around Delano, California, in 1966. Cesar Chavez's nascent, primarily Hispanic, group, the National Farm Workers Association (NFWA), joined the action, which lasted five years. When the two organizations merged, the resulting United Farm Workers (UFW) put labor rights and fieldworkers sharply onto the national agenda and focused attention on pesticide health risks to workers.

In 1975, the California Agricultural Labor Relations Act granted fieldworkers in the state rights to bargain collectively and protections for doing so, but the promise of labor rights ultimately went unfulfilled because growers were able to both mechanize and adjust the flow of migrants to undercut the union's position.[49] Nevertheless, for a protracted period, agricultural workers' rights were squarely and effectively on the agenda. UFW contract negotiators were successful in having the pesticide DDT banned in UFW contracts before the federal government did so in 1972.[50] UFW grape boycotts drew consumer attention to justice issues that remained at the core of the U.S. food system and taught an early generation of politicized food buyers to vote with their forks. In co-ops throughout the district, members would give the checker the UFW's number rather than their own, donating their dividend to the union. (See figure 4.1.)

California Agrarian Action Project

Many groups that present today as farm and sustainability advocates organized initially around a combined interest in labor justice and sustainability. For example, the California Agrarian Action Project (CAAP) organized at UC Davis and adapted the litigation model of the civil rights movement in behalf of agricultural laborers and family farmers. CAAP filed suit against the University of California, charging that the university research program violated the 1862 Morrill Act.[51] The group aimed to fight farm mechanization—just one of the conventional growers' response to initial UFW success—to protect both agricultural workers and family farms. The straddle between workers and growers soon proved challenging, and legal battles raised fundamental issues about agribusiness in California. The mechanical tomato harvester, introduced to address the labor shortage that resulted from the end of the bracero program, required a new variety of large, uniform tomatoes that were hard enough to endure a pummeling by the machine. This technology not only reduced labor requirements, but it also generated radical change in the water requirements, landscape, and scale of tomato farming. It became a symbol of both growers' mistreatment of workers and the declining quality of

Figure 4.1
Pete Velasco, Larry Itliong, and Cesar Chavez (left to right) at a United Farm Workers meeting in Delano California, c. 1968. Overcoming racial and ethnic divides, many of them deliberately deepened by growers seeking to undermine labor solidarity, was one of the greatest challenges and achievements of the UFW. *Source:* Walter P. Reuther Library, Wayne State University.

conventional foods. Author and Texas Agriculture Commissioner Jim Hightower's iconic volume, *Hard Tomatoes, Hard Times* (1972) summarizes the combined concerns. The CAAP case languished for a decade before a district court directed the university to develop plans to ensure that small farmers benefited from campus research. The decision was overturned in 1989, but the issues it raised did not disappear from California's agricultural politics, and the university's ultimate but partial victory did not end the challenge to its research priorities.[52]

California Rural Legal Assistance

The California Agrarian Action Project (CAAP) was represented in that fight by California Rural Legal Assistance (CRLA), one of the first and most durable legal programs supported by the War on Poverty. CRLA was founded in 1966 and was the first War on Poverty legal program to serve rural areas.[53] It played a major role during the period of UFW ascendancy and continues to be a major force in civil rights litigation in California.

Not all of CRLA's work relates to farmworkers or immigrants. It was a major force in the beginning of what became and endured as a Calfornia Hispanic civil rights movement to accompany the one gaining traction in the 1960s and 1970s in the South. CRLA has litigated to procure for U.S. citizens of Hispanic background the right to vote, fair wages and working conditions, and an education.[54] CRLA brought one particularly startling case that challenged the common practice of putting Spanish-speaking children into classes for children with mental retardation based on the results of IQ tests given in English.[55]

In the context of agribusiness, the greatest part of its contribution is that CRLA's presence raised the possibility that the law might be enforced. "[A]gribusiness and its spokesmen in public office were accustomed to ignoring [existing laws] with impunity." CRLA's string of legal victories is impressive: it won farmworker cases enforcing state law requiring employers to provide toilets and drinking water for farmworkers and another that restrained federal Department of Labor officials from deporting braceros in violation of their own regulations.

The CRLA's efforts on behalf of farmworkers were enhanced by passage of the California Agricultural Labor Relations Act in 1975, which established basic protections for fieldworkers and established legal guidelines and procedures for labor organizers and unions. It also created a state Agricultural Labor Relations Board to mediate and oversee labor relations. The agency had a proworker orientation under Governer Jerry Brown, but declined in its efficacy with the appointment of progrower board members under Republican Governor George Deukmejian.[56] Still, the new law resulted in some significant changes. Worker advocates used it to challenge the short-handled hoe as an unfair labor practice in court. Growers argued that the twenty-four inch implement resulted in more accurate, efficient fieldwork, but it was painful and debilitating to use because it forced workers to bend or stoop. The tool had been the cause of injuries and the focus of protests since the 1920s, and it was widely despised: in New Jersey agriculture, it was called *el brazo del Diablo,* the devil's arm.[57] The court seriously restricted, but did not eliminate, use of the hoe, but the CRLA continued to agitate and its use was finally prohibited in California by administrative order in 1975. Earl Butz complained that in California, Mexican farmworkers were "no longer allowed" to use their traditional hoe. This is not, Butz opined, "because the workers or the farmers want to change: but apparently because city people, driving by, feel more comfortable watching the workers use the kind of hoes that look good through car windows."[58] In spite of loopholes that allowed continued hand weeding, the result was a significant improvement in working conditions.

First Brushes with Food Security, Food Miles, and Peak Oil

Although the term *food security* has taken on a significantly different meaning in the twenty-first century, the term was first used in relation to national defense. In 1969, the secretary of the army recognized food provisioning as a military concern and commissioned a study that explored the concept of "food miles," raising implicit questions about an increasingly consolidated and centralized agricultural production system that required long supply lines as well as a far-flung distribution system. Stanford-based analysts estimated that the average U.S. food item traveled more than 1,300 miles between the place of production and consumption. Under two different nuclear attack scenarios, the report concluded that "transportation appears to be increasingly essential to production and may constitute a vulnerability."[59]

The U.S. Army's concerns with nuclear survival were soon overshadowed by Carter era oil shocks: rising fuel prices exposed the vulnerabilities created by an oil-based farm economy. Steinhart and Steinhart (1974) argued in the prominent journal *Science* that increased energy inputs into the food system had resulted, perhaps unexpectedly, in dramatic decreases in overall efficiency. Their data suggested that the potential for the "application of energy" to increase agricultural production in the United States had already been maximized and that further intensification would not produce any more significant gains in productivity (figure 4.2).

More profound, they identified deep and negative consequences of conventional U.S. agriculture, demonstrating that increased yields were accompanied by losses in nutritional and other values. "Faith in technol-

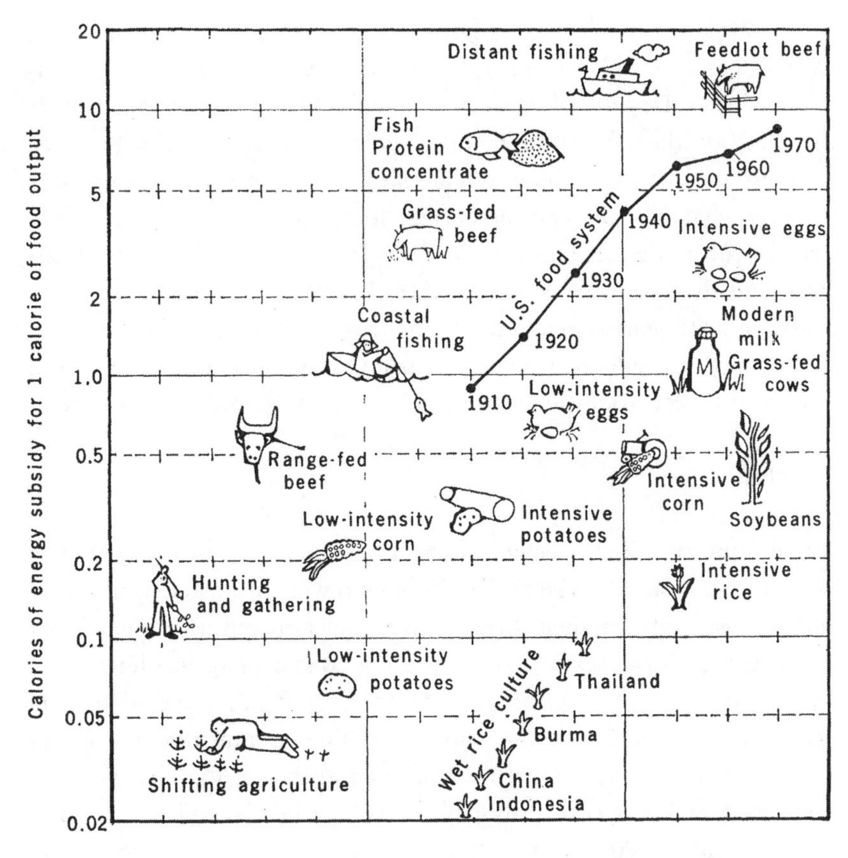

Figure 4.2
The energy subsidies for various food crops, the energy history of the U.S. food system 1910–1970, with comparisons for low- and high-intensity production of selected products. From Steinhart and Steinhart (1974), 312. *Source:* American Association for the Advancement of Science.

ogy and research has at times blinded us to the basic limitations of the plant and animal material with which we work," they concluded. "The cost of the increased yield has been the loss of desirable characteristics—hardiness, resistance to disease and adverse weather, and the like. The farther we get from the characteristics of the original plant and animals . . . the more care and energy is required."[60] Moreover, they noted, trends toward more intensive processing, storage, and distribution could exacerbate the problem of energy use.

A major and much praised book-length treatment of similar issues, more focused on oil dependence, appeared five years later. Now in its third edition, David and Marcia Pimentel's *Food, Energy, and Society* (1979, 2009) became and remains a widely read classic text. But even with growing concern about global warming, comprehensive steps that address what we now call the carbon footprint of modern agriculture are difficult to detect. This may be one place in which the nonscientists have had the upper hand. Lappé's (1971) *Diet for a Small Planet* was ahead of its time in assessing the energy costs of particular food choices, and her work produced results, channeling some of the concerns about energy into vegetarianism, at the time frequently regarded as quackery.

Conclusion

Three important points emerge from about a century of opposition to conventional food. First, many of those who were dismissed as unorthodox, or just "nuts," in their own time have been proven right. That ought to encourage those seeking to change the food system against huge odds. But the deck is still stacked against change. Second, those advocating alternatives, particularly those who are envisioning transformative rather than incremental change, need to understand that well-informed, well-connected, and frequently powerful critics have been working for reform since the late 1890s. Despite frequently enormous public outcry and demand for change, their efforts have not stopped the juggernaut of increasingly unhealthy, unsustainable, and unjust food. Indeed it is not clear when and whether it is helpful to reformers to claim a victory. Perhaps we would be better off, for example, if we had admitted in 1906 that what is known as pure food and drug legislation was weak and full of loopholes and that it got worse, not better, when the bill was amended in the 1930s.

Finally, the reformers in the district whom we describe in chapters 5 to 8 did not start from scratch. Learning and innovation in the Bay Area elaborates on decades of effort and learning elsewhere. Back-to-the-landers pulled Rodale's, Steiner's, and Balfour's books off the shelves and gradually adapted their insights to the crops in the region. They demonstrated commercial possibilities for more sustainable farming long before the term came into common parlance. In so doing, they followed paths of innovation not so dissimilar from the specialty growers that had transformed agriculture in California less than a century before. The back-to-the-landers were inclined to build alternatives to the system rather than boycott the grocery retail behemoth. Nevertheless, the first wave of innovation in our district was more the province of political and economic elites and focused on land protection. The story of our alternative food district begins with the land.

II

Waves of Innovation in the Bay Area Alternative Food Community

If we are correct in asserting that an alternative food district is operating in the Bay Area, we should be able to locate activists, entrepreneurs, consumers, and producers who are developing new relationships, institutions, and innovations that have resulted in higher expectations for food quality. Part II explores that hypothesis. In the next four chapters, we discuss innovations that raise the bar regarding the requirements for environmental sustainability, healthy food, and justice in district food.

Our narrative shows that important steps to protect the land required for the food district began early in the twentieth century. The common enterprise that defines our district began to take shape in the 1980s when chefs and sustainable growers established relationships that allowed them to distribute the alternative growers' superior product. We will also see that the growers' sustainability priorities overlapped fairly easily and early with chefs' French-influenced ideas of excellent food. By the mid-1990s, new producers, increasingly attentive consumers, and activists were collaborating on an even broader food quality agenda, emphasizing health, personal and public, in what we call a maturing district.

Addressing equity issues in the district has been more complicated and has taken longer. Ideas about affordable food and fairness for food and fieldworkers that were important in the Bay Area as early as the 1920s significantly shaped food institutions and conventions. However, those earlier priorities are different from ideas about food justice that emerged in the late 1990s as a new kind of public health issue became obvious. The health problem was not adulterants or toxins, but obesity and diabetes that have arisen from diets made up of conventional foods available in every school and home. Poor communities of color disproportionately experience those health issues. Starting in the late 1990s, community organizers began to define their own justice agenda and pushed their still-evolving understandings of food justice to the center of the district's shared agenda. But theirs has not been the last word. Growing attention to crafters and high-quality prepared foods in the district has been accompanied by retail and food service workers' projecting their own view of injustice onto expectations about just food.

The evolution of alternative food has been an iterative, cumulative process. Today's alternative was not specifically planned or envisioned by any individual or institution, and very few have addressed the issue comprehensively. Proximity has allowed the constant interaction and learning that has led to shared perspectives and enhanced expectations about alternative food. We argue that justice has now moved to the center of that conversation. The intersection of environment, food, and justice is creating opportunities for continuing innovation in the Bay Area alternative food community.

5

A Civic Culture of Parks, Planning, and Land Protection

Although many innovative industrial regions do not require a land base, a food district probably does. The Bay Area's experience in land protection, dating back more than a hundred years, anchors the district both symbolically and more concretely in possibilities for creative food production and processing. Our food story therefore begins in the coastal pastures of Marin County across the Golden Gate Bridge to the north of San Francisco. Long before there was an area designated as the "Bay Area," there was "Marvelous Marin."[1] At statehood in 1850, both San Francisco and Sonoma counties argued for control over the little spot of land between them. While neither was successful, that dispute suggests, correctly, that Marin was divided: the suburban or southeastern half of the county was connected closely to San Francisco's urban elites, while the county's agricultural western half was integrated into Sonoma's agricultural community to the north.

Marin's elite drew their economic strength and political power from "the city," to which many commuted, but they also took great interest in the rural parts of the county, which they considered more or less an extension of their backyards. Although they were not professionally involved in the agriculture that dominated the western part of the county, suburban Marin residents initiated what became a statewide culture of protecting land and natural resources. Once they had gained the experience of several decades of advocating for parks in the county, civic leaders were well positioned to tie post–World War II regional planning, growth control, and environmentalism to new methods of conserving county rural lands. A subsequent generation of back-to-the-landers built on that heritage, demonstrating a new path for agriculture that could sustain working farms and ranches.

Early Civic Environmentalists

The Marin civic leaders' first approach to land protection combined enthusiasm for national parks with a tendency to buy whatever land was threatened.[2] Marin County Congressman William Kent, a leader in efforts to establish the National Park Service (NPS), Save the Redwoods League, and the state park system, advocated for a national park on Mount Tamalpais, the county's most prominent geological feature. When that idea faltered, he donated land on its southeastern flank that became Muir Woods National Monument. Local hiking and conservation clubs—the Tamalpais Conservation Club (1912), the Tourist Club (1912), and the California Alpine Club (1914)—then worked to establish Mount Tamalpais State Park and saved the landscape from a highway and subdivision proposed in the 1920s.[3]

Garden club women in Marin began the Citizens Survey Committee in the early 1930s. They produced planning maps and a report designed to prepare the county for land use impacts from the pending Golden Gate Bridge. The group reorganized in 1934 as the Marin Conservation League, which has been at the forefront of Marin's civic environmentalism ever since.[4] The league identified Point Reyes Peninsula as a recreational and scenic resource and drove the county's early efforts to protect the area.[5] Following their lead, the county began to establish parks on coastal Marin beaches that were donated by the owners or acquired by public fundraising. The Marin Planning Commission pushed both the state and the county to protect additional bayside beaches and parts of the Inverness Ridge in 1943. Two years later, the first purchases for what became Tomales Bay State Park began.

Marin's land protection has always reflected the unusual wealth of those involved. It is nevertheless standard practice to puff the "selfless environmentalism and philanthropy of a few individuals" for saving the area's natural beauty, which is fairly routinely described as wilderness.[6] That ignores the agriculturalist occupants of the land and rarely acknowledges Marin property owners' considerable interest in minimizing the supply of housing and the demands on public services in the county while maintaining the surrounding beauty. Yet the consequences of Marin's land protection for the rest of the Bay Area have been significant, both positively and negatively. The protected lands provide recreational opportunities for the region in gorgeous landscapes; however, Marin's conservation has arguably intensified housing shortages in the county and shifted development pressure onto neighboring counties. With a growing portion of its land under protection, Marin real estate values rose steadily, and the county became an exclusive enclave by any standard.[7] At the beginning of the twenty-first century, Marin was one of the wealthiest counties in the United States and had the most highly educated population and the highest proportion of white residents of any metropolitan-area county in California.[8]

Deflecting the Postwar Boom

As state and federal government policies encouraged regional development after World War II, Marinites deflected the building boom that appeared unstoppable elsewhere. By 1960 San Jose had grown from a smallish prewar city of 68,000 and a lot of fruit orchards into a regional city of 204,000. Over the same period, Alameda County's once expansive farmland eroded as a system of largely segregated suburbs was created around Oakland.[9] Contra Costa agriculture faced similar pressures as the county's population quadrupled in the same two decades.

Marin County residents divided slightly on growth issues. Developers retained considerable heft until the late 1960s, but the county responded sharply to a 1959 Army Corps of Engineers report that predicted enormous population growth in the county and to the accompanying blueprint for accommodating the growth with transportation infrastructure throughout the North Bay. Before the decade ended, conservationists were making progress, and the extensive highway proposals were in the shredder.[10]

Protecting the Coastal Pastures

Even as highway revolts shook the developers, Marin's conservationists rattled the ranching community. When they first moved to protect the coast, the suburbanites began with the familiar: advocates sought a national park, asserting that Point Reyes was experiencing a "tremendous explosion" of development that the county and state could not handle. Interior Secretary Stewart Udall came rapidly on board; his department, in fact, had long aspired to a foothold on the California coast. Udall was convinced of the "short sightedness" of local officials who, he said, could not be trusted to protect what he described as the largest area in the United States near a population center that had been "unaltered by man."[11] Boyd Stewart, an early leader in both the West Marin ranching community and among suburban conservationists, dissented from that basic premise. The presence of ranchers was itself a demonstration that the landscape was not "unaltered." Stewart also noted that development pressure among ranchers was overstated: "Nobody wanted to sell." Many of the landowners had only recently escaped from being tenants.[12] They were not inclined to become tenants again, nor to exchange their coastal ranchlands for homes elsewhere.[13]

A New Coalition: Sustaining Farmers on Working Landscapes

Building support needed for federal legislation was beyond what the traditional conservation leadership could provide. Antihighway and growth control activists, planners, and the growing outdoor recreation movement eventually built a coalition with coastal ranchers.[14] Their compromise joined the emerging idea of open space protection and the even less familiar notion of a working landscape. Open space was an innovation at the time and remains a complex sell. It requires convincing park "purists" that slightly trammeled periurban lands merit federal protection. Similarly, maintaining working landscapes requires convincing wilderness advocates to accept farms and other cultural features as resources rather than blights to be expunged from protected landscapes.[15] Pursuing open space and working lands involved more than just a compromise: it required new expectations about protected landscapes and new institutions to support them.

A Pastoral Zone in the Seashore

The compromise sufficed to pass the legislation establishing not exactly a park but the Point Reyes National Seashore (PRNS). The new land protection ideas came to the Marin coast in a "pastoral zone," a roughly 21,000-acre region where all agreed that land would be privately owned and agriculture would remain.[16] However, the NPS arrival was destabilizing. In the early 1960s, the agency had limited experience in acquiring land, and its PRNS acquisitions were particularly maladroit.[17] It overpaid considerably for its first purchase, and the whopping sum ended the possibility that coastal ranching property could change hands within the ranching community. As both land prices and property taxes increased, even ranchers who did not want to sell began at least to consider it. The annual taxes on one ranch outside the seashore soared from $1,200 in the early 1960s to $22,000 in 1969.[18] Additional nominally willing sellers were forced forward, and the NPS condemned other parcels. Soon the agency had spent its original allocation of $14 million, largely, it turned out, to acquire land in the pastoral zone.[19]

Strapped for cash, the agency then proposed raising funds by selling off land for a subdivision within the seashore.[20] The development, complete with a country club, marina, polo field, gas station, and golf course, was intended to fund the acquisition of additional tracts in the seashore and other NPS units. The NPS scheme motivated the Save Our Seashore campaign in 1969, which attempted to raise additional funds to complete acqusitions on the coast. The first Earth Day, in 1970, pressured a previously reluctant Congress to approve an additional $43 million for land acquisition at Point Reyes. Shortly after, Congress established a second NPS unit, the Golden Gate National Recreation Area, south of Point Reyes. The process overrode the original pastoral zone elements of the seashore compromise and put virtually all coastal ranches between the Point and the Golden Gate Bridge in NPS ownership.[21]

Ranchers were not the only ones caught off guard by the NPS subdivision proposal. The Sierra Club, never fully committed to the working landscape notion, led a successful effort in 1974 to designate portions of the Point, including most of the pastoral zone, as wilderness or potential wilderness.[22] The designations included most of the pastoral zone but did not bar working dairies or aquaculture, however. The working pastures therefore remained an important part of the region's rural economy, supporting the creamery, veterinarians and similar. They also had an important political and visual presence, providing an agriculture-friendly precedent that shaped both growth control and open space protection from San Francisco to Mendocino County.[23]

Supporting Working Landscapes

The Williamson Act

The working landscape idea was soon built into the 1965 Williamson Act, which authorized counties to create agricultural preserves in which land became eligible for use-based tax assessment rather than its market (development) value. Voluntary enrollment required a ten-year commitment by farmers not to develop, sell, or subdivide their land.[24] The state, for its part, would make up the county's tax losses.

Counties in the district responded to the Williamson Act very differently. In Sonoma, it had little impact. The county supervisors respected the private property enthusiasms of the dominant agricultural community and worked to accommodate growth. Sonoma became the fastest growing of the nine Bay Area counties.[25] Slightly inland in Napa County, the Williamson Act came in time to prevent a "national vineyard." Its advocates had envisioned a federally protected wine-producing landscape akin to the pastoral zone in Point Reyes. The Williamson Act offered a more palatable alternative, and Napa landowners resisted what was widely viewed as an NPS takeover.[26] Figure 5.1 compares the impact of strategies adopted in Marin and Napa, which made concerted efforts to control and direct development, to Sonoma County, which resisted any growth controls until the late 1980s.

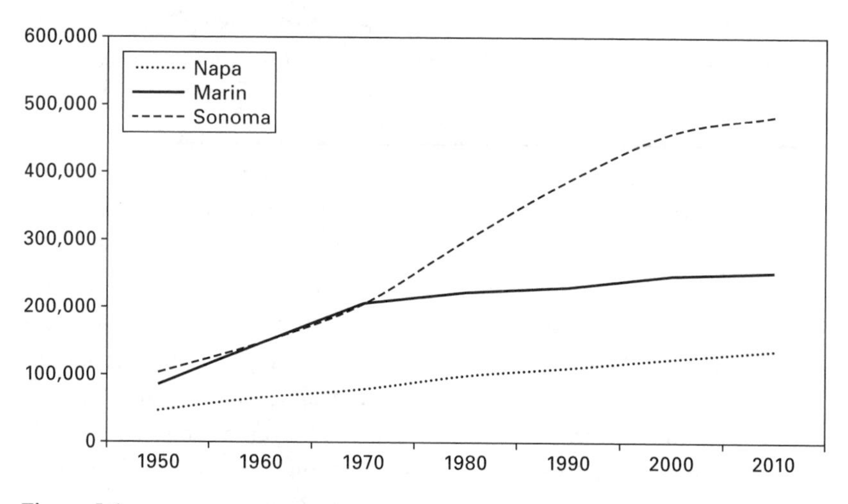

Figure 5.1
Population of Napa, Marin, and Sonoma counties, 1950–2010. Marin and Napa counties adopted stringent growth control measures, including agricultural land protection zones under the Williamson Act, in the 1970s, and they seem to have worked. Sonoma County started later and continues to grow more rapidly. *Source: U.S. Census*, various years.

The Marin Countywide Plan

Before building conservationist-rancher cooperation into county policy, the Marin Board of Supervisors commissioned two studies to determine whether agriculture in the county had a future. The resulting planning reports, *Can the Last Place Last?* (1971) and *The Viability of Agriculture in Marin* (1973), demonstrated that several factors—proximity to San Francisco markets, Williamson Act protections, and strong community support—could offset high land and production costs to sustain a ranching economy in the coastal pastures.

Emboldened, the Marin supervisors made agriculture the center of the 1973 *Countywide Plan.* The plan protected working landscapes in two of three permanent land use zones. What it referred to as the Urban Corridor along Highway 101 was slated for development; a central Inland Rural Corridor was reserved for agriculture and compatible land uses. The Coastal Recreation Corridor included the seashore and the Golden Gate National Recreation Area, dedicated to recreation and environmental protection (figure 5.2). In addition, the county adopted sixty-acre minimum lot sizes in the optimistic hope that it would discourage subdivision. The 1973 plan has become a sacrosanct element of district culture: essentially unchanged since it was adopted, it continues to define development in Marin.

Drought Relief and Regulatory and Price Supports

The Marin County government also backed agriculture in other ways. When runoff attracted attention from both local oyster growers and environmentalists working to restore salmon runs in Marin streams, the county subsidized ranchers' efforts to meet new requirements. Similarly the county financed a program "to haul water to West Marin's bone-dry ranches" in 1976 and 1977.[27] County planners and conservationists also turned out in force at hearings in 1977 in Los Angeles to support an increase in milk prices that would buffer their ranchers from the impact of the wildly fluctuating fluid milk market. The county would probably have continued those programs, but a statewide referendum intervened. In late 1978, Proposition 13, California's opening volley in what became a national tax revolt, capped property taxes and took the flex out of county budgets. It became clear that all the relationships,

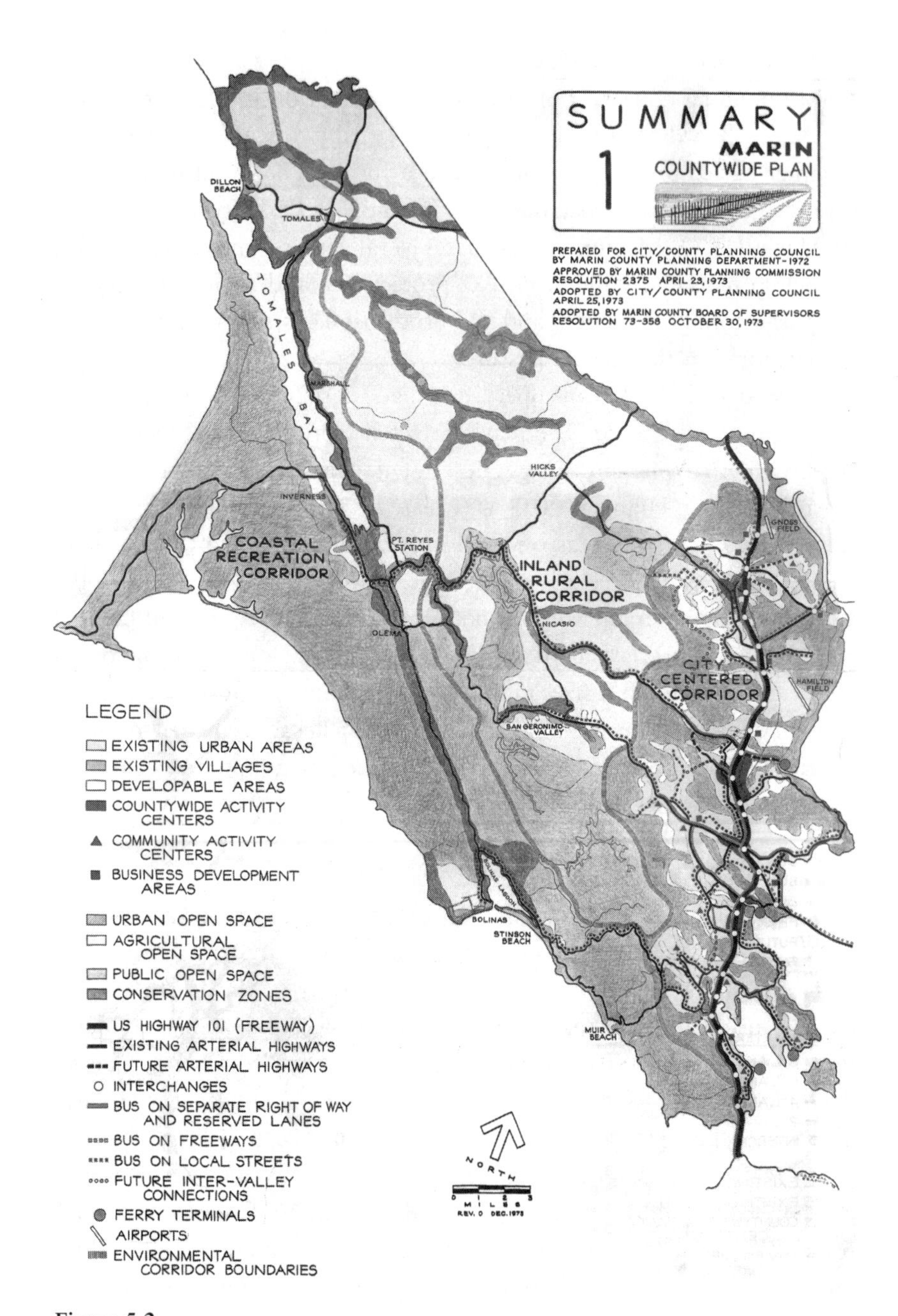

Figure 5.2
Summary map of the urban, rural, and coastal corridors defined by the 1973 *Marin Countywide Plan.*

innovations, and plans were insufficient. "Something more permanent" was needed.[28]

The Marin Agricultural Land Trust

The first change was an institutional innovation, the Marin Agricultural Land Trust (MALT). Land trusts and conservation easements, billed as voluntary, private, compensated land protection, proliferated in the Reagan era. A conservation easement allowed landowners to sell some of the bundle of ownership rights in their land while retaining others. MALT was created in 1980 to purchase the development value of ranches while allowing landowners to continue agriculture on their properties. The influx of cash might allow a ranching family to invest in the business or transfer land from one generation to the next by buying out nonfarming heirs.[29] In spite of the potential, ranchers did not rush to enter into binding contracts.

The slow start was not a problem because, although MALT began with great fanfare, it had little money. Taxpayer support for the working landscape concept came to MALT's rescue. Although MALT is glossed as a private organization engaged in private transactions, the Marin County Open Space District helped MALT fund its first easement in 1983. The Buck Trust, a private tax-exempt foundation, and the California Coastal Conservancy both made million-dollar grants to the organization, and in 1988, a bond act earmarked $15 million for agricultural easements in Marin, making MALT an important instrument of local public policy.

Within the decade, ranchers had warmed up to the easement concept, and MALT enjoyed the support of about 1,700 dues-paying members. The program and its accomplishments seemed so solid that its founders, Ellen Straus and Phyllis Faber, began work on a celebratory volume, *Farming on the Edge*. The book presents the collaboration of county, NPS, MALT, ranchers, and conservationists as a model for protecting natural and cultural landscapes by stabilizing periurban agriculture. The rapprochement, however, was never as cozy or secure as the widely circulated success story suggests. Not all the players accepted a continuing role for agriculture in the protected landscape. Even MALT was unclear on its priorities: it heavily promoted the nature preserve elements of the seashore and ignored the ranchers for almost twenty years. Nevertheless, the coalition basically halted residential development on Marin's coastal pastures.

That success encouraged another regional group, People for Open Space (POS), to study farm protection possibilities for the entire Bay Area. POS produced five dense reports documenting the existing scale, function, and status of agriculture in the region, as well as the importance of local food production for community, economics, jobs, habitat, water preservation, recreation and tourism, landscape values, and cultural heritage and history.[30] The final document, *Endangered Harvest: The Future of Bay Area Farmland* (1980), shifted expectations throughout the district where farmers and policymakers had long believed that time was running out for farming in the Bay Area. Figure 5.3 suggests some of the outcomes; protected farmland became a vital component of Bay Area open space.

Whether it was desirable or merely inevitable, aging farmers increasingly planned to sell their land to developers and retire comfortably. POS analysis demonstrated that farming declines were reversible and, moreover, that reversing them was critical to the future health and prosperity of the Bay Area. The report was turned into an art exhibit that toured California, attracting thousands of viewers, and ended up on long-term display in Oakland's Museum of California. That, plus several years of television and newspaper coverage, forums, and hearings spread the *Endangered Harvest* message and supported a vision of the Bay Area as a farming region with a future.

A New Path for Bay Area Agriculture

A second-generation problem soon confronted the district: What has been accomplished if we protect agricultural land but the farmers and ranchers cannot make a living? When the threat was residential subdivision, supporting ranches and establishing sixty-acre lot minimums seemed a reasonable response. But over time, subdivision receded as a threat. Aspiring country squires and dot.com millionaires were not daunted by the sixty-acre minimum. When working ranchers ran into financial difficulty, and the chaotic milk market virtually guaranteed that they would, there was no shortage of wealthy buyers who could afford a faux farm in the coastal hills.[31] However, hobby farming does not support veterinarians,

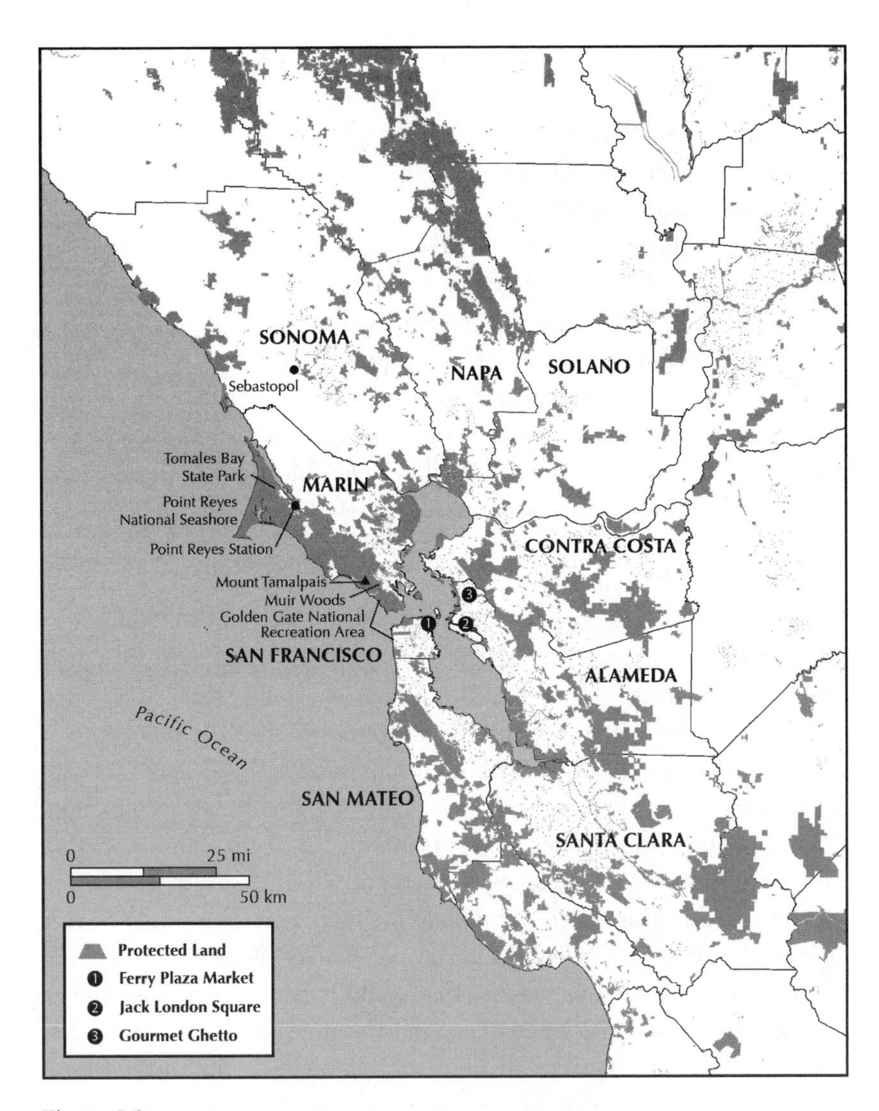

Figure 5.3
Map of Bay Area protected land. A century of conservation, agricultural, and open space activism supports a regional mosaic of recreational, open space, and agricultural land in the San Francisco region. *Source:* Greeninfo Network, California Protected Areas Database.

creameries, feed stores, and agricultural workers. Economic viability for the agricultural community became the next problem to solve.

As the POS and Marin County studies had argued, the Bay Area was well situated to respond to the challenge of ranch viability. The region's longstanding culture of food produced a tribe of both farmers and entrepreneurs who began to refurbish traditional practices and products in the coastal pastures and a phalanx of consumers who purchased their products. Initially motivated by their own personal values and a desire to stay on the land or to farm their own piece of ground off the conventional grid, growers came to the region, or came back to family land, or worked with and against their parents to push family operations in new and sustainable directions. They built viable individual businesses and created a dynamic rural community of innovators.

A New Kind of Producer

In the early 1970s, as Secretary of Agriculture Earl Butz was full tilt in his "get big or get out" campaign, coastal dairy ranchers were under growing price pressure from large feedlot dairies to the south. Some tried and failed to address the problems by getting bigger. In Bolinas, two long-haired agriculturalists, a school teacher and a Shakespeare scholar, began instead to explore new approaches to small-scale production. Bolinas, a small, rural town at the southern tip of the county, was an early haven for pot growers, communes, and kindred spirits well before the town became famous for removing highway signs that might attract an outsider or a tourist.[32] Bill Niman and Warren Weber were looking for ranch management techniques that could preserve the soil as well as products that emphasized the place and the producer, not a low price. They were part of a cadre of idealistic neophytes who worked with neighbors and fellow sustainability advocates to develop a new way of producing food. They shared insights, started support organizations, and began in new ways to fill the truck farming niche in the Bay Area.

Minnesotan Bill Niman arrived in Marin County in the late 1960s to teach school in Bolinas. A community of back-to-the-landers soon drew him to a small parcel of agricultural land. His wife tutored neighbor Boyd Stewart's granddaughter, Amanda, for which she was paid two calves.[33] Those calves became the start of Niman's meat operation. He was an odd duck in a diversifying pond: although the livestock was normal for the coastal pastures, Niman stood out from homesteaders who typically focused on vegetables and fruit, perhaps a few pigs, pot, and chickens. The Stewarts became a mainstay in his education about the meat business, and, for a time the Nimans and Stewarts ran cattle together.

Warren Weber had turned his back on all the opportunities afforded by a Berkeley Ph.D. in English and took up a youthful passion for farming. He began operations with five acres and a horse-drawn plow and successfully marketed branded greens, herbs, and vegetables. However, distribution was a problem. Weber responded by buying a truck and driving around to the food co-ops that proliferated in the district. He also became an architect of the learning and cooperating culture that characterized early organic agriculture in the district.[34] As soon as organizations were formed, he worked with them, and when they did not exist, he worked to create them. Although Weber is the face of diversified agriculture in Marin, he also diversified his land base and began working twenty acres in the Coachella Valley south of Palm Springs. That allowed him to create a year-round crop so that customers would not forget him and his brand when his Bolinas operation was not producing in the winter.[35]

A New Kind of Milk

Although coastal vegetable growers like Weber were first to experiment with sustainable agriculture in the district, ranchers largely dismissed them as organic gardeners for decades. Important as their institution building and demonstration of alternative possibilities has been, it is the ranchers, not the veggie folks, who drive our story, because it is the pastures that needed protection. The leader in that effort is not hard to find.

Straus Dairy is not a multigenerational family operation with deep roots in the district: the elder Strauses, Bill and Ellen, left Germany before the Holocaust to settle in California. They were not, as Ellen frequently stressed, inclined to move again under pressure. Ellen began to transition ranch management to something new after reading *Silent Spring*. First, she and her family began to eliminate pesticides from their ranch. The elder Strauses also became deeply involved in the growing environmental movement. Bill was initially the family's networker. He was a founder of the Tomales Bay Association, a local environmental group, and joined the Marin Conservation League, where he ultimately became a director.[36]

The Strauses were among the first ranchers to support the county's sixty-acre zoning. When they opposed development on a neighboring ranch, their opposition "split the whole community."[37]

Ellen then devoted herself to putting that community back together, and in the process, she came to personify the rancher-conservationist coalition in the district. Working with microbiologist and marsh ecologist Phyllis Faber, Ellen founded MALT, and she served on the boards for organizations that spanned the full ranch-environmentalist spectrum in the district.[38] A family legend insists that when the phone rang at the ranch, Bill would announce, "It's the environment calling."[39] But it was

Pasteurization and the Conventional Vision of Milk

Doom was pending on the Straus Ranch in part because of the heavily regulated but still unstable conventional milk market. Drinking fresh milk is a recently acquired taste for Americans. Well into the twentieth century, milk was dangerous. After Louis Pasteur discovered the organisms that sour milk in 1857, most states developed a dairy certification program to address milk safety issues.[a] Grade A dairies maintained the highest sanitary conditions for the production of milk: cleanliness of the facility, the cows, and the milking parlor and careful handling and cooling of milk were critical to a top rating. California adopted its first sanitation scorecard in 1905. A decade later, more stringent requirements in the Pure Milk Law brought an end to the Point Reyes Peninsula's famous farmstead butter operations. The higher, more costly standards caused farmers in the region to form cooperatives to pasteurize milk.[b]

Pasteurization ultimately displaced dairy certification programs in part because it enabled processors to expand, sourcing product from larger and larger areas.[c] In Milwaukee, for example, the number of milk distributors fell from 200 to 32 in six years after the city required pasteurization. By 1932, two companies, Borden's and National Dairy, handled over 20 percent of all the milk in the United States. The milk co-ops began regulating market access and pricing. Their rules were adopted as marketing orders under federal New Deal legislation.[d] However, the rules never controlled levels of production effectively. Pasture-based dairies struggled in glutted milk markets as the industry consolidated on factory-like feedlots in southern California.

a. Leavitt (1982).
b. Livingston (1993).
c. Leavitt (1982) and DuPuis (2002).
d. Woeste (1998) and DuPuis (2002).

their techie first-born son, Albert, who rejected a career in electronics to stave off "impending doom" on his family ranch.[40]

Conclusion

The first waves of innovation in our district secured a land base for agriculture near urban markets but did not ensure farmers a livelihood. To retain their homes, coastal ranchers acquiesced as urban elites imposed city notions of working landscapes onto rural agricultural communities. The seashore created havoc in ranch land markets and tax rates. Nonetheless, it linked urban recreationists to agricultural preservation and growth control in the coastal pastures. What is more, ranchers inside and outside the seashore soon realized that although protecting agricultural land from development was an essential first step, maintaining viable family farms was something far beyond that.

In part, Albert Straus had a chance to "stave off doom," as we show in chapter 7, because he lived in a community created by earlier innovators and mavens: Warren Weber joined the elder Strauses and Boyd Stewart in a cohort of institution builders and networkers. Nevertheless, creative as they were in developing new institutions and values that would protect their ranches, they worked in a culture that was already primed for their efforts. Congressman Kent, the planning advocates of the Marin Conservation League, the regional agricultural advocates in People for Open Space, and the generations of suburbanized Marinites who loved their marvelous county had all invested time, money, and energy in building relationships and institutions that would protect it.

The MALT-founding generation operated under conventions and values that became increasingly clear about the need for a growing role for supportive, well-heeled suburbanites in managing the countryside. When those steps were not enough, the next generation went searching for new ways of producing food. That led to a whole different set of relations with more urbanized parts of the district. Marin's ranchers and farmers soon became closely tied to entrepreneurs who were developing businesses across the Bay, specifically, in what many consider to be a separate universe altogether: Berkeley in the turmoil of the 1960s.

6

Radical Regional Cuisine

Solving the distribution problems for fresh, local, sustainably grown produce created the face-to-face interactions that comprise the foundation of the Bay Area's alternative food district. Innovations at the junction of two fairly distinct Bay Area food cultures gave the early district both momentum and a recognizable name: *California cuisine.* Preoccupation with food defines the first critical culture, and it began in the ethnic diversity of the Gold Rush. No matter how enamored they have become of Big Macs and soda, Americans have never totally ignored the tastes and pleasures of food, and particularly its importance in families and communities. Those factors never seriously eroded in San Francisco. Reasserting them as an appropriate focus of conversation and national social, economic, and political concern may be the district's singular contribution.[1]

The second element, now less familiar, is the residue of Bay Area food politics of the 1960s and 1970s. A network of frequently quite radical food processors and distributors—worker-owned co-ops, particularly bakeries, collectives, and buying clubs (occasionally known as food conspiracies)—provided the first distribution networks for alternative growers, and the survivors remain important.[2] Those food and political cultures intersected in high-end restaurants and at farmers' markets and led to a distinctive, politically engaged, regional cuisine. Briefly, as Reagan-era prosperity and then the dot.com boom unfolded, preoccupation with taste threatened to overwhelm progressive politics. Nonetheless, alternative producers' interactions with discerning chefs and environmentally oriented consumers redirected expectations from food aesthetics to environmental sustainability.

Both radical politics and food quality preoccupations are closely identified with Berkeley, California. Two downtown blocks, more or less, of Berkeley's Shattuck Avenue are frequently described as the Gourmet Ghetto. It has been frequently parodied as precious and overpriced, and with some justification. It is not accurate, however, to describe its priorities as simply displays of either wealth or connoisseurship. Regional cuisine in the district developed as simultaneously radical and haute and became, we argue, a critical proving ground for developing alternatives to the increasingly toxic conventional system. What we characterize as radical regional cuisine became imaginable as the alternative producers described in chapter 5 sought markets for their new products. We have been repeatedly told that it is hard to grow good food. Driving it around, explaining it, marketing it, and managing wholesale accounts are also difficult and require a particular skill set, or several of them. As in the coastal pastures, then, something more permanent was required. The solutions began to emerge in two settings that continue to be important in the district: quality-defined restaurants and farmers' markets.

The Political Roots of Radical Regional Cuisine

Baby Boomers in the 1960s and 1970s

The left politics of the Depression cooled during the McCarthy excesses of the Eisenhower years, but the baby boomers who reached college age in the 1960s pursued diverse social reform agendas. Their efforts included elements of the civil rights movement, the women's movement, the environmental movement, the antiwar movement, and farm workers' unionizing efforts. Period discussions of a generation gap reflected younger Americans'—particularly the privileged ones—discomfort with the prevailing culture and their experiments with communal and cooperative living, drugs, free love, and peace. In the district, that played out in a context shaped by the hippies, the free speech movement (FSM),[3] the Diggers, and the Black Panthers.

The reformers ranged from well-intended and sweetly naive to seriously revolutionary and occasionally intended to be or were accidentally violent. The Bay Area was less convulsed by the race riots that gripped other major urban areas in the 1960s, and the turmoil in our district is still typically glossed by reference to Bezerkeley, the free speech movement,[4] and the Summer of Love (1967–1968), a global phenomenon that drew thousands of young adults to the Haight-Ashbury section of San Francisco. Food values and institutions were surprisingly important in the antiestablishment chorus of the decade.

The San Francisco Diggers and the Black Panthers

The Diggers and the Black Panthers defined what was then seen as an extreme end of the food-relevant spectrum. Patterned on a seventeenth-century Protestant agrarian community led by British radical Gerrard Winstanley, the Diggers pursued a radical agenda with street theater, art "happenings," and occasional direct action. Digger challenges to capitalism included the Free Store (Trip Without a Ticket) where everything was free of charge, reflecting the Digger mantra, "It's free because it's yours." Most famously, they distributed free food every day to all comers in the Panhandle of Golden Gate Park. Their heyday was just ending as the Summer of Love attracted media attention.[5]

Bobby Seale and Huey Newton founded the Black Panther Party across the Bay in Oakland. Originally called the Black Panther Party for Self Defense, the group intended to protect black communities from police violence and brutality. Although its reputation remains steeped in guns, the FBI's surveillance and disruption of Panther programs, known as Cointelpro, was arguably as responsible for the violence as the group's own provocations.[6] The organization rapidly became a national phenomenon and developed a comprehensive set of social programs to improve community health and support black victims of oppression "pending revolution."[7] The Black Panthers' breakfast program for school children was among the first in the country, and just one of its many food activities (figure 6.1).[8] Frequent claims that their program shamed or inspired the federal government to establish a similar one—the School Breakfast Program, in which the free meal is served in public schools—and the common assertion that the Panthers were soon feeding upward of 10,000 children per day are difficult to assess.[9]

Alternative Food Access: Collectives, Conspiracies, and Co-ops

Digger and Black Panther goals were widely shared among political and religious radicals active in food conspiracies.[10] Some were 1930s-style

Figure 6.1
Food distribution was one of a suite of social programs organized by the Black Panthers, and commentators tend to minimize its scale. Here, Bobby Seale surveys groceries assembled in Oakland to be distributed by volunteers, c. 1972, probably as a part of a voter registration drive (photograph by Howard Erker). *Source:* Oakland Tribune Collection, Oakland Museum of California. Gift of ANG Newspapers.

buying clubs, groups of neighbors seeking cheaper alternatives to conventional supermarkets and food; others saw food as a platform and wedge for organizing a more comprehensive revolution. Simultaneously, in the early 1960s, the Berkeley Co-op became both more professionally organized and increasingly clear about its political priorities. Curl's summary of the Co-op's annual campaigns during that decade catalogues the food politics of the day:

> In 1962 [the Berkeley Co-op] instituted "free speech tables" near entrances for literature and petitions. In 1963 they debated milk contamination from a proposed nearby nuclear power plant; supported a local anti-discrimination housing ordinance; stopped stocking products boycotted by the Central Labor Council. In 1964 they increased minority employment; held a food drive "to aid persons suffering Civil Rights discrimination" in Mississippi; pioneered biodegradable detergent with a Co-op label. In 1965 they packaged meat with the better side down; lobbied for a bread and cereal enrichment law; educated on peanut butter additives. In 1966–67 they lobbied for a Fair Packaging and Labeling law; contributed to the United Farm Workers (UFW) co-op in Delano; instituted unit pricing on all shelves; lobbied for regulation of diet foods and for a unit pricing law; agitated against a phone rate increase; labeled all Dow products as boycotted because of their napalm production; assisted the legal defense fund for besieged integrated Southern co-ops. In 1968, they authorized centers to ban smoking; removed all nonunion grapes; withdrew from the Chamber of Commerce because of their consistent opposition to consumer legislation. In 1969 they battled against utility rate hikes; donated food to the Black Panthers children's breakfast program; demanded the "immediate termination" of the military occupation of Berkeley by the National Guard (ordered by Governor Ronald Reagan because of People's Park); posted statements at all co-op centers condemning the [Vietnam] war; closed in solidarity during a People's Park protest march and on Vietnam Moratorium Day.[11]

But even with all that, the co-op was insufficiently political for many in the district. The communes and food conspiracies pursued very diverse political agendas and visions, and many were affiliated with ashrams or other religiously oriented groups. In common, they sought cheap, nutritious food—the familiar bulgar, brown rice, whole grains, and tofu. Worker-run storefronts and then health food stores—with names like Seeds of Life, the Haight Store, and the Good Life Grocery—gradually displaced less structured efforts and were among the outlets that provided important distribution opportunities for the first alternative and organic farms in the area.

The People's Food System

The storefronts proliferated widely, and some soon loosely organized into a group known as the San Francisco Food Conspiracy that supported a more organized collective procurement and distribution center, the People's Food System (PFS).[12] They ran the Common Operating Warehouse, which had started distributing dried fruit (figure 6.2), but evolved into an explicitly political project using food distribution as a tool of "community organizing and political education."[13] The Common Operating Warehouse was not "just a network of stores. Its members saw it as an organization dedicated to radical social change."[14]

Overlap with the Black Panther agenda notwithstanding, it would be an error to conflate white middle-class radicals' priorities with those of poorer black customers. Quite the opposite, one participant underscores

Figure 6.2
People's Food System activities centered in the San Francisco Cooperating Warehouse, which supplied the network of small cooperatives and alternative neighborhood stores in San Francisco in the 1970s. *Source:* foundsf.org.

that PFS "volunteers, coordinators and collective members . . . were mainly young educated white men and women in their mid-twenties," dropouts who "worked part-time, received food stamps, unemployment and welfare or lived on savings." The white counterculture often "alienated sectors of the community most in need of inexpensive food," as one participant remembered:

> The stores drew a hard line on products, choosing to serve only natural foods. Small stores made room for kefir and cashew butter, but refused to stock canned tomatoes. . . . Hippies would drive up from all over Berkeley to save money on natural foods, but the Black people who lived in that neighborhood remained outside the process and continued to shop at Safeway and at the corner liquor store, paying extortionate prices.[15]

Collapse and Survivors

Conflict over political priorities caused dissension within the PFS community. The diverse procedures for decision making within PFS's constituent groups made these difficult to resolve. As the political agendas conflicted with the requirements of organizing and administering an expanding business, rifts among volunteers, store managers, and shoppers became fairly constant. A shooting at an emergency meeting in April 1977 left one participant dead and rocked the PFS.[16] Some briefly boycotted feminist produce distributor Veritable Vegetable, which was blamed for the incident. The violence did not bring down the PFS or the Common Operating Warehouse, but it signaled that an era was ebbing. A few organizations survived that chaos and endure to this day, vectors of the values of the 1960s that continue to be important elements of district culture:

Greenleaf Now a major West Coast distributor of fine produce, Greenleaf is perhaps the straightest of these durable firms. But it originated in a Hunga Dunga commune program that divided wholesale lots of food products among its members. Its founder, James Patton, became skilled at distributing produce from small-scale organic producers, purchasing the produce and reselling it to communes and natural food stores. Greenleaf remains one of San Francisco's most important vegetable distributors.

Rainbow Grocery Similar to Greenleaf, Rainbow Grocery began as a food-buying club in an ashram that first tried to avoid, and then to replace, the usual distribution channels. Rainbow withdrew from the PFS before the shooting incident and began taking inventory, hiring professional buyers, and "moving vitamins away from the door, where they could not be so easily shoplifted (the days of hippy trust and goodwill were evaporating)."[17] Still worker owned and still vegetarian, Rainbow has gone sufficiently upscale to have its own celebrity cheesemonger,[18] but it remains a fixture of food retail in San Francisco and uses its decades of experience and considerable clout to support local start-ups that share its values.[19] Most recently, Rainbow Grocery workers and those who worked there in the past have mentored social entrepreneurs at People's Grocery in West Oakland, which is discussed in chapter 8.

Veritable Vegetable Veritable Vegetable (VV) was organized as a predominantly female, lesbian, consensus-based collective. It started in 1974 when four members of a food club purchased a truck and began buying and selling produce from small and medium-size growers in the region. After the PFS imploded, VV continued to distribute organic, locally grown food. When suppliers asked VV to sell more produce in order to keep them in business, VV significantly expanded the size and scope of its operation. VV sees itself as "both a business and an instrument for social, economic, and environmental change," and, like Rainbow, it has played an enormous role in creating the relationships that sustain the district.

The Political Bakeries

Bakeries have long been the most visible element of the district's political food culture. During the 1970s, several dozen bakeries—one participant noted that when he arrived in Berkeley in the early 1970s there were thirty-two collective, cooperative bakeries operating in the area—presciently combined political messaging with environmental sustainability and health education. A 1983 collaborative recipe book chronicles the myriad uprisings in bread baking:

> The most basic of these is the grain growing from the earth, nourished by the rain and sun. . . . Bakers, with a little help from yeast and other leaveners, create another uprising, as dough rises to produce fresh-baked loaves, filling our senses.

The third uprising is the cooperative ethic of the bakeries we work in. There are no bosses, no employees. Instead we all do the work together, sharing the responsibilities and the rewards. Our businesses put priority on serving the needs of the community, not on making profits for a select few.[20]

Some of the 1970s bakeries have disappeared, and others have scaled up, but the intersection of good food and political commitment in the region's bakeries suggests the district's diverse approaches to combining politics and good food.

Your Black Muslim Bakery Black Muslim leader Elijah Muhammad's two-volume *How to Eat to Live* inspired Yusuf Bey to start a bakery in Oakland in 1971.[21] It was intended to build community and empower black people by providing healthy, natural foods to the local community. Bey's enterprise soon expanded to include a school, a beauty store, and a dry cleaning establishment. The bakery also became a center of political power in Oakland's black community. Bey ran unsuccessfully for mayor in 1994, but both the successful candidate, Elihu Harris, and later his successor, Jerry Brown, made political "pilgrimages" to the bakery. Over time, however, the bakery was implicated in diverse thuggery. Bey died in 2002, and his successor was murdered. Although the bakery remained to some an important "symbol of black enterprise and empowerment" in a neighborhood devastated by crack and violence, the operation closed.[22]

Rubicon Bakery Rubicon was organized in 1973 with a very different product and justice agenda. No dark whole grains or protruding pumpkin seeds in Rubicon products: the bakery still makes premium desserts for upscale markets. Rubicon was formed after Governor Ronald Reagan closed many of the state's mental hospitals, which worsened existing problems of homelessness and poverty. The closures were to be offset by a web of community-based programs, and Rubicon was formed when they did not materialize. Named for Julius Caesar's *alea iacta est* moment, the bakery provided jobs and support programs that help people to "cross their own personal Rubicon and transform their lives."[23] Its programs aim "to prepare very low-income people to achieve financial independence" by providing housing, training, and employment. Revenues from the baked goods support the programs.

Uprisings Baking Collective A workers' co-op opened the Uprisings shop in 1975. Most of its business was wholesale, and the operation took advantage of the diversifying alternative distribution system developing in the area, "delivering breads to farmers markets, health food stores and mainstream supermarkets." Uprisings is remembered for its "talking bread": in every loaf, a little fortune cookie–like enclosure spread the word about demonstrations, issues, and ideas. The bakery was both a political presence and a force for food reform. Ironically, it faltered when peasant-style whole grain and organic bread went upscale, and it closed in 1997.[24]

A Continuing Genre Although many of the bakeries of the 1960s and 1970s have gone the way of the Common Operating Warehouse, similarly political bakeries continue to maintain the tradition. Mission Pie, for example, operates in the Mission district of San Francisco, providing a neighborhood gathering place; access to organic, healthy food and training; and employment opportunities for youth at nearby high schools. The bakery was founded by an early supporter of a political farm, Pie Ranch, which includes an education and advocacy organization "working hard to grow literacy about food and farming."[25]

A Tradition of Fine Food

It is surprising how much is written about food systems, food politics, and food access as if the food itself—the taste, preparation, culture, sharing, and indeed, the passion for it—mattered hardly at all. Since the founding of San Francisco, food has been absolutely central, and it remains so.

Ethnic Traditions and Skills

The cuisine elements of the district began simultaneously with rapid development of the wealthy, ethnically diverse city of San Francisco. The Gold Rush was as much a driver of the cuisine as it was of the agriculture we discussed in chapter 3. A rapid influx of immigrants, each group and family with its own food traditions, farming and cooking skills, and frequently seeds or cuttings, turned the region into a global center of agriculture and also of cuisine.[26] By late 1849, San Francisco had its first French Restaurant, The Old Poodle Dog (figure 6.3).[27] Three years later, the French novelist Alexandre Dumas reputedly observed that San Francisco had more fine restaurants than Paris and that the Chinese and French cooks, colocated in Belden Place, still a major center for San Francisco diners, were experimenting with each other's traditions.

Gold miners were famously obsessed with food, and many had plenty of money. Luxury provisions were not confined to cities. Purveyors of fine food items followed the miners out to the mining communities, and a startlingly exotic spectrum of ethnic and European foods was quickly available throughout the mining frontier.[28]

When they left their diggings, many miners turned to farming and provisioning the growing city of San Francisco. The familial and cultural cohesion of immigrant groups translated into economic institutions and resilient cuisines that have remained quite stable. By the 1860s, immigrants from Switzerland, Portugal, Ireland, Italy, and Germany had transplanted the knowledge of both animal husbandry and butter production that made Marin a leader in milk processing and Point Reyes butter famous in San Francisco, while Portuguese and Italian communities supported traditional cheese makers.[29] Zionist women who had settled in Petaluma started keeping chickens for pin money and wound up making Sonoma County the national leader in poultry and egg production.[30]

French Food and Fast Food after World War II

Family traditions surrounding growing and eating food survived in the district despite two competitors that have dominated Americans' food imagination: Americans' historic reverence for French cuisine and their subsequent but opposite attraction to cheap, highly processed, convenient commodities. Soldiers returning from World War II in Europe were the critical vectors of both trends.[31]

Americans' regard for French cuisine as the touchstone of gourmet and also as so snooty as to be unpatriotic is not a recent phenomenon. Patrick Henry complained that Thomas Jefferson "abjured his native victuals in favor of French cuisine."[32] Jefferson and a growing phalanx

Figure 6.3
Front of the menu at the Old Poodle Dog, 1922. The name was probably a misappropriation of Le Poulet D'Or; it was among the first restaurants in the city and inhabited a number of locations. It was destroyed and then rebuilt after the 1906 earthquake, but was done in by Prohibition, although a restaurant by the same name limped along into the 1960s. *Source:* Poodle Dog Restaurant and Marcel Mailhebuau manuscript collection, California State Library.

of his countrymen were apparently undeterred by the canard. Several centuries later, the Francophiles began to expand beyond the upper class: in 1941, the first issue of *Gourmet Magazine* featured recipes with primarily French names, and French food was on its way to becoming a middle-class entertainment.[33] James Beard's major career as a writer, cook, and food personality began in 1940 with his best-selling *Hors d'Oeuvres and Canapés*. GIs, along with the increasing numbers of Americans who traveled in Europe (and Europeans residing in the United States), returned home with new insights into food possibilities. The Culinary Institute of America was founded in 1946 "specifically to train returning World War II veterans in the culinary arts," a clear sign that a culinary profession was developing in the wake of exposure to the European theater.[34] Craig Claiborne became the first male food writer at the *New York Times* in 1957, underscoring that cooking had outgrown the ladies' page.[35]

Julia Child's *Mastering the Art of French Cooking*, published in 1961, sealed that transition, bringing French techniques and sensibilities within reach of the increasingly affluent middle class.[36] Child's television program *The French Chef*, which began in 1963, spread the message even more widely.

Mac and Cheese, GIs and Square Meals

The spread of French food and technique in some American homes partially concealed a sharp general decline among Americans in shopping and cooking knowledge. GIs' exposure to chow line fare that included meat, a vegetable, and a starch had eroded their diverse family and ethnic cuisine traditions. That idea of a "square meal" opened the way for the astounding homogenization of American food that followed the war. Some inexpensive processed foods had become available during the Depression, providing meals that were suited first to hard times and then to wartime meat scarcity. General Mills's Bisquick was first marketed in 1931, and Kraft Macaroni & Cheese dinners (KDs in Canada) were introduced in 1937.[37] After the war, the floodgates opened on convenient, fast, and increasingly engineered and manufactured foods.

There is no question regarding which of these two trends dominated. In spite of postwar efforts to get Rosie the Riveter out of the job market and back into the kitchen, women remained in the job market or joined in droves during the women's movement, and convenience triumphed. Cooking receded in importance as a source of food and a locus of family life. Fast food was nailed into the American diet after the war, but fast is just the tip of the iceberg.[38] Along with its many imitators and competitors, McDonald's has reconstructed basic food products to meet its exacting needs. The McDonald's potatoes to which Americans have become accustomed—they are the most frequently consumed vegetable in the nation by far, followed by catsup, also a top-ranking vegetable in terms of consumption—bear little relationship to the nutritious "treasures of the Andes" potato that one author claims "saved" the Western world.[39] The convenient edible food-like substances that normalized after World War II have remade Americans as well: the Centers for Disease Control reports that the average American adult has added twenty-five pounds since 1960.[40]

Politics and European Food in the Early Gourmet Ghetto

The radical politics and the food priorities come together in our district, both symbolically and literally in the Gourmet Ghetto. For almost half a century, the district's rising expectations about food quality have been devised, nurtured, and marketed in the food enterprises that still come to the famous precinct. Although the neighborhood on Shattuck Avenue is frequently associated with Alice Water's famous Chez Panisse, she was actually a late arriver. Peet's Coffee is now a national brand, but initially there was a Mr. Peet who began on a side street, the Gourmet Ghetto's first enterprise oriented to European taste. According to the company's Web site, Dutch-born Alfred Peet was "appalled at the poor quality of coffee being consumed by Americans."[41] When he opened a European-style café in Berkeley in 1966, the place was already somewhat oriented to food: the Berkeley Co-op had an outpost nearby, and an excellent butcher shop, Lenny's Prime Meats, a survivor of the cut-and-wrap revolution, operated down the street. Peet's became a pioneer in sourcing fair trade coffee, but the store was and remains focused on taste. The place became a mecca for the first blush of Bay Area folks who cared about coffee.

The Cheese Board Collective

The Cheese Board became a collective in 1971, the same year Chez Panisse opened its doors. Together, they made the neighborhood recognizable as a site offering distinctive food. One of the district's enduring political bakeries, the Cheese Board brought fine cheese to the Shattuck Avenue neighborhood before even the proprietors knew what that meant. Elizabeth and Sahag Avedisian started the shop with no particular experience with either retail or cheese. They wanted a quiet establishment that would permit them to pursue research and other interests during the store's slow moments. In that they failed. When it first opened as a small cheese shop in 1967, the Cheese Board attracted Berkeley's large European population and "well-traveled locals." Four years later, the owners sold the store to the workers and it began operating in its new iteration as a collective.[42]

The flat business structure encouraged the worker-owners to innovate: they experimented with baking and started selling baguettes at a time when few Berkeleyites had even heard of a baguette. The Cheese Board evolved into a breakfast café and bakery, then a pizza place, and an eclectic social center.[43] By then the shop was at the center of three important subdistricts. The first is place specific, the much-discussed Gourmet Ghetto. The second is less familiar, more a community of shared interest and a heartland of the Bay Area workers' cooperative movement. The third, also a community of interest, is the already discussed assortment of politically motivated, frequently quite radical bakeries.

Early Chez Panisse Food Quality Priorities

Chez Panisse fit well into the Gourmet Ghetto's neighborhood values. Waters had made a defining visit to France just as Child went on television. Like so many before and after her, Waters fell in love with French markets. But she was less oriented toward the technique and tradition of Child's careful transcription of classic French practice and instead prioritized fine ingredients. Waters was also at home in the political climate of her time and place in Berkeley. She left for France just after Mario Savio climbed on top of a police car during Sproul Plaza demonstrations in 1964 that began the free speech movement. But she returned home to Free Speech Movement inner circles, paired with one of its leaders, artist David Goines.[44] While she honed her cooking skills feeding friends in the movement, her politics did not bar infatuation with quality ingredients. Indeed her own political path has been shaped by the challenge of obtaining and sharing those ingredients.

Legend has it that she started Chez Panisse hoping to make a living cooking for friends who had been gathering at her house to eat. The enterprise began on rather shaky financial footing and remained unstable for its first decade, even as it developed a reputation for increasingly breathtaking food. The food was challenging in part because sourcing options for quality ingredients were limited. Waters's food came from the Berkeley Co-op, local Chinese markets, and Bill Fujimoto's Monterey Market.[45] Staff foraged in the traditional sense—along the railroad tracks and around town for herbs, berries, and mushrooms, and neighbors bought lettuce and Meyer lemons to the back door of the restaurant in sacks.

Over time, Chez Panisse generated a national and international buzz that attracted new entrepreneurs to open related high-quality food shops. The neighborhood soon included a charcuterie (the Pig-by-the-Tail, opened in 1973 by an early Chez Panisse alum, Victoria Kroyer Wise), a purveyor of stunning chocolates (run by the Gourmet Ghetto's "other Alice," Alice Medrich), and Le Poulet, home to soon-to-be national sausage impresario Bruce Aidells.

Transitions in the 1980s

The Demise of the Berkeley Co-op

The aura of the Shattuck Avenue enclave of fine food grew and spread, threatening its complex balance of values. The closing of the Berkeley Co-op in the mid-1980s signaled the nature of problem. Beyond the much-chronicled political infighting, increasing prosperity left many Co-op customers disinclined to do the hard work of running their own grocery store. The same inclinations pushed Co-op managers to compete with both increasingly sophisticated upscale supermarkets and the next new thing in retail, discount stores, both of which successfully moved in on the co-ops' turf—providing health products, bulk packaging, and discounts.[46] The Berkeley Co-op responded with a high-end look and retail services that irked the still-committed members who remained committed to affordability and reliance on member volunteers. When the

Co-op finally closed, it was, tellingly, replaced by a distinctly upscale grocery retailer. Food aesthetics could have displaced the Cheese Board–style combination of food and fairness, and for some the connection was perhaps lost in "bourgeois piggery."[47] However, the quality-oriented restaurateurs were soon pushed to a more complex understanding of food quality based in sustainability.

Industrial Districts and Regional Cuisine

A reimagining of regional cuisine helped the Gourmet Ghetto change track. Elizabeth David's 1962 classic, *French Provincial Cooking,* maps unexpectedly well onto the industrial district framework. David puts distinctive local ingredients, not fetishization of haute cuisine, specifically at the heart of a common food enterprise. "Country and seaside restaurateurs," she observed, "began to realize the possibilities of attracting tourists by advertising some famous local dish on their menus."[48] Nonetheless, those possibilities require pursuing shared interests, David notes, and relationships among proximate, related producers that keep "a flourishing tradition of local cookery" alive: "The cooks and the housewives must be backed up by the dairy farmers, the pig breeders and pork butchers, the market gardeners and the fruit growers," working to adapt "the old methods to changing tastes and altered conditions without thereby standardizing all the food."[49]

Chez Panisse and Regional Cuisine

The regional cuisine idea became central to making Chez Panisse "better than it was."[50] The process perhaps started when the restaurant began to celebrate a "new sense of place, of where we lived and ate."[51] The October 7, 1976, menu marks a significant step in this direction. Jeremiah Tower, the early, mercurial Chez Panisse chef, was inspired by a recipe for "cream corn soup à la Mendocino" to create a regional feast.[52] The menu was written all in English and featured, in addition to the corn soup, Tomales Bay oysters, Ig Vella's Sonoma cheese, and fruits and nuts from the Alemany Farmers Market. It was, Tower relates, as if "a switch had been thrown." Thereafter, Chez Panisse became the heart of a region like those David described, articulating a California cuisine deeply flavored by both the alternative growers and the political environment of the Gourmet Ghetto.[53]

Building a New Distribution System

The Farm-to-Restaurant Project

Tower's reorientation intensified food distribution problems. Regional eating required finding ingredients that paralleled the kind of chickens, breads, and cheeses that were available to French regional cooks.[54] Waters's first approach was agricultural: in 1977 and again in 1980, she unsuccessfully sought to source high-quality ingredients by growing them on farms directly connected to the restaurant.[55] Waters was soon looking for the "hippies raising goats up in the hills of Marin and Sonoma [who were] beginning to learn to make chevre as beautiful as the small-farmstead cheese of France. But of course they never advertised. They had to be found . . . farmer by farmer, artisan by artisan."[56] Chez Panisse took financial risks and made commitments that positioned it as a major investor in alternative food growers and processors.

Greens, Mudd's, and Regional Cuisine

Waters's quest for fine ingredients was not solitary. Two other district restaurants, Greens in San Francisco and Mudd's in San Ramone, were

Laura Chenel and the North Bay Cheese Community

Laura Chenel introduced goat cheese to the Bay Area and most of the rest of the nation. She learned to love the animals and the cheese at a time when both were viewed, even among foodies, as odd and probably undesirable.

Chenel was raised in Sebastopol, California, where her father taught school and her family ran a restaurant and a turkey ranch. She had her first European sojourn as an exchange student in Holland, came home in 1967, and studied cheese making with Ig Vella, master of the famous local jack cheese and David Viviani of the Sonoma Cheese Factory. She then returned to Europe and studied with Jean-Claude Le Jaouen, who wrote *The Fabrication of Farmstead Goat Cheese* (1987). When she came home again, she supported her cheese explorations by waitressing. Once she finally got the cheese right, she did what aspiring small producers still do to establish their product and reputation: she tried it out on Alice Waters. As she had done for Bill Niman's meat, Waters put Chenel's chevre and her name on the menu, where it became a permanent fixture.[a]

a. Starchefs.com (n.d.).

equally motivated to seek high-quality produce.[57] When Greens opened in San Francisco in 1979, it was not an attempt to recreate a French regional tradition. Greens is affiliated with the San Francisco Zen Center. The restaurant was built by Zen Center carpenters, served Zen Center farm produce, and provided an opportunity for Zen Center students to "extend their Buddhist practice into the workplace." The Zen Center farm that supplies some of the Greens produce—the work, in part, of the Alan Chadwick who was so important in developing the garden at UC Santa Cruz—is a place for meditation. It is no longer necessary to be a Zen Center student to work at Greens, but it was at the beginning.[58] In fulfilling its religious mission, Greens "brought vegetarian food out from sprout-infested health food stores and established it as a cuisine in America."[59]

Mudd's embodied yet another aspiration. Like Greens, the name is not a manifestation of European priorities nor does it reference an earthier tradition in food. Mudd's was the brainchild of Virginia Mudd, a one-time school teacher in San Ramon.[60] In the 1970s, she envisioned a teaching garden that would sell its produce to local restaurants. Ahead of her time, few customers materialized, so she developed both the garden and the restaurant herself. Growing your own vegetables, however, is more like a farmstand than a common enterprise, and it did not encourage a community of growers and consumers. Mudd moved on in the early 1990s and left her gardens for the city, with the intention that the land would remain a community garden.[61]

Sibella Kraus, Foraging, and the Farm to Restaurant Project

Chez Panisse Café line chef Sibella Kraus embraced the task of finding product and producers and connecting them effectively to restaurants. Her arrival at Chez Panisse was probably as pivotal as Jeremiah Tower's debut a decade earlier. Tower himself asserts that among all the changes taking place at the restaurant in the early 1980s, "innovation really began after Sibella Kraus came . . . and planted those little lettuces in Alice's garden." Kraus's Farm to Restaurant Project created the relationships and collaborations that were essential to regional cuisine. Kraus infused the project and the district with her particular priority on supporting farmers on the land and moved the chefs beyond their preoccupation with aesthetics. On Kraus's watch, Chez Panisse and the district became

For the Climatologically Challenged

California enjoys a particularly favorable, occasionally twelve-month growing season that encourages dissenters to grumble that eating seasonally and locally is not possible elsewhere. They miss an important point: people from Greenland and the Sahara, among others, have "eaten local" for millennia. A Mediterranean climate helps, but those who used to think that wine could be made only in France have gotten accustomed to wine from California, New York, Oregon, Washington, Chile, New Zealand, and even Michigan. L'Etoile in Madison, Wisconsin, and an evolving food district in Vermont's Northeast Kingdom have traversed a path to seasonal cuisine.

Odessa Piper opened her restaurant, L'Etoile, in 1976 when she was twenty-four. She promptly descended into near bankruptcy. But she struggled long after "any reasonable person would have closed up shop a dozen times or more," investing her college savings, her first marriage, and her health. "It was very important to stay in business," she relates from the safe distance of thirty years on, "in order to tell the story of . . . seasonal, 'regionally reliant' cooking."[b] The low temperatures in Madison in December, January, and February and March average between 5°F and 12°F.

Piper honed her approach to food not in France, but by farming suboptimal ground in a New Hampshire commune and later in the Kickapoo River Valley of Wisconsin. When she opened L'Etoile, she had Julia Child's book in hand and was literally surrounded by the Dane County (Wisconsin) Farmers Market. Piper treated a percentage of her food budget as "research and development"—an investment in products and relationships required for regional cuisine at a time when a small-scale distribution infrastructure for high-quality product was nonexistent in the Upper Midwest.

She nevertheless committed L'Etoile to local, seasonal eating. In her telling, that effort was constrained as much by lack of space and spare cash to buy and store ingredients as by climate. Piper therefore collaborated with the nearby Harmony Valley Farm to plan the varieties and amounts of vegetables she would need to get her through the winter. She also invested in constructing a root cellar on the farm. In the dead of winter, Piper served about 60 percent local produce and about 30 percent of that fresh.[c]

Similarly, when Tom Stearns appeared at the 2011 EcoFarms conference, he reported that it was 30 degrees below zero back in the Hardwick, Vermont, area where his High Mowing Organic Seed Company is located. Promoted as "the town that food saved," Hardwick was a mining center that had lost momentum first when the granite market collapsed in the 1920s and 30s. The dairy-based economy that replaced it fell under the same pressures that afflicted Bay Area coastal dairies. But an influx of Internet-savvy food entrepreneurs and agriculture enthusiasts beginning around the turn of the twenty-first century have developed climate-appropriate, financially innovative businesses.[d] Jasper Hill Farm, for example, is cashing in on the nation's growing appetite for artisanal and farmstead cheese. The operation began in 1998, the same year the region lost five dairy farms, and has become a center for aging and marketing cheeses from nearby crafters. It is home to a 22,000-square-foot facility blasted into the same granite that had failed the region earlier as an economic base.[e] All this is possible at 30 degrees below zero.

b. Odessa Piper, telephone conversation with Sally Fairfax, October 11, 2010.
c. Trubek (2008).
d. Hewitt (2010).
e. See "Vision of Jasper Hill" at http://www.jasperhillfarm.com/index.php?option=com_content&view=article&id=79&Itemid=149 (accessed February 19, 2012).

increasingly associated with "advocacy for farmers markets, for sound and sustainable agriculture, [and] for Slow Food . . . as well as the Garden Project at the San Francisco county jail."[62]

Serving a regional cuisine required discovering, encouraging, and connecting hundreds of moving parts, each with its own peculiar constraints. Fortunately Kraus was more about connections than cooking. She admits that she never handled the line very well. She frequently notes that one needs a very calm temperament in a professional kitchen, which requires a kind of energy that is probably more appropriate to saving a drowning child. Inspired by the People for Open Space (POS) "Endangered Harvest" exhibit at the Oakland Museum,[63] Kraus shifted emphasis, volunteered at POS and then happily traded out of a shift at the café to supervise a project that Waters and POS jointly supported.

As the project started, a number of small farmers had demonstrated that they could support themselves producing sustainably grown products, and a few chefs had demonstrated a market for beautiful, delicious food. Nevertheless, it was frequently difficult for the chefs to access the products for which they were looking. Even after Chenel's chevre found a niche on the Chez Panisse menu, it was a challenge to get the cheese out of the hills. Kraus's early solution illustrates the distribution problems: when Full Belly Farm's fresh corn was headed to Chez Panisse on a bus, Kraus would call Chenel and tell her to get her product down to the Greyhound station. Kraus could then collect both shipments in the afternoon.

Much of the produce that is familiar today was not available as Kraus set out in the early 1980s. She located heritage products that had survived the conventional system's rush to standardized, transportable commodities. She credits UC's cadre of county-based Small Farm Advisors for encouraging small-scale growers to diversify.[64] But she is alone among those with whom we have spoken in describing seed savers as essential to California cuisine. She found families, hobbyists, and assorted enthusiasts who were passionate about seeds: a vein of melon lovers in the Midwest, people "mad about" all different kinds of potatoes, a fanatic garlic breeder, conventional tomato growers who raised heritage varieties for their family, and networks of immigrant families savoring strains of rapini that most gourmets had never heard of. These producers were not, according to Kraus, initially inspired to cultivate unique food crops for fancy restaurants. They were protecting their ethnic and family food traditions.

With care, some could be talked into growing their specialties for the restaurants Kraus was organizing. But after the seed was saved, sprouted, grown, and harvested, distribution—especially distribution of small amounts of product—remained a major barrier to meaningful trade. Communication was one of the primary challenges: it was difficult for restaurants to find out what was available from whom in time to put it on the menu. To notice only the most obvious difficulty, farmers get up to pick crops just as chefs are coming home from work and going to bed. Helping them to communicate about what was at its peak in the field and what was needed in the kitchen was not a simple task.

Beyond that, "schedules and routes of established delivery systems were," Kraus recalls, "the limiting and defining factor." Veritable Vegetable had already begun to shape the market: the collective offered "farm pick-up service on twice weekly routes running through the Central Valley to the L.A. area and returning . . . [plus] a delivery service, used primarily by natural food stores and restaurants." VV also agreed to serve as the "main centralized drop-off and pick-up point" for Kraus's project, but it was not positioned to discover and promote new farmers and new products nor equipped to handle extremely perishable products like baby squash or squash blossoms.[65] Moreover, in Marin, Napa, Sonoma, and Contra Costa, despite a growing number of small farms, there was no delivery service at all.[66]

Even packaging was a problem. The normal unit for conventional produce is the boxcar, broken into pallets—fifty-six boxes of forty-eight heads of lettuce on each pallet—but none of the alternative producers grew that much at one time and none of the restaurants needed that much. What is more, the baby lettuce heads that were stylish then did not travel well in standard wire-bound lettuce crates. Styrofoam lugs designed for grape clusters were better. However, the heads still arrived damaged, and it was clear that an innovation was required. When Kraus and Warren Weber decided to try cutting the lettuce before it headed up and shipping it in bags, a whole new food industry was born.[67]

The farmers and restaurants also had different priorities. The chefs had specific goals: they wanted tree-ripened fruit and fresher and more interesting produce, and they wanted to be introduced to "the best of what was out there."[68] For that, Kraus observed that chefs "were willing to pay a fair price." Farmers wanted reliable customers who would pay on delivery, and they appreciated both help in moving product when it was at its peak and feedback on what customers, including chefs, might want them to plant. They also appreciated ideas for new products, as when chefs began curing sun-dried tomatoes in olive oil, a product that was then commercially available only as an import.[69]

Nonetheless, supporting sustainable farmers was not on chefs' agenda when the project began. "Local" did not figure into understandings of food quality, and as Waters notes, when she was starting, "I was never looking for sustainable farmers or organic food. I was really looking for taste."[70] Kraus's project encouraged the chefs to support their suppliers, develop "a sense of investment in the producer as well as the product," and "view the support of local growers as an investment. If you want the product," Kraus insisted, "you want to keep the farmers on the land, so you have to understand what they need. In the long run," she argued in her project report, "encouragement—especially loyalty to specific 'proven' growers—will serve their needs."[71]

A festive dinner at Greens concluded the project and opened the way for the relationships between chefs and farmers that have become commonplace in alternative food. Many of the producers and chefs met face to face for the first time at Greens. The idea of a common enterprise was codified in Alice Waters and Warren Weber's long-term contract for supplying lettuce to Chez Panisse. They agreed that Star Route Farm and Chez Panisse "will continue to work in an open and communicative way to understand each other's needs and problems in fulfilling this contract," and signed "in good faith."[72]

The initial 1983 dinner became an annual event, A Tasting of Summer Produce, for creating relationships and educating restaurants about farmers' growing and harvesting schedules and delivery problems. As public interest in the restaurants and the wonderful product exploded, Kraus expanded beyond growers and chefs and invited the general public to join the festival at the Oakland Museum. The event began shifting around in the calendar to showcase different products and was replicated in Los Angeles.[73] The project instilled the idea of a responsibility shared between rural producers and urban chefs, chefs who then brought the concepts of sustainable, organic, and local into more general expectations about food quality.

Chez Panisse became both a symbol of and the headquarters for rethinking expectations about fine food. But sustainably produced food is expensive, and as the Co-op closed its doors, critics decried elitism in the emerging food culture. That, however, misses a fundamental point: California cuisine, and Chez Panisse in particular, made a point of "uncoupling good eating from fanciness."[74] Still, Waters has observed that the proximate Cheese Board kept her restaurant from becoming too expensive or exclusive and recalls that the "merry collectivists" of the Cheese Board celebrated their "authentic food and authentic politics" by bursting naked through the restaurant's dinner one evening, "the very embodiment of ecstatic, anarchic nature, if not anarchosyndicalism."[75]

Decriminalizing Farmers' Markets

Although the upscale restaurant was an important beachhead for bringing chefs, discriminating consumers, and sustainable growers together, it was too narrow to become the base for what has been frequently called a "delicious revolution." Sustainable production achieved a far broader audience in California's farmers' markets, which became important to food retail, community building, and education. Kraus did her foraging in farmers' markets that were not quite legal. Before relationships between farmers and consumers could become almost literally a product on offer at California farmers' markets, years of industry-encouraged regulations had to be rolled back. Before exemptions were established through a state-level direct marketing program in 1977, it was illegal for farmers to sell their products until they had been graded, sorted, and packaged, meaning that it was effectively criminal for farmers to sell directly to consumers.[76]

The renaissance of farmers' markets in the 1970s and 1980s reflected a convergence of energy, uniting anti- hunger activists, farmworker advocates, small farmers, urban and rural community developers, and church groups. The process was too attenuated to allow us to say that a coalition was built or that a great victory inspired activists of diverse priority to see a common cause. Nevertheless, all the elements of alternative food—environment, food, and justice—were visibly and effectively involved in the process of promoting the direct sale of fresh fruit, vegetables, meat, and dairy.[77]

By the 1970s, a number of diverse groups had identified farmers' market as potential tool for their antihunger work. Church-based activists such as the Quakers and Oakland's Interfaith Hunger Coalition sponsored urban farmers' markets designed to serve the poor. Minority farmers—frequently Asian immigrants—supplied Stockton's farmers' market, which was located under a freeway overpass and was organized and patronized largely by farmworkers and local food cooperatives. Organic farmers in search of an outlet for their product began the Covered Market in Davis in 1975.

While isolated farmers' markets flew under the radar, the issue of legality became pressing in the summer of 1975. Peach growers gave away truckloads of their crop near the capital in Sacramento to demonstrate that they could not afford to process it for sale: low prices made the expensive packing and sorting requirements untenable. According to farmers' market lore, Governor Jerry Brown was dismayed to learn that the roadside fruit stands and U-pick operations were illegal and appointed a committee to investigate. By the following spring, several bills to clear a path for farmers' markets had all been defeated. Retailers, and some farm and labor interests, opposed what they saw as a fundamental change in food distribution.

Brown's secretary of agriculture, Rose Elizabeth Bird, finally bypassed the reticent legislature and granted exemptions from the packing regulations. Ironically her actions relied on the same authority over crop marketing programs that were earlier intended to benefit large growers and co-ops. The U.S. Department of Agriculture, sensing a groundswell, began to offer guidance for establishing farmers' markets through its own small direct marketing program.[78] Congress passed the 1976 Farmer-to-Consumer Direct Marketing Act, which enabled the cooperative extension service to work with local groups to develop farmers markets' and other forms of direct marketing. The act provided impetus and funding for market organizers to overcome state and local prohibitions.

Conventional food industry lobbyists nevertheless continued to oppose California legislation, and farmers' market legislation did not pass until 1985, well after a wave of new farmers' markets had been established under Bird's exemptions. Moreover, the law that passed allows California farmers' markets to operate only under unusually restrictive regulation. The law requires farmers to sell their own crops directly to consumers, limits the products that can be sold, and limits who can organize and operate a farmers' market. Although intended to curtail farmers' markets, the direct producer-to-consumer relationship has arguably worked to the markets' advantage, helping them to become, and to remain, an important alternative to conventional groceries and supermarkets.[79] As in the rest of the nation, California farmers' markets began to multiply: from 15 before Bird and Brown's agitation, to 140 in 1990, and 443 in 2002. In 2011, there were an estimated 754 farmers' markets in the state.[80]

Currently about 15 percent of all U.S. farmers' markets are located in California, and these markets have been the salvation of many small farms.[81] Because of the higher profits per unit that come with the elimination of middlemen, direct sales increase the viability of farms that could not produce enough volume to interest wholesalers.[82] Scholars estimate that gross returns for farmers at farmers' markets are more than twice those of conventional sales. Seventy-seven percent of California farms with revenues of less than $50,000 per year or fewer than fifty acres participate in farmers' markets or other forms of direct marketing. A single individual or family owns almost nine out of ten farms that market directly.[83] Nearly a quarter of organic farmers in California report that farmers' markets are their primary marketing outlet, and more than half engage in direct marketing (only 8 percent of all California farms participate in direct marketing). Although organic farms constitute only about 2 percent of all farms in the United States, they represent nearly 20 percent of farms participating in farmers' markets.

Conclusion

Farmers' markets and ingredient quality-oriented restaurants developed in tandem with the burgeoning sustainable and organic farming movement. The three are clearly linked in a distribution system that enabled California cuisine and crystallized consumer interest in safe food and food quality. The connections among them have become so routine that it is necessary to remember that the growers and chefs celebrating The Taste of Summer did not start with shared priorities: chefs had to learn that sustainability is an important element of quality.

The rush of attention to California cuisine underwrote a growing conversation about food quality. In a brief time, innovators who had initially focused on simply prepared, beautiful food turned to encouraging investment in greenhouses for alternative growers and educating consumers about sustainably grown heirloom produce varietals that had disappeared over the past century. Alternative producers prospered, supported by growing numbers of demanding farmers' market customers who were willing to pay a premium for such product.

Nevertheless, these core values of taste and sustainability developed as the Berkeley Co-op was closing. The chef and the farmer replaced Cesar Chavez at the center of the discourse about fairness, and the fact that careful artisanal production required careful artisan workers was largely overlooked or ignored. Thus, the district's revitalization of the food conversation was initially narrow, even with sustainability in the mix, and anarchosyndicalism to the contrary notwithstanding, it emphasized the preferences of prosperous consumers. But that was just the first phase.

7

Maturing the District

In the Bay Area, the conversation about what it meant to produce and eat good food did not remain narrow for long. It rapidly became contentious and complex. New sectors—dairy and beef—joined the alternative enterprise, and new buyers, particularly schools and hospitals, emerged. As these businesses and related institutions thickened the district, others expanded and began operating in multiple locations. Through this process, many of the distinctions between urban and rural values and priorities eroded. Alternative processors—artisanal cheesemakers and cheesemongers among them—worked specifically to build and knit together the district. Producers increasingly started identifying themselves collectively and in association with a particular geography in the style of a European appellation. The district itself became more recognizable to news media, consumers, and government officials.

Protecting farmers on the land remained a priority for the new generation of alternative food advocates, but their efforts were shaped less by early environmentalism than by intensifying health concerns. As the conventional dairy industry deployed the engineered hormone rBGH to increase milk yields, Straus Dairy went in a different direction, gaining organic certification and opening an organic creamery.[1] Nearby, several generations of alternative beef producers struggled to get their synthetic hormone, antibiotic-free, and, later, grassfed product to market. These new products made healthier, sustainably raised milk and beef available to growing numbers of prosperous consumers.

We tell that story in the first part of this chapter. Our narrative is less serene than its pastoral setting might imply. For example, a growing oyster sector subject to stringent state health regulation repeatedly challenged dairymen on water quality issues. More broadly, scaling up became highly controversial across the alternative sector. As businesses grew, their connections to land, markets, and district norms changed. Some businesses grew by partnering with neighbors, creating space for smaller operators to start without having to do everything simultaneously. Some evidence suggests that the scaling-up process improved opportunities for workers, but alternative food was not definitively on the side of equity and justice. Renewed controversy about the short-handled hoe—and some organic growers' defense of it—suggested the opposite: that organic producers were not so different from their conventional counterparts.

Scaling up exposed and complicated the question of the relationship between alternative food and justice, and debates over how much growth, and what kind of growth, were consistent with alternative norms and values became common. One perspective considered even small expansion as a fall from grace and branded scaled-up enterprises "industrial organic."[2] Fundamental issues about the meaning of alternative food swirled in charges and countercharges large and small. Within the district, Bill Niman's Niman Ranch, once a celebrated local alternative beef producer, ran into push-back. As Niman became involved in a national market, he ran afoul of district values prioritizing small-scale, artisanal, and local in preference to expanding operations in order to increase access to meat raised without antibiotics or synthetic hormones. Straus encountered controversy over amended organic pasture rules that were intended to curtail the industrial producers but seemed to ignore climatic differences and prioritize happy cows over increasing the availability of organic milk.

Our second topic takes place somewhat removed from both public view and many of the personal relationships formed in the early days of the district. Institutional buyers—hospitals and schools particularly—became leaders in nutrition, food, and health education. Kaiser Oakland and Chez Panisse addressed issues of scale from two very different perspectives. Kaiser Oakland and Healthcare Without Harm, two institutions not generally associated with alternative food activism, together promoted connections between institutional food buyers and alternative producers. Just as restaurants had in earlier days, large buyers needed guidance to connect with alternative producers. Programs linking the two addressed the distribution mismatches and introduced a new array of possibilities for small-scale producers to participate in new and larger-scale markets. Farm-to-school programs for linking those sellers to schools were first piloted in in the late 1990s and have become important throughout the nation. The school lunch took on a particularly high profile in the district in 1994 when Alice Waters began to connect her interests in cooking, sharing, and education to growing concern with health at the Edible School Yard kitchen and gardening project at a Berkeley middle school.

Finally, we turn to three programs that express some of the ambiguities about justice goals and the means to achieve them that arose as the district matured. Food Not Bombs (FNB) continues 1960s radicalism in district food politics, gleaning and sharing leftovers in the tradition of the Diggers and the Black Panthers. At the other end of public perception, the Ferry Plaza Farmers Market (FPFM) is the frequent anchor for allegations that alternative food in the district is snooty and too expensive. However, for all its clarity, FNB's political intensity has had limited impact, while some market sellers regard FPFM as essential to financing fair labor practices. Between those two antipodes, Marin Organic gleaning programs, which distribute non-commercially harvested crops from organic farms to school children and the elderly, well express the district's learning as health moved to the center of food quality.

At the end of the 1990s, expanding access to healthy food was a concern but not a consistent focus of effective attention. Nevertheless, as a new generation of innovators found their feet, they focused on human health and began to develop new ideas of food quality. Unfortunately, good meat and farmers' markets' leftovers by themselves cannot right the wrongs of a society that seems bent on subsidizing cheap, unhealthy food while denying basic rights to agricultural and other food system workers. It therefore would be easy to overstate what was accomplished.

Haute Healthy Food

Straus Family Dairy

The Straus Dairy led coastal milk producers into the alternative arena. After two years of transition, the dairy was certified organic in 1994, institutionalizing Ellen Straus's long commitment to pesticide-free production. Shortly after, her son Albert opened the Straus Family Creamery to process organic, European-style milk products. His timing was perfect. Four days after the creamery opened, Congress ended a moratorium on the use of rBGH. As conventional dairies began using the synthetic hormone, alternative producers rushed their rBGH-free and organic milk to market.

The Petaluma creamery that had processed virtually all West Marin and Sonoma milk since the early twentieth century banned rBGH milk.[3] They did not, however, help Straus's transition to organic. For unclear reasons, the creamery refused to accept organic milk, so Straus was forced to sell what he could not process at his small start-up creamery not as fluid milk but far cheaper "nonfat milk solids." As organic milk became more familiar, the Petaluma co-op made things easier for the growing cohort of organic dairymen by accepting their surplus milk. But innovator Straus's early losses were tremendous as he traversed a relatively uncharted field.[4] Five almost separate sets of rules define organic dairies. First, all feed must be 100 percent certified organic. Second, the animals must be raised in a pasture-based system that provides access to active grazing. Third, no hormones or hormone implants can be used to increase production or speed growth. Fourth, antibiotics cannot be used prophylactically to prevent spread of disease among crowded animals or clinically to treat a sick animal. Fifth, all of these conditions must have been in place for a minimum of a year before milk can be sold as organic. Compliance with all of these rules must be carefully documented and reported.[5]

The pasture requirement for organic dairy certification went some distance toward restoring the natural advantage provided by the coastal pastures: Confined animal feeding operations (CAFOs) are difficult to frame as pasture based unless the farmer lies, which is not unprecedented.[6] However, even Marin's coastal grasslands cannot support cattle year round: the State Water Quality Control Board does not allow cattle to graze on coastal pastures during winter rains.[7] Even counting the silage that the dairy harvests for winter, Straus has to buy about 40 percent of its feed, including most of the protein. During the transition year in which cows must eat organically before the milk can be sold as organic, Straus paid about $100,000 extra for organic feed with no price premium to compensate.[8] Nevertheless, the organic feed issue was a relatively straightforward challenge compared to treating sick animals without antibiotics. A course on homeopathic medicine was helpful, but Straus soon focused on prevention. Herd health actually improved: the cull rate dropped by about 10 percent.[9]

Straus's most famous, and widely imitated, innovation is a digester that turns manure into electricity. Over twenty years of strong press interest, the lead story on his digester has evolved from energy savings, to water-quality protection, back to energy, on to sustainability, and then to global warming. Straus was focused on manure management, but when his electricity meter began to spin backward, others took notice. Digesters have since became so common that they are no longer as newsworthy, but Straus's innovation and tinkering in this area allowed other families to shorten their trial-and-error process.

Processing Healthy Haute Food

Organic milk must be processed in an organic creamery. There are advantages to avoiding the conventional framework: Straus could distribute and market products without government setting the prices.[10] At first, that was small comfort because financing a processing plant was a major problem. The federal Small Business Administration, which is supposed to support new enterprises, did not help the Straus family. "When it gets down to it," Straus observes ruefully, "they don't like new business . . . they wanted 150 percent collateral." Straus got started in a reconditioned kitchen at a former drug treatment facility.[11]

Although he is outside the main foodie scene, connoisseurs consider Straus's minimally processed products to be superior to most other dairy options. Alice Waters, according to legend, encouraged Straus to make fine butter. But that process also required tinkering, including resurrecting equipment from small operations that had folded as conventional processors consolidated. Straus proudly proclaims, "We make as much butter in a week as most butter plants make in an hour," using a 1950s churn that resembles a front-loading washing machine. The Straus operation also abjures the salt that lengthens the shelf life of butter. But even that apparently slight change required learning: Straus butter has less moisture and may burn if it is overheated. Discerning cooks who use it, have had to adjust.

Glass bottles are Straus's signature. Although pressure from retailers eventually forced Straus to offer some plastic containers, the company still emphasizes the bottles, which are made from 40 to 50 percent recycled glass and are each used approximately six to eight times.[12] Constant tweaking has reduced bottle wash water consumption from 12 gallons to half a gallon per minute, and the wash water is used to hose down the creamery floors.[13] Straus has yet to solve the problem of plastic, particularly for yogurt. Heating flexible plastic, as when incubating yogurt in a plastic cup, releases chemicals that can act as hormone disrupters, so Straus yogurt gets incubated in a stainless steel vat and is cooled before it is put in the cup. Most consumers are unaware of the issue, but Straus is not happy using plastic at all (including cornstarch-based plastic, which he believes creates a market that simply encourages transgenic corn production without assuring reliable biodegradation).[14] He aspires to plastic-free packaging, but for now, he relies on it as a stopgap because "the company does not have the resources or space to deal with returnable yogurt containers."[15]

Working for a Cheese District: The New Normandy North of the Bridge

Although Straus's experimenting gave the alternative milk community a jump start and he has invested enormous effort in helping other dairies convert to organic and in protecting organic standards, knitting folks together was his parents' specialty, not his. A different cadre of innovators has continued that relationship building process. Local and national recognition of the alternative food district is frequently associated with the flowering of artisanal cheese, which began with Laura Chenel's chevre in the 1970s and took flight in the Cowgirl Creamery in the 1990s.[16]

Sue Conley and Peg Smith, who share a collective identity as "the Cowgirls," arrived from the East a decade too late to be flower children. Smith was drawn to the district by early press attention to California cuisine and the revolutionary idea of a female chef—Waters at Chez Panisse. She applied immediately, and unsuccessfully, for a job at Chez Panisse. Rejected, she cooked all over the district until she landed a position at Waters's increasingly famous restaurant.[17] Conley was not initially attracted to the food scene—she intended to be a journalist—but was soon drawn in. She signed up for classes at San Francisco City College Hotel and Restaurant School.[18] In a surprisingly short time, she was running her own restaurant, Bette's Oceanview Diner, where an hour's wait for breakfast was common and an unusual tipping policy, in which the tip was included in the check, approximated fairness in the almost uniformly exploitative restaurant industry.[19]

Like Waters and Kraus, Conley too pursued interests beyond the kitchen, working with partners to develop a family of businesses: a carry-out restaurant, a bakeshop, a catering facility, and a line of baking mixes so that fans could reproduce Bette's scones and pancakes at home (in less time than it took to get a table at the restaurant). This diversification allowed both Conley and Bette's employees to grow professionally, moving between the enterprises to develop new skills. When Conley left Berkeley to follow her partner, a state park ranger, to the coastal pastures, Marin Agricultural Land Trust (MALT) founder Ellen Straus interested her in the ranchers' struggle. Conley honed in on the local milk and explored the possibilities for premium cheese.

Integrating their urban and rural worlds and contacts, Conley and Straus combed the district, displaying Straus products to specialty food markets and restaurants. In 1994, Conley began distributing Bay Area cheeses as the products of an identifiable region.[20] She taught local restaurants how to construct a cheese course while she learned the business. But she was not making an adequate living when Napa winemaker Robert Mondavi, a renowned networker in his community, recruited Smith to be executive chef at a biannual wine event in Bordeaux. Smith hired Conley, and the two immersed themselves in European approaches to marketing regional specialties.

They fell in with Jean d'Alos, a Bordeaux cheesemonger and *affineur* (one who manages the *affinage,* or aging process, for small cheesemakers in a region), who runs a cheese shop in Paris that promotes products of the Bordeaux region. The Cowgirls also studied Neal's Yard Dairy in England where Randolph Hodgson had learned how to distribute cheese for small producers and was midcourse in resurrecting the moribund British farmstead cheese community. The Cowgirls' business model was born somewhere between Bordeaux and London.

Conley and Smith headed home to open their own creamery and aging facility, intending to create a European-style appellation that would protect farmers on the coastal land north of the Golden Gate bridge. They brought with them urban associates and food sensibilities, plus a passion for paying the real cost of high-quality food. That passion grounds their unusual commitment to the fair treatment of labor, which developed over decades of knocking around in the frequently brutal world of high-end restaurants.[21] As they have since demonstrated, a fairly priced product can direct the resources of regional and global cheese markets to sustaining the land and rewarding workers with fair wages and viable careers.

Selling the District in the Countryside

The food district notion got off to a slow start. Growers have to find consumers who are willing to buy a fairly priced product, and farmers' markets can be an important venue for drawing together and educating those consumers. State law prefers nonprofits as farmers' market managers, and in Marin the likely candidates, the Marin Agricultural Land

Neal's Yard

Alternative agriculture is frequently associated with the reuse of historic buildings, and Neal's Yard is a prime example. An alley in London's Covent Garden, Neal's Yard had served as a fruit and vegetable market for several centuries before urban congestion got the upper hand in the 1970s. The seventeenth-century buildings were threatened with redevelopment until the area was listed for protection in 1973.

Rejuvenation centered on returning food-related retail, and Randolph Hodgson opened Neal's Yard Dairy in 1979. Entering a British cheese market famous for its "rubbery, plastic-wrapped" cheddar, Hodgson encouraged artisanal producers to maintain traditional methods and quality and linked them to urban customers.[a] He also mentored his customers. Some were enthusiasts, but more were neophytes, though willing to be educated by knowledgeable sales staff about their cheese heritage.

As the cheesemaking business became both economically viable and even somewhat glamorous, young people who had abandoned their farm roots returned home to learn the traditional methods from their parents and join the business. New producers also learned the artisanal methods and entered the once nearly abandoned field. In the process, their revitalized skills and land base, and the heritage breeds of livestock that depend on them, have also achieved a new relevance and stability.

a. Kapoor (2000).

Trust and the local Farm Bureau, both declined, asserting that there were no farmers in Marin, only ranchers. Three years passed before it became apparent that another new organization was required: an ad in the local newspaper lured potential sellers to a founding meeting of what became the Marin Growers' Group, which formed to start a West Marin farmers' market. When it opened in Point Reyes Station in 1995, the new market provided a distribution venue for the region's small organic growers. It also connected the town to its agricultural history and, optimistically, its future. The Growers' Group continued to meet, later developing both a grown-in-Marin label and a county organic certification program.

Another Creamery in the Countryside

Starting a farmers' market was simple, though, compared to opening a creamery in Point Reyes. The Cowgirls wanted to open shop in a refurbished barn, envisioning that it would provide a center for developing, displaying, and marketing local agricultural bounty. In their vision, a takeout grill and rotisserie that also served ice cream would attract the public, and the creamery would produce artisanal cheeses. Although the county was supportive of agriculture, planners were suspicious of food processing. And although the barn was not new construction, the conversion was a fight. Competing food purveyors encouraged the county's hesitation, and not everybody was on board with agriculture enterprises that might attract tourists needing parking places to a protected landscape that many regard as wilderness.[22]

Financing was also a problem. Like Straus, the Cowgirls went to the Small Business Association, but when the loan fell through at the last hour, the Cowgirls refinanced their homes and turned for help to neighbors and friends. After several frustrating years, the creamery was approved with no rotisserie, sixty extra parking places, a complicated and expensive septic system, and very little operating capital left.

Nonetheless, at the opening party, urban and rural strands of the district came together on the rancher's turf: well into a rainy February 1997 night, restaurateurs from the "farm-to-restaurant" crowd and beyond mingled with local producers, artists, and assorted notables. The Cowgirls became familiar fixtures in community activities as Ellen Straus brought Conley onto the MALT board of directors, hoping to ensure that a fuller range of agricultural interests, including row croppers and processors, gained a voice in its operations. The creamery Barn became an important element of district life, and civic events began to frequently feature a platter of Cowgirl cheeses on the table or in the raffle (and still do today).

Building Networks: Cooperative Extension and Institutional Thickening

Cowgirl Creamery was at the cutting edge of something that became a groundswell. As the state withdrew support from the Small Farm Program that had encouraged small producers in earlier days, the director of UC Cooperative Extension (UCCE) in Marin and Sonoma counties put her energy into building community around diversifying agriculture. Ellie Rilla began with a series of workshops and community discussions that gave direction to both public and private efforts. UCCE has contributed some of its own funds, but its more important role has been in steering public attention and directing county and private funding to support alternative agriculture.

For example, Rilla wrote grants that initiated support for a permanent coordinator of organic and sustainable agriculture in the county. She raised funds for a county ombudsman to facilitate effective county planning and health department attention to agricultural projects, and she took a lead in guiding county planners and regulators in learning about agriculture. UCCE workshops and training have supported the often challenging generational shifts in district farming families and Rilla has cultivated connections with and amonga younger and far more diverse generation of about-to-be farmers. Frequently working one-on-one with new owners, UCCE advisors have also helped to keep family farms in agriculture and to secure a place for recently arrived landowners in the agricultural community. In addition, Rilla has had considerable success pitching stories about local agriculture to the region's newspapers and pushing the coastal agricultural community into closer relationships with the urban areas of the district.

The Marin Growers' Group Becomes Marin Organic

Even as the institutional webs in the district grew denser, an effort to reinvent the Marin Growers' Group to encompass organic certification uncovered remaining rifts in the local agricultural community. Marin's county agriculture commissioner believed a county certification was essential, but somehow the word got around that the mortality of a leading local advocate was major concern, and the goal was to have the whole county certified "before Warren Weber croaked." Some thought that that was a worthy target, but others worried about the possibility that growers would be required to certify. A different segment believed that Marin Growers should go "beyond organic," to include wildlife protection, water conservation, and "community based & socially just economic practices" in the mix. Others stated flatly that if the group tried to mess with labor issues, they were leaving. The Cowgirls provided refreshments and UCCE raised funds to pay for a facilitator. At length, an optional county organic certification was created, with a Marin-specific label, Marin Organic (MO). The label MO first adopted suggests some "beyond organic" aspirations, but behind it, some tough issues remained unresolved.

Figure 7.1
The first label designed for Marin Organic, which is no longer in use. The border suggests the early organizers' range of aspirations for the certification program. *Source:* Marin Organic.

The New Cheeseheads

Rilla and the Cowgirls also collaborated to offer workshops to train dairy ranchers in cheesemaking. The cheese option gained momentum as dairymen came under pressure from oyster growers and environmentalists to reduce runoff due to concerns about the water that flowed down from coastal pastures into shallow Tomales Bay during and after big rainstorms. Ranchers who sold cheese, which added value to the milk their cows produced, could reduce the size of their herds and any associated manure-polluted runoff from their fields.

Worries about runoff arose less from federal water quality standards than from the state version, the Porter-Cologne Act, and the National Shellfish Sanitation Program. The sanitation program requires that pollution events affecting the oyster-producing area be "predictable and manageable."[23] The California Department of Health Services attempts to "manage" the events and any accompanying elevation of fecal coliform levels in Tomales Bay by halting oyster harvest after major storms.

When water quality issues in the bay arose in the 1990s, they were not new, but the oyster-dairy connection was. A biannual State of the Bay Conference began discussing real estate development threats to the bay in 1988.[24] Simultaneously, a growing phalanx of Tomales Bay oystermen redefined their product to take advantage of the regional emphasis in district restaurants. Oysters that had long been sold shucked in jars were upgraded and reframed to serve the district's expanding high-value, on-the-half-shell trade. In that changing market, and after decades of coexisting productively in and along the bay,[25] the oystermen and the dairies came into conflict. And in 1998, an El Niño year, they came to loggerheads.

Storms had kept the harvest closed for much of the winter, and the following May, the Department of Health Services connected 110 illnesses to consuming Tomales Bay oysters. With little evidence regarding the actual cause, oyster growers and environmentalists eyed the cattle on adjacent hillsides. They seemed unlikely culprits, as the virus struck well after the rainy season, and therefore any runoff that might contain manure had ended. But as tempers flared, studies that had been in progress for several years finally emerged. A State Shellfish Protection Act authorizes convening a Technical Advisory Committee (TAC) to report on such disputes, but those hoping for a clear cause and an obvious solution were disappointed. The TAC report was incomplete: the analysis implicated the dairies without assessing the possibility of contamination from residential septic systems, recreational boaters, or wildlife. Dairy owners who believed the TAC study was unfair and inadequate continued to argue with oyster growers, who found allies among environmentalists. In addition, the county assessed bay-side septic systems. The county's report noted that most properties had very limited space for on-site sewage disposal systems: most of the relevant land was in fact below the high-tide mark, and several gray water dischargers drained directly into the bay.[26]

Gradually the ranchers and oystermen cooled off and established new organizations to address the issue. With Robert Giacomini in the lead, eighteen dairy owners formed the Tomales Bay Agricultural Group. The Marin Community Foundation funded on-ranch testing of fecal coliform loading. With the Tomales Bay group involved, ranchers became more receptive to allowing researchers on their property, and best management practices were developed targeting high-traffic areas in dairy operations. Two new septic-related groups—the nonprofit East Shore Planning Group and the county-designated Onsite Septic Policy and Technical Advisory Committee—analyzed residential wastewater, and the National Park Service (NPS) began developing plans to upgrade restroom facilities for boaters. Finally, building on more than a decade of biennial discussions, diverse stakeholders including community groups, agricultural, maracultural and environmental interests, and public agencies joined in 2000 to form the Tomales Bay Watershed Council. In 2004 the group adopted a watershed stewardship plan and raised more than $1 million in grants earmarked for local watershed monitoring, restoration, and septic system overhaul.[27]

The issues that had pitted cows against oysters were resolved in the producers' joint recognition of their dependence on both the working landscape concept and the marketing of place. Like the winemakers to the north, district oystermen and dairymen market their specialty products as a direct link between consumers and environmental quality. "Fortunately that stuff is behind us now," notes one oyster grower. "The community is maturing."[28]

An interest in preserving Tomales Bay water quality was not Robert Giacomini's only incentive for looking into cheese production. His children had moved away from the family ranch, but they were intrigued by the possibility of a new family business producing specialty cheese. All four Giacomini daughters attended Conley and Rilla's first class with their parents. As a family, they decided to cut the herd to reduce runoff and use the smaller amount of milk to make a more profitable product. The Petaluma milk co-op went easier on Giacomini than it had on Straus, accepting whatever milk the family could not use in its start-up. Soon their Point Reyes Farmstead Cheese Company was turning milk, still warm from their cows, into "Original Blue." The product was such a success that the family expanded the line and built an event space to bring chefs and food personalities to teach classes on the ranch. While some fear that West Marin will turn out like Carmel, a precious tourist town south of San Francisco, there are probably limits on the Point Reyes approach to posh: one has to walk very close to the barn to get into the stylish facility.

The Lafranchi brothers' new creamery in Nicasio is more modest. After a century on a ranch established by their immigrant grandparents, the family was struggling to smooth out the irregularities of the fluid milk market. They ran a vegetable stand and a pumpkin patch, and rent part of their pasture to a mobile chicken operation managed by a neighboring alternative beef operation, but as the district gained momentum, they saw the cheeses of their Swiss homeland as a path to financial stability. The Lafranchis have also been advantaged by entry into a thickening collective enterprise. They too began with a Conley/Rilla cheesemaking class, and the county's planning ombudsman eased the approval of their new small creamery.[29] The Lafranchis now bill themselves as "California's only certified organic farmstead cheese makers" When a tourist biking through Nicasio asked if the new firm was "getting ready to give the Cowgirls a run for their money," the horrified Lafranchi cheesemonger responded that the whole point was to work with the Cowgirls. "We are part of a district that works together," she intoned with gravitas. "We would not be here without the Cowgirls. To help us get started, they lent us the molds that we use to make our cheese."

Not all of the new cheesemakers are multigenerational ranchers trying to stay on family land. In 2007 the ombudsman helped Marcia Barinaga secure county approval for a creamery in Marshall. The former journalist and *Science* writer was the first to buy a ranch encumbered by a preexisting MALT easement. While she leased the land to neighbors and MALT worried that it might go out of agriculture, UCCE wrapped the new owner in community support and advice.[30] Barinaga attributes her cheesemaking consistency, unusual for a beginner, to her scientific background. But she is deeply aware that she has become part of an established community of relationships, support, and knowledge and that she could enter an existing market without having to develop a context for her cheese business.

Building Networks for Alternative Beef

All this cheese talk could obscure the fact that many innovative ranchers are abandoning milk entirely.[31] When prices for conventional beef took a lower-than-usual dip in 2000, alternative beef production began to spread through the region. And when Michael Pollan's "Power Steer" appeared in the *New York Times Sunday Magazine* in 2002, the possibilities for healthier meat took off. Building a supply chain and scaling up has been challenging, but it is now possible to get grassfed burgers in a new breed of California fast food chains.

Bill Niman, In and Out of the District

Minnesotan Bill Niman had laid the foundations much earlier. He arrived in Bolinas in the 1960s. By the 1970s, Niman was soon selling pasture-raised meat on a very small scale. In 1978, Niman added a business partner, Berkeley journalism professor Orville Schell, and more land to his growing operation.[32] Congress helped the Niman-Schell start-up when it created the Golden Gate National Recreation Area, extending NPS land acquisition all the way to the bridge. After some resistance, the partners accepted an NPS buyout, grateful to get out from under an 18 percent mortgage. The NPS deducted a substantial fee for a leaseback that gave them access to their pasture for several decades. Thus, the NPS appears in the Niman-Schell story in a role similar to MALT's in other Marin producers: the cash infusion meant that the two could invest in their business without the burden of paying a mortgage. A newcomer without multigenerational ownership of coastal pastures could scarcely hope to acquire land, notes Niman, without such intervention or a small fortune.

For most of Niman-Schell's first decade, sales were so meager that they avoided the regulations that can create complex barriers to entry for small meat producers. By preselling animals on the hoof, they were able to use on-ranch slaughter services and a nearby butcher shop, neither of which required federal inspection.[33] Their first retail sale, however, required USDA-inspected processing. Neighbor Stewart trucked the cattle to Rancho Veal in Petaluma. A few local restaurants were serving Niman-Schell product when the young team got a big break: Niman-Schell was the first branded meat named on a Chez Panisse menu. That encouraged calls from the growing cohort of chefs adopting the Waters model.

As demand for their meat grew, Niman-Schell tried to scale up by arranging for their neighbors to grow beef according to their protocols. Although they had demonstrated the possibilities for branded beef raised without antibiotics or added hormones, they could not talk any neighbors into joining their operation. UCCE tried as well, inviting coastal beef producers to a meeting at which "natural beef" mogul Mel Coleman invited them to join the Coleman brand. The meeting was well attended, but when conventional beef prices rebounded just a little, interest evaporated.

Niman-Schell found partners elsewhere and kept growing. When Schell moved on, Niman renamed the operation Niman Ranch, restructured the debt, and brought in new partners.[34] To succeed in the meat business at any scale requires what is known as "balancing the carcass" (i.e., selling all of it, not just the high-end cuts). At a larger scale, Niman had enough volume of end meats to turn into hot dogs and other value-added products. He then contracted with a local processor and sent hot dog samples not to Chez Panisse but to Trader Joe's (TJ). At a time when the chain did not sell any meat at all, TJ's ordered 10,000 pounds. Niman recalls that before the Trader Joe's order, their largest batch was 200 pounds. But, they pulled it off, and the resulting "Fearless Frank" was a TJ staple for years.[35]

Niman's new partners bought him out during a dispute about protocols in 2007. By that time, all Niman Ranch cattle were finished at a midsized feedlot, and only a few remained in Marin. Internet wholesaler Freshnex began distributing Niman Ranch meats, as did the McDonald's-owned Chipotle chain. Although Niman Ranch's feedlots were a cut above the industrial version in terms of size and impact, the product was no longer at the forefront of sustainable practice and no longer small scale or local. As healthier grassfed beef became more readily available, Niman Ranch lost traction in the district, and Chez Panisse shifted sources. Shortly after, the firm lost its spot at the Ferry Plaza Farmers Market.

Grassf-fed Beef

While still a vast improvement over conventional beef,[36] Niman's product was feedlot finished and did not offer the health advantages of grass-fed cattle.[37] Although grass- and forage-fed beef is standard in many places, particularly Latin America, grassfed producers in the United States are up against Americans' preference for the heavily marbled meat that results from fattening cattle for slaughter—known as "finishing" in the industry—on a mostly grain diet. Grassfed production has waxed and waned in the United States, usually in relation to grain prices. Recent interest in the district and elsewhere reflects different consumer interests—in human health, environmental sustainability, and farm animal welfare—that have proven more stable. But even with growing demand, moving the product to the plate remains far more difficult than distributing vegetables.

Despite those challenges, Point Reyes rancher David Evans has built his own vision along the path first blazed by Niman. Evans's grandfather, fourth-generation Point Reyes dairyman Alfred Grossi, had just paid off the mortgage when the NPS condemned his pastoral zone ranch. Devastated and coping with low milk prices and debt to upgrade manure management facilities, he quit the dairy business. Grossi's daughter decided to stay on the ranch as an NPS lessee and try beef. She began at a time when antibiotics and hormones were the new tools of the trade, and she jumped right in. A generation later, Evans found himself rejecting his parents' approach and reweaving older skills into his operation, which he started on leased land.

Back when local gossip pegged Niman as an upstart, Evans worked with and for him, managing Niman's ranches in Bolinas and Petaluma. He learned the importance of a "branded product" there, but his own personal aha! moment came while reading *Stockman Grass Farmer* magazine. Editor Allan Nation described emerging niche markets for grassfed

Lines and Shadows on Labels

Product labels are a minefield for consumers and alternative producers. Processors of products containing genetically modified organisms (GMOs), such as corn, have opposed requiring labels that would state that fact. But they have also successfully urged Congress to prohibit producers of products containing no GMO inputs from putting that on their label.[b] In 2004, conventional beef producers hit an ethical low, urging the USDA to block an alternative beef producer from testing his cattle for mad cow disease. The producer was trying to export beef to Japan, which had closed its borders to cattle that had not been tested for bovine spongiform encephalopathy (BSE). The USDA claimed it was authorized to prevent use of the BSE test kits, and the District of Columbia Court of Appeals supported the conventional industry by agreeing.[c]

Labeling disputes around definitions are more common. The USDA's Food Safety and Inspection Service (FSIS) requires all meat and poultry labels, conventional or alternative, to contain specific information, and FSIS must verify that any "voluntary label claims"—like "natural" or "grassfed"—are true. FSIS relies on signed affidavits and rarely conducts in-field audits to verify these claims. Although that practice has raised complaints, the degree of controversy is small compared to the debates over what terms should mean. USDA's Agricultural Marketing Service standard requires cattle to be 100 percent grassfed but says nothing about confinement feeding or use of antibiotics and hormones. In contrast, FSIS standards allow meat to be labeled as partially grass fed, for example, 80 percent grassfed. That standard includes most feedlot beef, and its attendant hormones and antibiotics. The definition fracas motivated grassfed producers to form two national trade associations amid anxiety over grass-fed having similar problems to organic after the federal government became involved. Today only one remains: the American Grassfed Association, which issued a comprehensive standard in 2009.[d]

"Natural" may be even less informative on a label. FSIS states that "all fresh meat qualifies as natural" if artificial additives or preservatives are not added to the meat after slaughter. "Natural" implies that hormones or antibiotics have not been used before slaughter, but it means no such thing. Producers like Niman Ranch that fall in the middle—finishing beef on grain in feedlots without hormones or antibiotics—have had to use multiple labels to reflect that fact. USDA efforts to define "naturally raised" to describe the middle ground have not achieved consensus. Ironically, grassfed and organic products may create a false sense of security among those who choose them. The critical consequences of CAFO reliance on antibiotics and hormones arise in the waste stream rather than from eating the beef itself. Grassfed consumers, vegetarians, vegans, and even ovolactarians are equally at risk from antibiotic resistance among waste and foodborne pathogens that results from using antimicrobials to speed growth in conventional food animals.[e]

b. In July 2011, the United States dropped twenty years of opposition to labeling of foods containing genetically modified organisms (GMOs) at the U.N.'s Codex Alimentarius Committee, which establishes what food-related standards are permitted under international trade agreements. Since then, labeling GMO foods as such is permitted but not required. For insight into likely impacts, see Steve Suppan, "The GMO labeling fight at the Codex Alimentarius Commission: How big a victory for consumers?" Institute for Agriculture and Trade Policy Think Forward Blog, July 11, 2011, available at http://www.iatp.org/blog/201107/the-gmo-labeling-fight-at-the-codex-alimentarius-commission-how-big-a-victory-for-consum (accessed September 2011).

c. Creekstone Farms Premium Beef v. United States Department of Agriculture (2008).

d. Gwin (2006), available at http://www.americangrassfed.org/about-us/our-standards/ (accessed September 2011).

e. We have said nothing about claims that food processors make about their products. The food industry has learned that health claims on packaging sell food even when the claims are of dubious merit (Silverglade and Heller 2010).

products and wondered why, with foodies and chefs like Alice Waters around, was no one producing grassfed beef in the Bay Area. "I stopped in mid-sentence," Evans recalls, and thought, "This is the path to Oz." Evans had goals for his community as well: pastures that were productive in beef and had environmental benefits might, he thought, reconcile ranchers with the "new bohemians" who had come to West Marin in the 1970s. By his fourth season, Evans and his partner were so confident in the quality of their beef that they planned to market five head, a number he recalls seemed huge at the time.[38]

Michael Pollan's "Power Steer" article changed Evans's plans, however. In it, Pollan followed the life of a conventional steer from birth to steak, crystallizing consumer concerns about conventional beef. Days before it was published, Evans had purchased a $10 advertisement on eatwild.com, a Web source for grassfed livestock products, and Pollan mentioned the Web site in his article.[39] "I had 30 phone calls the day after the article came out, from all over the U.S.," Evans reports. He issued a press release that was picked up by several major news networks and leaped into the field. Marin Sun Farms (MSF) sold twenty-five head of cattle that year.

Selling that many meant a challenging 500 percent ramp-up for MSF, and scaling up processing is, even today, a major problem for small and midsized beef producers. To be sold, meat must be processed at a USDA-inspected facility from slaughter to packaging or at a state-inspected facility in one of the twenty-seven states that have them.[40] Nonetheless, such facilities are limited: four companies control more than 80 percent of processing in the United States, which is consolidated in a few extremely large plants. Many small and midsized processing sites that might have handled grassfed and other livestock for local markets have closed for complex reasons. Regulations are challenging, the sheer operational costs can be staggering, even for a small facility, and finding skilled, efficient workers is difficult, especially when demand for processing services is seasonally uneven.

The only USDA-inspected cattle slaughter facility in the Bay Area is Rancho Veal in Petaluma.[41] Evans used it to slaughter, but took his "cut and wrap" (i.e., turning the carcasses into packages of cut beef) to a family operation in San Francisco. Because that facility was not set up for small, specialized orders, Evans first had to work side by side with the butchers and then haul the packaged meat to rented freezer lockers north in Sebastopol, where it sat until he loaded it up again for weekly deliveries, often back to San Francisco. The resulting supply chain "made no financial sense," Evans admits, but he pieced it together and made it work out of necessity while building his brand and customer base.

When Evans later tried to solve his problems by building a butcher shop in Point Reyes, he quickly ran into the same planning roadblocks that had all but confounded Straus and the Cowgirls. Although the county was even less excited about a butchery than it had been about creameries, the new ombudsman helped Evans coordinate between the planning and health departments, and a local foundation supported the technical assistance he needed to achieve regulatory compliance and permitting. Still, the county would not allow a butcher shop as a stand-alone: to get the permit, Evans had to open a twenty-five-seat restaurant as well.[42] For cash flow, he followed his alternative colleagues' by then well-trod path to San Francisco and began selling at the Ferry Plaza Farmers Market.

Storms that flooded out his butcher shop in 2005 turned out to be a blessing. They required Evans to rebuild, which allowed him to refocus on fundamentals. Bill Niman's scaling-up troubles provided another piece of luck. Niman Ranch had gone far in refurbishing the butchering skill base in the district, and when it moved out of its Oakland processing facility, the newly unemployed workers did not pound the pavement looking for new jobs. Niman's celebrity butcher, Oscar Gedra, called Evans. Evans did not think he could afford Gedra's Niman-level salary, but after a trial week, he discovered that "the extra pay he [Gedra] wanted was easily worth it." As quickly as he could afford to, Evans hired on Gedra's relatives and other Niman staff.

Evans also benefited from Bay Area customers' willingness to learn and experiment with food. Unlike Straus butter, which requires no particular skill to appreciate, grassfed beef was easily overcooked and arguably not very tasty at first, and producers had to simultaneously improve their product and teach their customers to prepare and prefer it. Bay Area consumers were willing to pay extra for grassfed, initially for a less-than-perfect product, in part because district chefs provided support and leadership. As Evans learned how to finish his cattle on grass, meat quality improved and more restaurants climbed on board: Evans saw restaurant doors "opened for me. Chez Panisse, Acme Chophouse. I got them all."[43] Still important educators and arbiters of taste, restaurants today account for about 60 percent of MSF's gross sales.

Scaling-Up Issues

While such success obviously presents opportunities, expanding a business is not easy. Moreover, district innovators learned, as Niman had earlier, that even when scaling up is successful, it presents conflicts with other district priorities. When Straus, Evans, and the Cowgirls began expanding, each in different ways, their growth revealed important inconsistencies in district expectations about quality food. Expansion at Straus Creamery, one of the district's more successful alternative operations, was challenged by wild swings in organic milk supply. The creamery first lost traction, ironically, during a shortage of organic milk: as demand shot up, large organic processors sought partnerships with smaller producers. When one of his partners found a better deal, Straus lost 25 percent of his milk supply. Straus, who had walked the neighbor through his transition to organic, was frustrated by the sudden change, which has follow-on ripple effects in the district. The Cowgirls' Red Hawk and Mount Tam cheeses disappeared from market shelves for a time as they and Straus cut back production while locating new milk sources.

They soon brought new dairies into the growing web of alternative producers, but it was not long before problems of oversupply shifted the organic milk market. Although some blamed the glut on dairies that fudge pasture requirements to produce feedlot organic, Straus believed it was due to typical cycles of agricultural supply and demand.[44] As the 2008 economic recession intensified, Straus scrambled, reforming his network with a much smaller dairy and developing new product lines for the food service industry: ice cream and frozen yogurt bases and "barista milk" for specialty coffee drinks. Straus Creamery continued to grow, albeit at a slower rate than previously.[45]

These shifts and growth have complicated Straus's position in the organic community. Although new organic rules had restored some of the advantages that pasture-based dairies enjoyed in the early twentieth century, the oversupply issue eventually led to efforts to restrict the market by requiring year-round pasture access for organic dairy cows. A subtext of these regulations was to curtail "industrial organic" production and protect the price premium for organic producers. Straus opposed these changes, believing that they might cause conflict with California water quality regulations that prohibit grazing during the coast's spectacular winter rains. Organic purists accused Straus of colluding with the bad guys and running his own feedlot. A compromise resulted in language that all could live with, but the fundamental conflict between availability and access versus limited supply and price premiums remains.

Evans's business model for scaling up, like Straus's, required that he partner with ranchers in his community. Those ranchers had rejected both Coleman and Niman, but what a difference the "Power Steer" article made. These partnerships have allowed MSF to work with wholesale customers requiring ever larger volume and achieve economies of scale without resorting to feedlots and growth hormones. Evans too has, however, run into issues with expectations about local artisanal production. Specifically, California's rules for certified farmers' markets require that sellers offer only what they have produced on their own land. Even without the feedlots, some markets have treated Evans as a meat aggregator rather than a rancher.

The Cowgirls first scaled up by opening a store in the refurbished Ferry Plaza building. That operation looked promising enough to lenders that they were able to obtain their first true loan from a bank. The Cowgirls then opened a store in Washington, D.C. Although they chose the location because it is near their families, the opening of the store appeared as if the Cowgirls were going national, and to some, it seemed to be a fundamental break with most understandings of "local." Nevertheless, the Cowgirls find it harmonious with their basic goals: they are replicating the Neal's Yard model in another location, using the established heft of their brand to develop and support a new community of artisanal cheesemakers and good jobs in the Middle Atlantic region's nascent quality food sector.[46]

This effort is similar to an international program in which the Cowgirls became involved at the invitation of the Global Environmental Institute and the Blue Moon Foundation. Through the program, the Cowgirls trained food entrepreneurs in developing countries in marketing, distribution, and retail of their local products within their regions. Ayesha Grewal, for example, had an internship to visit the Cowgirl Creamery and returned home to open a creamery in the Himalayas to serve local products to the increasing numbers of tourists who pass through her once-remote area.[47] Whether replication of the Neal's Yard model will work in a Himalayan setting remains to be seen, and questions about how the Washington, D.C., operation fits into diverse notions of "local" remain.[48]

Mike and Sally Gale, early grassfed innovators who have intentionally avoided expanding their operation, are at the opposite end of the spectrum. The Gales do not want to become employers in the California agribusiness tradition and rely largely on their own labor. After a troubling experience with a conventional beef broker, they began personally selling cattle they had raised and processed completely on their ranch. They had been advertising on the eatwild.com Web site for two years before the 2002 "Power Steer" trembler. Calls from retailers and restaurants that wanted grassfed beef by the cut were tempting, but selling high-end cuts would stick the Gales with marketing challenges and low prices for the rest. The Gales's approach to beef does have limits: not everyone is prepared to buy a chest full of frozen beef or to cook unfamiliar cuts. Nonetheless, some experts argue that this is the most cost- and time-effective way for small-scale ranchers to operate.[49] The Gales's ranch is profitable because it is diversified. They host weddings, art workshops, retreats, and other events on the ranch, and their heritage apples are pruned so Sally can harvest them alone. In addition, Mike's retirement income from his first career provides a useful cushion.

Labor, Scale, and Fairness

The Gales are rightly concerned about California's agricultural labor traditions, but we regard scaling up as essential to fairness. The Gales's practice concedes that a very small operation cannot support workers fairly. Nor does it support workers in numbers sufficient to have an impact on economic opportunity in the district. Nor, as the operation also demonstrates, can it achieve enough economies of scale to make its product accessible to more than a few diligent customers. While the North Bay's artisanal cheeses will not likely feed the masses either, they provide some evidence that as district firms have scaled up, they have moved to provide better wages and benefits to their workers (table 7.1).

Table 7.1
Wages, health, and retirement benefits in North Bay cheesemaking facilities, 2010

	Number of full-time employees		
Compensation characteristics	Fewer than ten	Ten to thirty-nine	Forty or more
Number of facilities in the size category	7	5	4
Average entry-level hourly wage	$9.75	$9.75	$13.55
Average highest hourly wage	$12.50	$22.20	$30.00
Health benefits	40%	100%	100%
401k plan available	No	Yes (40%)	Yes (75%)
Head cheesemaker salary	$35,000	$40,000–50,000	$50,000–80,000

Source: Rilla (2011, 18).

Nonetheless, while wages are a central element of workers' rights, they are not the only concern. One of the downsides of the county's growing support for agriculture is that officials appear to have become reluctant to pressure ranchers to meet state and county standards for worker housing. Housing is more than a place to live, however. Because most employers deduct housing costs from pay packets, the price of housing determines the rate of pay. Any deduction is illegal unless the worker has agreed to the arrangement in writing. Housing for workers is clearly related to larger, and chronic, affordable housing issues in Marin. Nevertheless, workers' housing is structurally different and has changed over the past forty years. Most obvious, shifting toward row crops and processing that require more workers has pressured limited on-ranch housing. In Marin, the issue broke into the news in 1990 when a Spanish-speaking priest at the local Catholic church inquired into housing conditions on seashore ranches. The priest was soon ousted by the rancher-dominated congregation, and the issue subsided. However, it did not go away.[50]

In 1992 workers on one ranch sought legal help to address wage theft and substandard housing problems and prevailed with help from Legal Aid.[51] In 1995 Johnson Oyster Company took a different approach: in the midst of a larger discussion of affordable housing in the county, Johnson laid off all his workers to pressure the NPS to expand his lease. A septic system inadequate for overcrowded worker housing had created sanitation problems.[52] When the issue reemerged again in 2004, attorneys assisted workers in obtaining $235,000 in compensation for unpaid work and substandard housing on yet another ranch.[53] A 2008 literary gathering in Point Reyes Station (figure 7.2), which intended to celebrate the town's agricultural landscape, did so without adequately acknowledging dairy workers, or their housing, as part of that landscape. Two more years passed before the *Point Reyes Light* reported that a public-private partnership would finally "put its toe" into a solution: a substantial grant from the Marin Community Foundation would support building six housing units. Even after more than twenty years of abuses, West Marin's supervisor emphasized that the county was concerned not to disrupt the established "working for rent" model, and wanted to give ranchers an incentive to improve their housing stock.[54]

The short-handled hoe returned to the workers' rights agenda in 2003, splitting the state's rapidly expanding suite of organic producers on

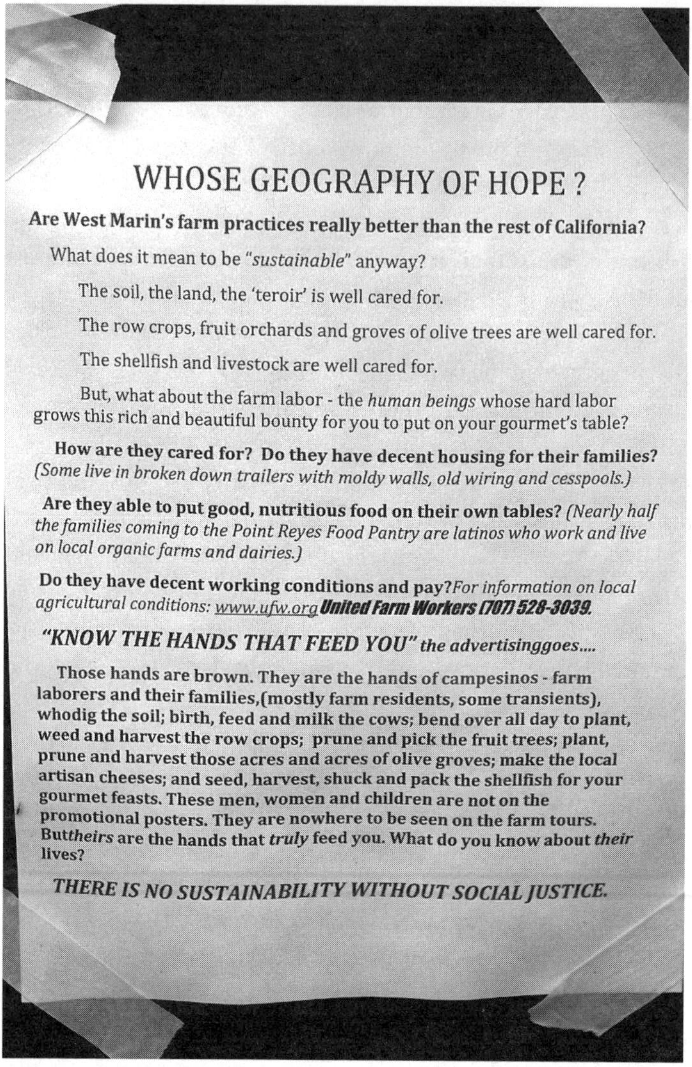

WHOSE GEOGRAPHY OF HOPE ?

Are West Marin's farm practices really better than the rest of California?

What does it mean to be "*sustainable*" anyway?

The soil, the land, the 'teroir' is well cared for.

The row crops, fruit orchards and groves of olive trees are well cared for.

The shellfish and livestock are well cared for.

But, what about the farm labor - the *human beings* whose hard labor grows this rich and beautiful bounty for you to put on your gourmet's table?

How are they cared for? Do they have decent housing for their families? *(Some live in broken down trailers with moldy walls, old wiring and cesspools.)*

Are they able to put good, nutritious food on their own tables? *(Nearly half the families coming to the Point Reyes Food Pantry are latinos who work and live on local organic farms and dairies.)*

Do they have decent working conditions and pay?*For information on local agricultural conditions: www.ufw.org* ***United Farm Workers (707) 528-3039.***

"KNOW THE HANDS THAT FEED YOU" the advertisinggoes....

Those hands are brown. They are the hands of campesinos - farm laborers and their families,(mostly farm residents, some transients), whodig the soil; birth, feed and milk the cows; bend over all day to plant, weed and harvest the row crops; prune and pick the fruit trees; plant, prune and harvest those acres and acres of olive groves; make the local artisan cheeses; and seed, harvest, shuck and pack the shellfish for your gourmet feasts. These men, women and children are not on the promotional posters. They are nowhere to be seen on the farm tours. But*theirs* are the hands that *truly* feed you. What do you know about *their* lives?

THERE IS NO SUSTAINABILITY WITHOUT SOCIAL JUSTICE.

Figure 7.2

The Geography of Hope Conference began in Point Reyes Station in 2008 to celebrate author Wallace Stegner's 100th birthday. The now-annual event "explores the relationship between people, land, and community through the voices of authors, environmentalists, farmers, and artists." The 2009 event focused on agriculture and featured Star Route Farms's Warren Weber, peach farmer and writer David Mas Masumoto, and *Farmworker's Daughter* author Rose Castillo Guilbault. Before the conference, persons unknown plastered the town with yellow, bilingual broadsides (with the original Spanish translated into English), commenting on farmworker exploitation in sustainable agriculture. See "Geography of Hope" at the Point Reyes Books website: http://www.ptreyesbooks.com/goh (accessed February 2012). (Photographs by Richard James).

Figure 7.2
(continued)

workers' rights issues. The organization that spearheaded the first organic standards in the nation, California Certified Organic Farmers (CCOF), opposed legislation that would have closed a loophole in the earlier hoe ban. The district's leading family farm advocacy group, Community Alliance with Family Farmers (CAFF), also opposed the legislation.[55] Although most in the alternative community strongly supported it, conflicts of interest between small family farmers and their employees and a worst-case vision of industrial organic played out in the district and the state.

While weeding fragile row crops is a serious challenge for organic farmers, CCOF and CAFF appeared to have lost their founders' concern for workers.[56] Complete alignment may be impossible: family farmers and agricultural workers are trapped one on top of the other at the bottom of a barrel where conventional agribusiness has long worked to divide potential opponents. But as Getz, Brown, and Shreck have noted, CCOF, which had placed itself in opposition to conventional agriculture, "willingly served as the face" for a conventional antilabor agenda, leading them to argue that "the organic movement, in and of itself, holds little promise for California's farm workers."[57] It is not clear to what extent organic certification was ever intended to promise California's farmworkers anything beyond a workplace free of exposure to synthetic chemicals. But as that incident showed, conflicts between organic producers and farmworkers on the issue of justice are a familiar and continuing theme in the alternative community.

Nevertheless, while scaling up may advantage workers, access for all to healthy food is also central to our understanding of alternative. The district's response to Niman's growth evinces serious disconnects in its food quality priorities. Bill Niman made hormone- and antibiotic-free meat products available to a large, growing audience, and Niman Ranch continues to do so. That firm has also generated numerous skilled jobs in Oakland and initiated the renaissance in the city's former meat processing sector. When Niman left the company, the company employed almost sixty people: twenty butchers, twenty packers, seven drivers, and ten office staff. How does one balance those accomplishments against the firm's shift from exclusively pasture-raised meat to sourcing from feedlots that use neither antibiotics nor added hormones? Is expanding access to healthier food the top priority, or is it greater comfort for cows plus healthy food for some? Niman Ranch lost stature at a time when expectations for justice were ill formed in the district.

Beyond the Pasture: Institutional Buyers and the District's Broadening Agenda

Institutional food buyers put a very different slant on scaling up. The food purchases of hospitals, schools, universities, and hotel chains are significant factors in shaping the market. That impact, plus the fact that the food they serve has an important role in both public health and nutrition education, illustrates the more positive side of scaling up. Two iconic Bay Area institutions, Kaiser Permanente and Chez Panisse, have been at the forefront of shifting institutional buyers toward healthy and sustainable food.[58]

School meals were a focus of concern long before Alice Waters became interested. But her work through the Chez Panisse Foundation has emphasized cooking, gardening, and sharing food in classrooms, and her involvement has crystallized broad concern regarding what children learn about food from eating school meals. Coming from a health care and wellness perspective, Kaiser has joined with Healthcare Without Harm and the Community Alliance with Family Farmers (CAFF) to bring healthy food into hospitals in ways that link institutional food buyers with local, alternative growers. Both of these efforts are part of a broad public learning about food and health. That public learning has become part of the platform on which justice advocates are now innovating additional approaches to food, health, and community leadership.

School Food

School food directors are arguably the most important institutional buyers in the nation.[59] The almost undecipherable political and administrative environment they inhabit is well treated elsewhere.[60] We relate a small part of it here for three reasons. First, for those who believe diet-related health problems result from bad personal choices, the role of school meals in encouraging bad eating habits provides a disturbing counterfactual. Second, racial and economic discrimination that has long been central in federal meal program regulations appears related to diet-related health problems among urban minorities. Finally, the school food issue provides an update on the discussion in chapter 3: conventional food systems arguably reach their nadir in America's school cafeterias. The early twentieth-century power shift from growers to processors and distributors is clearly on view in school cafeterias, with unhealthy lunches as one result.

A Brief History

After 150 years of diverse efforts to provide school lunches to needy children, most results are fairly dismal. The Children's Aid Society in New York City started the first U.S. school lunch program in 1853, intending to improve health among poor and immigrant children. Unlike the Black Panther breakfast program, the society's example did not inspire government action. Fifty years later, two volumes published contemporaneously with *The Jungle* demonstrated the connection between inadequate nutrition and failure in school. *Poverty*[61] and *The Bitter Cry of the Children*[62] demonstrated that hungry children are "in very many cases incapable of successful mental effort, and much of our national expenditure for education is in consequence an absolute waste."[63] Both volumes urged government attention to hungry school children.

When Congress finally did act, it pursued very different priorities: its school meals program was designed to absorb price-depressing agricultural overproduction. The USDA ran the food distribution, which was folded into a New Deal program that paid unemployed female workers to prepare and serve the meals. Both were discontinued during World War II. The school lunch that returned after the war was supported by odd bedfellows. The surplus distribution element was justified as a defense measure: too many malnourished young men were declared unfit to serve as soldiers. But while southern Democrats enthusiastically supported the part that subsidized farmers, they would not allow it to become a wedge for federal involvement in school desegregation. The program's home in the USDA minimized its connection to education, hence to desegregation efforts in schools, and also ensured that nutrition education has never been a central program element. Congress further directed that local governments would pay most nonfood expenses and manage the programs. Because that left important parts of a school lunch unfunded, Congress also allowed local Parent-Teacher Associations to supplement the programs' budgets.[64] That effectively inclined the lunch program toward becoming an upper-middle-class transfer program and minimized participation by schools and school districts dominated by poor and minority students.[65]

The resulting structural racial and economic discrimination was exposed in a study by the Committee on School Lunch Participation, which included representatives of the National Council of Catholic Women, the National Council of Jewish Women, the National Council of Negro Women, Church Women United, and the Young Women's Christian Association. The committee was hardly a radical organization. But the facts its members gathered spoke for themselves, and their report was unveiled at a press conference held the day after Martin Luther King Jr. was assassinated in 1968. At that raw emotional moment, the report's findings—"the greater the need of children from a poor neighborhood, the less the community is able to meet it"—caused a storm. In the backwash, a new "hunger lobby" was born, intent on turning the surplus management program toward feeding poor children.[66] Richard Nixon pledged to feed every poor child a free lunch within a year.[67]

When that goal foundered on cost issues, more odd bedfellows emerged. Liberal senators worked with the hunger lobby to protect the program from funding cuts. They advocated privatizing food services, but unintended consequences soon followed.[68] A two-tiered system emerged: unappealing program meals were available to poor children, while popular fast foods started being sold on school grounds and in vending machines. Participation in the lunch program plummeted: many poor children eligible for free meals dropped out from embarrassment, while wealthier classmates switched to the "competitive" foods. Nobody got a healthy lunch, and the funds available for the program, which were tied to participation, declined sharply.[69]

Since then, continuing reform efforts have proceeded in a complex environment that discourages even the stout-hearted. For example, Congress soon directed that school meals must meet minimal nutritional standards. Nonetheless, implementation has set up a counterintuitive roulette: trying to meet the caloric minimums while not going over the fat maximums is difficult. Adding fruit is expensive, and adding bread can put the meal over the sodium maximums. So the standard recommendation is to add sugar.[70] Industry food scientists have engineered their school offerings to meet the standards. But the concoctions are "not really food," as Poppendieck notes, pondering what a child learns about healthy eating from a Hot Pocket customized to meet the school requirements while resembling a regular Hot Pocket.[71] Complex debates about who should eat what for breakfast, lunch, and over the summer, who should pay, and who should manage the programs have been dominated by players with little interest in providing or teaching about nutritious food. The result is a war on surplus, a war on poverty, followed by wars on spending, on fat, on waste, and ultimately a war on food itself. Bentham's object of wholesome horror had been achieved.

The Edible Schoolyard

The perception that fixing the resulting morass requires national action is understandable. Reform would undoubtedly be easier if the continuing coalition of strange bedfellows was on board with the task. Yet ample evidence has emerged from every corner of the nation that concerted local and regional efforts can significantly improve outcomes. In our district, the Berkeley Unified School District collaborated with the Chez

Panisse Foundation in a unique and increasingly well-known effort to seek peace for Berkeley's school children in the food war zone. The Edible Schoolyard, a large school garden and associated cooking classroom, is confined to the Martin Luther King Middle School (MLK), but it soon drew Waters into a long-term effort to improve the school lunch program in the school district as a whole and to encourage similar efforts in other schools and regions.[72]

Waters's initial involvement appears almost accidental. In 1994, the story goes, she complained to a reporter that the grounds at a school near her home were a mess, and she was summoned almost immediately to the principal's office. Waters began working with him to create first a garden, and then a kitchen, and then a curriculum that integrates both into the science and humanities classes in grades 6 to 8 at the MLK school. The program, now in its fifteenth year, includes teaching and gardening staff paid for by the Chez Panisse Foundation, and an acre of garden, a greenhouse, a water catchment system, and enviable kitchen facilities as well. In 2005, the foundation began working with affiliates to develop different models throughout the country.

The resources that the Waters's foundation provides the MLK program allow it to demonstrate possibilities and raise expectations for school garden and education programs across the country. That connection has made the MLK program a handy target for the "good food is elitist" crowd: one particularly questionable article in the *Atlantic* asserts that Waters "thrust thousands of schoolchildren" into an experiment based on "a set of assumptions that are largely unproved, even unexamined" about the worth of learning in a garden setting.[73] The idea that school gardens contribute to hands-on teaching is venerable, however, and the related literature is voluminous. That should be no surprise: as the United States slipped from rural to urban after the Civil War, industrial expansion, and immigration, many reformers, including John Dewey and Maria Montessori, examined and embraced school gardens for diverse purposes, whether as parks, as an opportunity for experiencing nature that would "ameliorate the 'negative' effects of urban life," or to "recapture America's agrarian past." Others saw gardens as a "route to improving dietary and health practices" among the urban poor and immigrant populations, and regarded them as a place to teach a variety of school subjects, most frequently science.[74]

Two world wars refocused gardening on food production, and education elements lost ground as gardens were converted to playing fields and parking lots at the end of World War II. Nevertheless, those gardens are now coming back, again for diverse reasons that include pedagogy: a garden is a useful teaching tool in many fields, including math, ecology, literature, and nutrition. Anecdotal and statistical evidence demonstrates that students also learn about nutritious food and where it comes from in school gardens, and they share a sense of accomplishment while working with their classmates and teachers.[75] The gardens have drawn parents and volunteers into public education as well.

Berkeley's School Lunch Initiative

Somewhat later and apart from the Edible Schoolyard, Waters also worked in her hometown—like so many other parents in other places—to improve school lunches. Her efforts build on more than a decade of community effort. The Berkeley Superintendent's Group that began in 1997 as a "forum for the increasing number of parents concerned about the school meal program" produced the School Lunch Initiative.[76] Waters jumped into a related school district bond measure campaign to upgrade school facilities that passed with 83 percent of the vote and ultimately provided $11 million for new kitchen and cafeteria construction.

Implementation came slowly. In 2002 and 2003, the school district removed vending machines. The next year, the school board adopted a food policy that encouraged staff to procure food from school gardens and local farmers. It then won USDA funding to introduce swipe cards in cafeterias (the swipe cards conceal who is paying what for their meals and reduce the stigma for poor children). In 2005, the Chez Panisse Foundation provided funding that allowed the district to hire celebrity chef Ann Cooper, a former Kellogg Food and Society Fellow, as the director of nutrition services for the district.[77] On her watch, and not without controversy, nearly all processed foods were eliminated, and fresh, organic foods were added to the daily menu. The food service program remained within budget, and Waters contributed her expertise to the design and construction of a kitchen and school cafeteria at the middle school site. Initially intended to serve that one campus, it soon became the central kitchen where all of the food served in the school district is prepared.[78] Grassfed hamburger and hot dogs allow critics to call Waters an elitist and to mock her insistence that Americans should spend enough to ensure that children are eating healthy food.[79] Nonetheless, there is no pink slime in the meat—a small step toward raising expectations about what is normal and available to all.[80]

Waters emphasizes sharing food and understanding nutrition as important elements of school life and education. Her aura may obscure efforts by many others, but it highlights the importance of distribution to an alternative food system. Having taken major steps to connect small farmers to high-end restaurants, it was perhaps easier to envision a similar system that connected them to schoolchildren. And although a federal shift toward programs that teach about nutritious eating would be helpful, accomplishments in Berkeley and elsewhere demonstrate that significant progress can be made without it.

Healthcare Without Harm and Kaiser Permanente

Healthcare Without Harm (HCWH) has taken a different approach to distribution and scale, but like farm-to-school programs, it has helped small farmers access hospitals' enormous purchasing power. It cut its teeth on toxic materials little regulated in daily hospital use: mercury, flame retardants, phthalates, and pesticides.[81] HCWH then turned to unhealthy food, encouraging medical practitioners to contact food processors directly regarding rBGH in milk and antibiotic resistance arising from antibiotic overuse in the food chain.

HCWH's real innovation grew out a FoodMed conference the organization put together in Oakland, where it found support at Kaiser Permanente. The Oakland gathering drafted the Healthy Food in Health Care agreement that helps hospitals develop language for their general food purchasing order. The contract and electronic catalogue outline the relationship between a large customer and a large distributor. Typically the catalogue describes available product by weight and flavor (e.g., "a 6 ounce chicken breast with tarragon") but says nothing about how or where it was produced. The HCWH agreement encourages contracts that allow purchasers to specify local or organic products and seek alternative sources without penalty if the distributor cannot provide what the hospital orders. Two hundred fifty health care facilities, including many multilocation systems like Kaiser, have signed on. Hospitals have learned to push large distributors into the market for fresh, local products in ways that support regional farmers.[82]

Kaiser was already on board at that first gathering. Oakland cardiologist Preston Maring had focused Kaiser's attention on the importance of access to fresh fruits and vegetables in preventing chronic disease, and Kaiser adjusted its messaging and its food service to position its facilities as nutrition and health educators. It followed with an organic farmers' market outside the Oakland Emergency Room to underscore the point and provide access. The effort has expanded to Kaiser hospitals from Hawaii to Georgia. To implement their agreement, Kaiser collaborated with the Community Alliance with Family Farmers to create a distribution system that connected farmers not to high-end restaurants but to buyers who required huge quantities of food.[83]

Ambiguities in the Maturing District

Class and Access at the Ferry Plaza Farmers Market

Ferry Plaza Farmers Market (FPFM) is perhaps the most obvious manifestation of the ambiguities of the maturing district. Farm-to-Restaurant project manager Sibella Kraus, having spent some time working at Greenleaf and as a columnist for the *San Francisco Chronicle*, founded San Francisco's Public Market Collaborative in 1992 to ensure that a farmers' market would be included in the Ferry Building reconstruction that followed the 1989 earthquake.

As in the 1940s, a one-time event in 1993 demonstrated the potential for a permanent market: one hundred farmers and sixteen restaurants set up in a parking lot near the Ferry Building and generated the same enthusiastic response that the Alemany "one-time" market had in the early 1940s. The popular event amplified the district's growing infatuation with regional cuisine, but the collaborative also emphasized broader expectations, such as ensuring that reasonable rents would allow farmers "to make a reasonable profit" selling at "affordable and reasonable prices." It also asserted that education was "fundamental to the preservation of regional agriculture and a healthy population." Accordingly the collaborative formed a nonprofit, the Center for Urban Education about Sustainable Agriculture (CUESA), to manage the certified farmers' market now located at Ferry Plaza. The farmers' market continued operating in a nearby parking lot, growing steadily as investment in the neighborhood upgraded the infrastructure and aesthetics of its surroundings.

Kraus left CUESA in 2000, when her education priorities conflicted with others' intentions about the market that would surround the Ferry Building. The first property manager did not share the educational vision, preferring a market that echoed the shops and restaurants in the building, which represent the best of Bay Area cuisine.[84] A new CUESA director focused the nonprofit's educational mission on a cooking school, which never happened, and a less formal program showing a few well-heeled market shoppers and some less-well-heeled sample seekers how to cook with the market's seasonal produce.

In spring 2003, a decade after Kraus's first demonstration market, the market relocated to the plaza that surrounds the Ferry Building. It has become a mecca for foodies, culinary professionals, and tourists. Its gourmet orientation and spectacular setting make it easy to typecast as the domain of the wealthy and elite. However, the market looks very different from the perspective of Swanton Berry Farms, seventy miles south on Route 1 just north of Santa Cruz. There, Jim Cochran has run a small highly diversified operation since 1983.[85] In 1987, the farm became the first organic strawberry farm in California; and in 1998, it became the first organic farm to sign a contract with the United Farm Workers. In 2005 Swanton Berry Farm began offering stock bonuses to career employees. For Cochran, selling directly to customers, at his stand on Route 1 and at FPFM, where he can charge a premium price for his berries, is critical.

Table 7.2 compares the cost of produce at three San Francisco purveyors and demonstrates that while organic is more expensive than conventional and that the conventional produce at FPFM is, as at Whole

Table 7.2
Cost per pound for organic and conventional produce at three San Francisco retailers.

	Safeway	Whole Foods	Ferry Plaza Farmers Market
Conventional	$2.10	$3.10	$2.85
Organic	$4.45	$3.84	$3.24

Source: CUESA (2009)

Foods, more expensive than at Safeway, the organic produce at FPFM is far less expensive than either Whole Foods or Safeway. Moreover, at FPFM, most of the price differential goes to the farmer, not to fund a chain of stores or a corporation.[86] Cochran is clear that the price premiums that come from the Ferry Plaza Farmers Market underwrite his legendary labor practices in an industry not renowned for fairness to field workers.

Gleaning and Sharing

As FPFM turns toward the high end of district expectations, a spectrum of gleaning and sharing programs underscores the broadening understanding of and commitment to fairness.[87] Gleaning programs also emphasize the magnitude of waste in the U.S. food system. Those who assert that addressing hunger requires green revolution–style production intensification tools rarely mention it, but overproduction has been and remains the major problem in the agriculture sector. Somewhere between a quarter and a third of the edible food in the United States is wasted.[88] If you include the food crops ready to harvest that remain in the field, the figure goes up to about 50 percent. On a household basis, 14 percent of what makes it through the supermarket into the American home never gets eaten; about 15 percent of that is thrown away, never opened, still within its sell date, which costs the average family about $590 per year. Ten percent of food in fast food chains is thrown away, and smaller fast food outlets like convenience stores discard about 50 percent of their product. Anthropologist Timothy Jones estimates that reducing food waste by half could reduce adverse environmental impacts by 25 percent through reduced landfill use, soil depletion, and applications of fertilizers, pesticides, and herbicides.[89]

Efforts to recover those losses began in the district in the early 1980s but aimed at fairness and access, not efficiency. Daily Bread, a group of Berkeley activists, and Food Runners, primarily restaurant professionals, began to gather excess and leftover foods from restaurants, farmers' markets, and gardens to distribute to shelters and needy people. Both resemble Food Not Bombs (FNB), which began in Cambridge, Massachusetts, in 1980 and moved to San Francisco seven years later. Protesters at the Seabrook Nuclear Power Plant in New Hampshire held bake sales to raise money for their legal defense fees. When they invited hungry folks from a homeless shelter to join them at the protest, they discovered, as had the Diggers and the Black Panthers before them, that food could draw a crowd to political organizing and education. The collective created the iconic poster, "I'm waiting for the day when schools get all the money they need and the Air Force has to hold a bake sale for a bomber."

The FNB San Francisco era began in 1987, when cofounder Keith McHenry moved west. Between 1988 and 1995, two mayors, progressive Art Agnos (1988–1991) and conservative Frank Jordan (1992–1995), targeted FNB for distributing food in city parks without a permit, resulting in 1,000 arrests and citations.[90] Amnesty International threatened to adopt as "prisoners of conscience" any FNB volunteers who were convicted. Today FNB describes itself "not as a charity" but among the fastest-growing "revolutionary movements" in the world. Its mission is not food but progressive political change.[91] At district farmers' markets Daily Bread gatherers are indistinguishable from FNB gatherers.

Marin Organic (MO) is the most recent entrant into the gleaning effort. One of the most farm-centric of the district's thickening organizational scene, it began as the Marin Growers' Group. It has recognized the importance of connecting to urban centers and justice issues, and it is a leader in national efforts to address hunger and nutrition issues through school lunch programs and by connecting farmers to hungry elders. MO acts as a broker or distributor for county farmers with product that is either excess or insufficiently beautiful for the high-end organic market. It has knit together relationships between farmers and school food directors, senior centers, and community food banks. In making those connections, it resembles the Farm to Restaurant project of the 1980s: staff call farmers to find out what is available and call food directors and other recipients to learn what is needed. Then MO invites volunteers to pick vegetables that it cleans and delivers in a truck donated for that purpose. Marin Organic also coordinates cafeteria signage, field trips to farms, and similar programs designed to introduce school children to farmers and their food. The Kraus project was entrepreneurial, but the distribution problems MO resolves are similar. Working with school food service directors seems far more complex, but MO has the advantage of spreadsheets and the Internet. It is now coordinating a national gleaning week program intended to encourage the practice throughout the country.

Conclusion

As the district matured, it became clear that early innovators had developed and demonstrated a path that others could follow: infrastructure and public acceptance of alternatives allowed small operators to enter the market without having to invent the whole supply chain. UCCE brought county planners and diverse private funders on board, and new businesses built on both the successes and failures of the first wave of innovators. The momentum behind the idea of a common enterprise enabled the district to resolve many internal rifts. Oyster growers, dairymen, and environmental and organic purists squared off over details that seemed minute, and they learned to resolve at least some of their differences.

At the same time, issues inherent to small-scale, quality-based responses to large-scale, highly subsidized price squeezing practices emerged and were less easily resolved. Niman stubbed his toe on district preferences for small, local artisanal production even as his efforts expanded access to meat produced without antibiotics or hormone implants. However, as health issues increasingly mattered, two critical kinds of institutional buyers, hospitals and schools, put a different face on scale. They emerged as health and nutrition educators made expanding access to healthy food a priority. But they also developed and demonstrated ways that new relationships, communications, and distribution systems could significantly expand markets for alternative food and opportunities for small producers. Innovating in the rather obscure arena of group purchasing organizations (more typically referred to as GPOs), the reformers aimed to make healthier alternative food available to institutional purchasers.

At first it seemed that prosperous purchasers might avoid the worst of the conventional systems' decline if they shopped carefully. Indeed, the UFW was among the most effective in asserting that ethical purchases could help farmworkers as well. But personal health soon intersected with public health even for the wealthy. A lot of the ambiguity in the district can be summed up with reference to the all-American burger. Consumers can substantially reduce their personal fat, calories, and cholesterol intake by choosing grassfed beef or becoming vegetarian. Either option also reduces the energy consumption that concerned Lappé and the Pimintels. But you can spend or abstain only so far: feedlot products are a public health hazard even if you spurn them entirely.

The issue was not routinely discussed by the general public in terms of class, but a conflict between producing affordable healthy food for all and producing elitist food, arguably even tastier, more sustainable food for the wealthy, was clear. And although the conversation had expanded from aesthetics to include sustainability and then health, justice remained ill defined, and ambiguities about the main priority allowed dissension about scale and price to persist. At that point, new participants from the most affected minority communities created yet another round of innovation in the district.

8

Food Democracy and Innovation

Given the district's history of radical food politics and food-based political organizing, justice issues were surprisingly unformed and inconsistently pursued during the 1990s. That began to change early in the new century. Expressed geographically, one could say that the center of district innovation has moved from the Gourmet Ghetto to Oakland as diverse new voices and activists have asserted their own prerequisites for the U.S. food system. They are defining a new and more comprehensive justice agenda for the district that is reshaping regional practice, investment, and food discourse. The new voices are generally young, and they come largely from communities hardest hit by a major public health crisis: theirs will be the first generation of Americans living shorter lives than their parents. They are also pursuing their agenda during an unprecedented national and international conversation focused on America's "productionist" food policies, which emphasize increasing the total production of cheap food with little or no regard for social and environmental consequences.

Health issues have forced growing numbers of Americans to acknowledge that the Western diet has narrowed around low-quality, highly processed foods and beverages with dire public health consequences. Increasingly malnutrition in the United States is associated less with those who are underweight and hungry and more with overweight and obese people. The human suffering associated with this malnutrition is enormous, and the public policy problems, current and pending, are potentially overwhelming. Diabetes is probably the most discussed consequence: medical experts predict that by about 2050, 20 to 34 percent of Americans will be suffering from type II diabetes, and at a far earlier age than suggested by the name it once went by: adult-onset diabetes.[1] But diabetes is just one part of the health problem. Hypertension, heart disease, stroke, and some cancers are among the diet-related diseases that are also on the rise and increasingly concentrated in poor and minority communities where fast food take-outs and liquor stores are often the most readily accessible, and sometimes the only, food retailers.

This dismal situation has become a topic of intense national debate. The recent explosion of information available about food systems includes popular works by Schlosser, Nestle, Pollan, Spurlock, and others discussed in chapter 2. While some camps still point to improved personal choices as the route to ending the obesity epidemic, any serious response to these food-related health issues requires recognizing that critical parts of the problem are structural. Scientists and scholars have documented the role of food processors in creating and promoting food, snacking, and consumption habits designed to exploit humans' biologically hardwired food preferences for fat, sugars, and salt.[2] While stoking those addictions is profitable for the industry, it has created a cascade of health and social problems.

The severity of these interrelated epidemics has spurred activism among community leaders who are now determined to engage food debates that had largely bypassed them for decades.[3] The most basic way to address frontline communities' most pressing problems, convenient access to affordable healthy foods, was often simply not available to those at the greatest risk from diet-related illness. In many cases, efforts to address this have started with promoting urban agriculture and alternative retail, some of the same strategies discussed in previous chapters. Nevertheless, food justice groups in the district and elsewhere have focused far greater attention on empowerment, self-determination, and social entrepreneurship.

More recent still, young and low-wage workers throughout the food chain have joined the discussion. Relying in part on the ethical consumer–ethical producer notion that took root during the United Farm Workers boycotts, a separate but increasingly coordinated array of new institutions and leaders, "invisible no more," is defining food quality in terms of fair wages, safe working conditions, and health care for food system workers.[4]

Solutions to access issues are complicated by the fact that many health threats stem in large part from foods that most people consider completely standard in supermarkets, schools, restaurants, and homes. Food system labor issues, which have been contested over several centuries, are further complicated by the particular toxicity of immigration issues. But innovative advocates have expanded the conversation—and raised expectations—for change in the district, where food democracy remains a work in progress.[5]

Our primary focus in this chapter is on Oakland, historically and currently. Our version of the city's complex food situation highlights both the critical intersection between urban planning and food systems over time and evolving perspectives on the concepts of "local" and "community." The fertile alternative food scene there reflects a new-style Gold Rush–like period: new ideas and actors from throughout the nation and globe are arriving, rooting themselves while retaining previous links and contacts, and starting to drive the district's evolving understanding of justice.

Exploring Oakland requires additional background. The complexities of food and place in San Francisco and Marin County are well treated elsewhere; our abbreviated version here emphasizes that planning and environmentalism had very different impacts on the east side of the Bay, where large portions of Oakland were devastated by white flight and infrastructure construction. Many of Oakland's food access issues are related to the resulting destruction of place and community. Three elements of Oakland's history are particularly relevant. First, anticolonialism, black nationalism, and the Black Panther era were an important part of black Oakland's response to the devastation of place; second, the results of the subsequent, highly unusual blip in U.S. history, generally known as the War on Poverty, when eradicating poverty seemed both possible and an appropriate national priority, were disappointing in Oakland; and finally, the environmental justice response to mainstream environmentalism grounded Oakland's discussions of food justice.

With a bit of additional East Bay history in hand, we address three topics roughly in chronological order. Although most Oakland food justice groups began (and some remain) with urban agriculture, the various legacies of Black Power and community activism are not entirely subsumed in the garden. We begin by looking at contemporary food and agriculture advocates in the city who often define their activities in terms of self-determination, leadership development, and social entrepreneurship. Similarly, efforts to address food access and grocery retail problems proceed in a crowded arena of nonprofits and for profits that is not entirely harmonious. However, most Oakland food justice advocacy groups are less than a decade old and charting new territory. They are hopeful that the city's new Food Policy Council may provide better coherence and coordination, as well as an effective means of engaging with local politics and policy.

Second, we look at a new community of food processors, microenterprises, and food business collaborations in the district. Food professionals coming into Oakland looking for cheap space, capital, markets, inspiration and even raw materials often bring years of district business experience with them. Others in Oakland's new generation of activist entrepreneurs are starting from scratch in food carts and trucks, incubators and pop-ups with ethnic specialties or dreams of a small business. This eclectic community appreciates delicious food, but prioritizes social responsibility even as it creates issues of gentrification. Their appearance alongside Oakland's food justice advocates highlights important issues about the relationship between nonprofits and for-profit businesses. That relationship and both sectors' interactions with local governments remain to be sorted out. Nevertheless, the Oakland community is pushing the district to include a fuller spectrum of food democracy possibilities.

Finally, we discuss new approaches to food system labor, exploring district involvement in the development of workers' organizations that build on the UFW boycott legacy of ethical producers and consumers. We begin with a union farm and a labeling scheme, but we focus on the district's involvement in unionlike organizations. The North Carolina–based Agricultural Justice Project is extending the familiar labeling approach into workers' issues and is an important adjunct to the district's "beyond organic" debate. The Coalition of Immokalee Workers and the Restaurant Opportunity Center have developed a nonunion institutional form, the worker center, to focus on low-wage food system workers. We look to two district institutions—Bon Appétit Management, a Palo Alto–based food service firm, and Young Workers United, a food service workers organization in San Francisco—for two different windows into how worker centers work throughout food supply chains.[6]

Food, Place, and Community in West Oakland

Urban Planning

While Marin County leaders focused on the land protection that enabled early alternative farms and other food enterprises, Oakland activists were having an entirely different experience. That different history of the West Oakland neighborhood and the far larger region known as East Oakland is important in the food justice story. Growth accelerated in Oakland after the 1906 earthquake: many displaced San Francisco residents settled permanently in the East Bay. With its easy access to downtown, as well as to the city's bustling transcontinental railroad and port facilities, worker housing sprang up in West Oakland in the late nineteenth century.[7] World War II marked its heyday, with industry surging and shipbuilding at the forefront. War workers from throughout the country arrived in West Oakland, and diverse Spanish-speaking immigrants joined a growing majority of African American residents in East Oakland.[8] West Oakland had its own commercial and cultural center along Seventh Street, while East Oakland, always more diffuse, featured commercial districts along several major thoroughfares, including Fruitvale Avenue, Foothill Boulevard, and East Fourteenth street (subsequently renamed International Boulevard). Both areas were home to diverse and thriving working-class communities. However, events of the next few decades undermined both their diversity and their prosperity.

An exodus of industry in the 1950s and 1960s compounded white flight as major manufacturers sought cheap land, easy regulation, and tax incentives in new suburbs like Hayward and Fremont. When state engineers and rapid transit planners looked for a route through Oakland in the 1970s, they aimed at the struggling centers of black culture and community. West Oakland leaders had little traction with the state and regional bureaucrats, who wielded eminent domain with impunity. The I-80 and I-580 freeways shown in figure 8.1 were built, starting in the 1950s, to facilitate new suburbanites' commutes through Oakland into San Francisco. The neighborhood's Bay Area Rapid Transit (BART) station, with the largest commuter parking lot in the city, also obliterated housing and businesses in 1971, while providing little benefit for the working-class residents whose own commute led them toward buses.

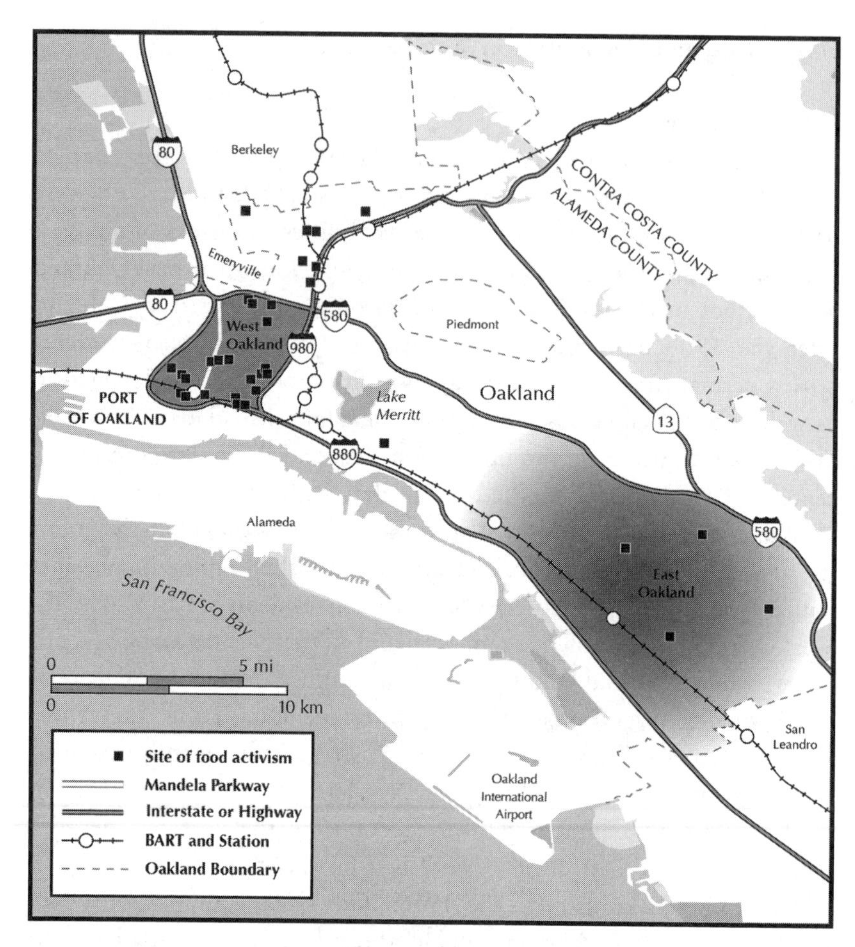

Figure 8.1
Map of Oakland showing the large portions of West Oakland that were cut off from the rest of the city by twentieth-century highway construction. They have become centers of food activism where urban gardens and farmers' markets provide food to underserved neighborhoods, along with community and school gardens (not shown). *Source:* Adapted from McClintock and Cooper (2010).

The two-tier Cypress Freeway that opened in 1957 cut West Oakland in half, isolating large parts of it from the rest of the city. Destruction of Oakland's black neighborhoods continued through the 1970s as whole districts were removed to facilitate construction of the Acorn Plaza housing project, Oakland's Main Post Office, and the West Oakland BART Station. By the 1980s, West Oakland was almost entirely black and had lost much of its commercial base. One of the poorest neighborhoods in the city, it lacked basic amenities, including access to fresh food.[9]

The Cypress Freeway became a focus for frustration among Oaklanders who believed, with ample justification, that they had had no meaningful opportunity to participate in the decision making that led to its construction.[10] An older generation of black leaders who had tied their aspirations to racial integration was displaced by younger organizers' more militant calls for self-reliance. Oakland's black nationalist and Black Power activists drew explicitly on Third World models of colonial subjugation to explain the plight of their city and embraced both community empowerment and self-determination to craft locally defined solutions.[11] The Black Panther Party was born of the neighborhood's growing anger.[12]

West Oakland is also home to the West Coast's first containerized shipping facilities. Since 1962, the port has grown to include a thousand acres of marine terminals, rail facilities, and maritime support areas. It is not, however, an unmitigated blessing for West Oaklanders. The city's air pollution problems reflect "growing port activity, the densely populated regions where most ports are located, and the prevailing onshore wind patterns that accumulate, rather than disperse" the pollutants, creating a "perfect storm of threats to public health." Port development has provoked intense but largely ineffective opposition over time. To settle a lawsuit brought by West Oakland Neighbors, the port agreed in 1997 to a plan that included a $9 million program (out of a $1.4 billion budget) designed to reduce air emissions from marine terminal equipment, tugboats, local buses, and trucks.[13]

A Blip in the 1960s

West Oakland's destruction played out, ironically, during an unusual period of national attention to hunger and poverty.[14] Between John

Kennedy's 1960 presidential campaign and Ronald Reagan's election in 1980, global and domestic upheavals allowed a new direction on policy to help the needy. The turn from Bentham's "wholesome horrors" was brief, and gains in addressing poverty were perhaps less than both the period's mystique and the reaction against it suggest.

Kennedy's campaign initially focused largely on white, rural communities, especially in Appalachia. But the civil rights movement soon forced a more inclusive view. Several years after the 1963 March on Washington and passage of the 1965 Voting Rights Act, Martin Luther King called for a new "economic bill of rights" to address structural economic disparities at least as crippling as Jim Crow laws. King was assassinated in 1968 even as the Poor People's Campaign planned a "Poor People's March" styled on the earlier one. Although the march took place, it was probably the racial violence that erupted in many U.S. cities soon afterward that jolted leaders and citizens alike into considering the relationships between racism, poverty, hunger, and malnutrition.[15]

The Office of Economic Opportunity (OEO) that was part of the War on Poverty that emerged as the centerpiece of Lyndon Johnson's Great Society was far from King's economic bill of rights.[16] The OEO did not directly address the consequences of "an economy still permeated with the legacy of Jim Crow," and its programs generally allowed existing local power structures to control the implementation and distribution of resources.[17] After intense efforts to achieve even limited goals through the OEO programs, Oakland's black leadership concluded that it could not serve as "an effective arm of the larger Black liberation struggle."[18]

Politics around hunger did, however, shift during this unusual period. In 1967 Congress deviated from normal practice and directed not the U.S. Department of Agriculture (USDA), but the Department of Health, Education and Welfare (HEW) to study hunger and malnutrition.[19] In the following year, CBS's news weekly show *60 Minutes* aired a disturbing analysis, titled "Hunger, USA," prepared by former War on Poverty staff at the Citizens' Crusade Against Poverty.[20]

Richard Nixon's incoming administration took up the cause with vigor, convening the White House Conference on Food, Nutrition and Health in December 1969. The president's opening address at that conference explicitly rejected the "unworthy poor" notion and identified hunger alleviation as a moral imperative for government action: "Until this moment in our history as a nation, the central question has been whether we as a nation would accept the problem of malnourishment as a national responsibility. That moment is past. . . . I not only accept the responsibility—I claim the responsibility. . . . Our national conscience requires it."[21] Nixon proposed the Family Assistance Program (FAP), which would set a minimum annual income—or "negative income tax"—guaranteeing all families sufficient funds to buy food. But the proposal ran into hostility from both parties, and as the urban riots subsided, it quietly disappeared.[22]

The proposed FAP was threatening to the USDA's constituency because it would curtail government purchase of surplus food and allow participants to buy food in the supermarket like everybody else. Agribusiness preferred that USDA continue purchasing surplus food for distribution to the poor. The existing pattern allowed bureaucrats to direct benefits to specific producers. USDA allies also argued, somewhat incredibly in retrospect, that Americans were simply not eating enough. Senator George Aiken of Vermont identified "under consumption" as one of two major problems facing American agriculture, and President Kennedy also urged addressing the twin dilemma of hunger and surplus food by encouraging people to eat more.[23]

Although the FAP never saw the light of day, it was serious enough to persuade substantial segments of agribusiness that their world had changed. Convinced that declining rural populations and congressional redistricting boded ill for their programs, the USDA and its clients moved to create programs within USDA that would benefit nonrural constituencies.[24]

The 1960s and 1970s also saw the start of a durable coalition of progressive groups that is still known as the "hunger lobby." Their efforts led to a reinstatement of the Depression-era food stamp program; after years of pilot and demonstration programs, food stamps got a new life (renamed the Supplemental Nutrition Assistance Program in 2008). Although frequently described as a cornerstone of government aid to needy Americans, the program did not, even at the peak of national attention to hunger and poverty, move very far from supporting agriculture. As in the Depression, participants were required to purchase stamps "equivalent to their normal expenditures for food." Thereafter, they could use subsidized stamps "to obtain a low-cost nutritionally adequate diet."[25] Although the hunger lobby led the fight for food stamps, it resisted efforts to take hunger programs out of USDA. It still does; despite all the nutrition- and education-related contraindications of USDA management, the hunger lobby views human services as politically weaker than USDA and fears the program would disappear if it were moved.[26]

Attention shifted away from hunger and poverty issues as the Vietnam War and then Watergate heated up,[27] and political support for systematically addressing poor people's issues declined substantially.[28] Repeated funding cuts refocused public attention back on charity.[29] Poor people may not starve, but the same system that encourages unsustainable overproduction also ensures that channeling the least desirable leftovers from that system to poor and hungry people remains the nation's primary response to hunger in America.[30]

Environmental Justice and Environmentalism

The environmental justice movement has strongly influenced Oakland food justice advocates' responses to these issues. While Rachel Carson's concern with environmental poisons and health did much to precipitate the modern environmental movement, wild land and wildlife preservation and conservation groups initially defined its agenda. Up to and following the first national celebration of Earth Day in April 1970, their focus remained largely on pre–World War II wilderness, parks, and recreation issues.[31] Embracing nature and getting back to the land were regarded by some at the time as a less divisive alternative to the unrest that accompanied the civil rights movement, Vietnam War protests, and the women's movement that emerged at about the same time. Even as they turned to address air and water pollution, Big Ten[32] environmental groups, whose board members and paid staff were (and remain with a few exceptions) largely white and male, did not regard minority communities' issues as environmental.[33] Even after elaboration of the National Environmental Policy Act in 1969 institutionalized requirements for public comment on "major federal actions significantly affecting the quality of the human environment," people of color were not conspicuously involved in shaping the environmental agenda.[34]

Unacknowledged by mainstream environmentalism, people of color nonetheless waged many of the epic environmental battles of the second half of the twentieth century, including those in Louisiana's Cancer

Alley[35] and on PCB dumping issues in Warren County, North Carolina.[36] Environmental groups failed to understand routine, disproportionate exposure to toxic pollutants as an urgent environmental issue, and that failure fueled growing tension between environmental organizations and fence-line communities.[37] This lack of support for an environmentalism based on civil and human rights to pollution-free lives led minority communities to develop a separate environmental justice (EJ) paradigm that included racial factors as central elements of analysis. In that setting, minority leaders defined and pursued their own priorities.[38]

Much of the EJ movement took inspiration from 1979 litigation that suggested that disparate experience of environmental hazards might be a civil rights violation.[39] It gained traction in 1987 with the release of research conducted by the United Church of Christ demonstrating that health hazards associated with industrial waste treatment, storage, and disposal had a disproportionate impact on people of color. Even when economic status was considered, the racial makeup of the surrounding community was the single most significant variable in the distribution of waste site locations. Four years later, in 1991, the first National People of Color Environmental Leadership Summit attracted a diverse group of new leaders to Washington, D.C., to "begin to build a national and international movement of all peoples of color to fight the destruction and taking of our lands and communities" and to draft a set of principles of environmental justice.[40]

The EJ movement was heir to the moral force of the civil rights movement, and the principles, approaches, and remedies to environmental hazards it has pioneered have been influential in defining food justice. Nevertheless, EJ came of age in the wake of the Reagan revolution, a time of growing hostility to activist courts and the government. The litigation strategy that had shaped the successful assault on Jim Crow laws and then developed environmental law was running out of steam by 1990, and despite a very few successful cases,[41] by and large EJ advocates could not persuade courts that disproportionate experience of environmental harm constituted racial discrimination.[42]

Environmental Justice and Food Justice

Food issues were not central in early environmental justice debates. The 1991 Principles mentioned food just once, in the preamble.[43] And although the Second EJ Summit in 2001 included farmworkers' organizations and discussed their issues, summit documents still framed food issues primarily in connection with foodborne diseases.[44] A 1994 *Journal of the American Medical Association* article is generally acknowledged as the turning point when medical professionals became aware of an important obesity epidemic.[45] Lead author Katherine Flegal noted that that while obesity rates "had been pretty stable for about 20 years," analysis of new data showed a "noticeable and surprising increase."[46] The paper set off alarms, and further investigation demonstrated that obesity and related health problems of diabetes, stroke, high blood pressure, and cancer were disproportionately concentrated in poor and minority neighborhoods.

That led to discussion of "food deserts."[47] Interestingly, food desert conditions were an issue in and beyond the Bay Area long before the term became common. In the early 1980s, public interest attorney Lois Salisbury had identified the pattern and was fighting urban grocery store closings in San Francisco: "Decades ago the supermarkets moved into these neighborhoods and ran the smaller markets out by underselling them. Now they're pulling out. You have to remember that these are neighborhoods populated by the poor, the aged, immigrants, the handicapped, people who can't just jump in a car and drive to some other store." Salisbury was quoted in a 1984 *Mother Jones* article, "The Supermarket Shuffle," reporting that large grocery retailers were abandoning the inner cities throughout the country, and food prices in the mom-and-pop stores that remained were 18 percent higher than in the suburbs.[48] Nonetheless, as Amartya Sen pointed out, food access reflects the distribution of entitlements, and people with little or no money cannot buy food even if it is for sale next door.[49] With diet-related illness rates soaring and increases in morbidity and mortality concentrated in poor areas where healthy food was unavailable, food access quickly emerged as a survival issue and an urgent priority for community organizers.[50]

New Concepts and Vocabulary

These new food organizers did not turn to France or Julia Child for inspiration. In Oakland, many food justice advocates built on local history and Black Power advocates' vocabulary of colonial subjugation to ground new calls for community self-determination. In an era of increasing globalization it is not surprising that many food justice terms have similar international roots, underscoring both that local food systems inevitably exist within a global context and that many food justice activists consciously address global issues through their work with local organizations.[51]

The concepts of food security and insecurity, food deserts, and food sovereignty all originated outside the United States.[52] The term *food security* was used within the U.N. system for years after shortages following an enormous sale of U.S. wheat to Russia in 1972 raised food prices all over the world.[53] It was taken up in the U.S. hunger lexicon following the Rome Declaration on Food Security adopted by the U.N. Food and Agriculture Organization (FAO) at the World Food Summit in 1996. The declaration defined food security as a right to "safe and nutritious . . . adequate food and the fundamental right of everyone to be free from hunger." The summit also declared it "intolerable that more than 800 million people throughout the world, and particularly in developing countries, do not have enough food to meet their basic nutritional needs." The Rome declaration identified poverty as "a major cause of food insecurity, and [that] sustainable progress in poverty eradication is critical to improve access to food."[54] The declaration should not be confused with the FAO's own definition of *food security,* however, which uses similar language without the word *right* (allowing a closer alignment with U.S. policy). The FAO considers food security to exist when "all people, at all times, have physical, social, and economic access to sufficient, safe and nutritious food to meet their dietary needs and food preferences for an active and healthy life."[55]

Many activists regard the concept of food security as preferable to the term *hunger* in policy discussions because it shifts focus from an individual experience to the social context in which people have, or do not have, reliable sources of food for themselves and their families.[56] Nevertheless, they also point out that according to the FAO definition, food security simply exists (or not) without reference to how it might happen or who is responsible for achieving and maintaining it. Food First analysts note, for example, that because the term "says nothing about where that food comes from or how it is produced, Washington is able to argue that importing cheap food from the U.S. is a better way for poor countries to achieve food security than producing it themselves."[57]

The concept of food sovereignty is more complex, more contested, and closer to the Oakland activists' concerns. Originated by the global peasant organization La Via Campesina,[58] the term addresses the destabilization that industrialized countries' food policies cause in food systems around the globe. In particular, food sovereignty targets neoliberal trade and food aid that the northern nations deploy both to absorb their own overproduction and redirect global South production toward export of products for the North. The policies undermine food production for local consumption in both the global North and South. Food sovereignty advocates similarly condemn the United States's aggressive exports of proprietary and centralizing agricultural technologies, which have also eroded local production systems in the global South. Accordingly, food sovereignty asserts rights beyond those dealing strictly with food, including self-determination regarding food, that is, the

> right of peoples to define their own food and agriculture; to protect and regulate domestic agricultural production and trade in order to achieve sustainable development objectives; to determine the extent to which they want to be self-reliant; to restrict the dumping of products in their markets; and to provide local fisheries-based communities the priority in managing the use of and the rights to aquatic resources. Food sovereignty does not negate trade, but rather, it promotes the formulation of trade policies and practices that serve the rights of peoples to safe, healthy and ecologically sustainable production.[59]

This understanding of food sovereignty was introduced at the 1996 World Food Summit as well. While not taken up in the Rome Declaration, it informed protests at the 1999 meeting of the World Trade Organization (WTO) in Seattle. The "battle of Seattle" introduced many Americans to the role of free trade in global hunger, and the profile of food sovereignty has risen steadily since.[60]

U.S. food advocacy groups are mixed in their attitudes regarding the term *food sovereignty*. Direct service providers generally avoid the term, framing their antihunger efforts in the context of religious or humanitarian values and emphasizing a strictly local focus. The Community Food Security Coalition,[61] a national organization that does incorporate food sovereignty terminology, describes food security as "a condition in which all community residents obtain a safe, culturally acceptable, nutritionally adequate diet through a sustainable food system that maximizes community self-reliance and social justice."[62] The group also aims to create a "socially just, ecologically and economically sustainable food supply,"[63] but it stops short of asserting the community self-determination that is the core of La Via Campesina's basic human right to healthy food.

Recent food justice scholarship incorporates self-determination issues as one facet of an increasingly comprehensive approach. Gottlieb and Joshi, for example, deploy the term to describe efforts to address all aspects of the food system that might be considered unjust, framing it as ensuring "that the benefits and risks of where, what and how food is grown, produced, transported, distributed, accessed and eaten are shared fairly."[64] Food justice activists we have encountered generally have a more specific perspective. Their attention is focused by the health problems linked to limited access to nutritious food, and their agenda emphasizes skill building, leadership development, and self-determination, priorities that resonate with their specific history. Oakland food activist Brahm Amahdi emphasizes self-determination, saying that those without food are being denied a basic human right and that food justice requires that communities provide their own tools and solutions to address the disparities they experience.[65]

Food Justice in Oakland

Although international influences are important in the district, the politics around food access and related justice are also intensely local and reflect the histories, cultures, environments, and economics of particular places and people. Although a citywide lens might seem adequate for assessing Oakland's food needs, it would not pick up the geographical specificity of its food justice groups. While it is easy to overstate the distinctions, the differences between the east and west areas of the city are very much part of the city's reality, in which partnerships must be built across all sectors of the alternative food community.[66]

West Oakland is a rectangle bounded by MacArthur Avenue and two freeways, I-980 and I-880, as shown in figure 8.1. The population is predominantly African American and Latino. The last supermarket left the area in the 1990s. People with access to cars can shop at grocery stores located in other neighborhoods, but the rest frequent one of the forty to fifty liquor stores, corner stores, or fast food outlets that dot the area, where prices for what is available are generally nearly double the prices found in supermarkets elsewhere in the region.[67]

The borders of East Oakland are harder to identify. The region lies between Lake Merritt and San Leandro and is home to the largest Latino population in the city, as well as sizable numbers of people of other ethnic backgrounds. East Oakland has attracted recent redevelopment and includes some relatively affluent regions (Mills College is located there), plus professional sports stadiums and Oracle Arena.[68] Other East Oakland areas are characterized by the drugs, gangs, and sex trades that are glamorized in East Oakland's distinctive style of West Coast hip-hop music. Diabetes is hardly the area's only problem: International Boulevard is a major center of sex trafficking.[69]

The two regions are regarded and regard themselves very differently. One observer who grew up in West Oakland asserts that while West Oakland has a history that gives the neighborhood collective meaning for residents, East Oakland tends to be merely a place to live. The devastating highways that border and barricade West Oakland may have promoted its recognition as a distinctive, cohesive place, but in his memoir about growing up largely in a Chinese restaurant, Ben Fong-Torres observes that to a "Chinatown family, East Oakland was another world. . . . East Oakland people seemed to be better dressed, to have better cars, to speak perfect English. I never imagined we'd live there."[70]

That geography is reflected in the operations of Oakland's food justice groups. When they were founded, some food justice organizations focused on East Oakland, others on West Oakland, and they have tended to remain geographically separated. Trying to build a geographically more comprehensive program would risk both encroaching on one another's territory and growing beyond one's institutional capacity. Funding organizations have tended to reinforce this by supporting nonprofit institutions in distinct areas. For example, the California Endowment, a wellness grant-making program that was created when the nonprofit health insurer Blue Cross morphed into a for-profit corporation, identified so many food justice and related organizations in West Oakland that it focused its Building Healthy Communities program in East Oakland, in part to avoid redundancy and help spread resources. The Building Blocks Collaborative is a public-private health and well-being program that has focused generally on West Oakland, in collaboration with other organizations.[71] There is nonetheless basic agreement on the primary problem in both neighborhoods and on how to fix it: by confronting the pervasive lack of access to healthy, affordable food. Dozens of organizations great and small—with very different ideologies and with and without major external funding, community, or city government support, or local or broader public recognition—are working valiantly to bring healthy, nutritious food to underserved populations.

Most of Oakland's food justice programs began with urban gardens, farmers' markets, and variations on similar familiar models. Several organizations grow, sell, or distribute food that is subsidized, to make it affordable and to help build a constituency for fresh produce that may have become unfamiliar to residents in West Oakland's neighborhoods. Many of these small nonprofits are also intent on developing and retaining resources in their community and cultivating community, leadership, and entrepreneurial skills while promoting nutritional and cooking skills. How subsidized operations that depend on volunteers and outside funding will fare with that kind of complex agenda remains to be seen. Furthermore, community in many Oakland groups includes meanings that are race based. That is neither inexplicable nor unprecedented, but it does complicate white participation. Additional issues arise from the fact that relationships between food justice nonprofits and city government are often uneasy or nonexistent, although this may be changing as a food policy council gains traction in Oakland.

Mo' Better Food

The first seeds of Oakland's food justice groups were probably planted by David Roach, the son of part-time Texas sharecroppers. Roach studied business at Morehouse College and came to Oakland as a schoolteacher. His efforts have emphasized "healthy economics." The Familyhood Connection, the nonprofit group that Roach started with Bryan Phillips and Tony Dorsett in 1994, organized a conference that asked and explored who would supply healthy food to the African American community. Although Roach presumed the answer was African American farmers, attendees at the conference did not know where there might be any. His organization took on the conference name Mo' Better Food and the mission of connecting these farmers to Oakland schools, markets, and restaurants.[72] Roach's path became complex and remains so, but he probably precipitated the discussions that gave early shape to the Oakland food justice scene.

Social Entrepreneurs in Oakland: Urban Gardens and Urban Agriculture

Food justice work frequently involves some form of garden. Gardens are figuratively and literally low-hanging fruit: the self-provisioning opportunities they offer can rapidly improve access to healthy fruits and vegetables, while providing opportunities for skill building and product for small-scale processors and distributors. Gardens also have a visible impact, transforming neighborhoods, and when ramped up into urban agriculture, they make important contributions to urban waste management and food supply. Finally, gardens have the potential to draw both school children and community volunteers into learning about nutrition, cooking, and the pleasures of sharing a meal. Oakland's food justice activists thus pursued a well-trodden path to the garden and had many models and experienced garden activists to draw on.[73] Among the numerous food justice groups supporting, designing, building, and managing gardens in Oakland, Will Allen and Growing Power have been a significant inspiration and model.

Oakland's Food Justice Gardens

The largest and one of the oldest food justice gardens in Oakland is OBUGS (Oakland Based Urban Gardens), started in 1998 by two friends intent on using their horticultural skills to help Oakland youth and school children. OBUGS began with a single garden at Lafayette Elementary School near Jack London Square and now supports or manages seven gardens. It also runs cooking classes and science activities at the intermittent Mandela Farmers Market where OBUGS gardeners can sell the produce and flowers they have grown.

City Slicker Farms runs, as the name suggests, at a larger scale. It was founded in 2001 by a group of West Oakland neighbors who were concerned about the lack of healthy food in their area. Sonoma County transplant Willow Rosenthal established City Slicker's first farm and stayed to build an organization that brings a production and urban agricultural emphasis to urban gardening. Although many gardeners can and do consume what they grow in their own back yards, the program priority is on high-yield urban agriculture that is intended to bring quality produce to extremely low-income people. City Slicker runs a

Urban Agriculture: Will Allen's Multicultural Coalition for Real Food

Growing Power's extraordinary success as an ethnically and conceptually diverse food justice organization may arise from its eclectic roots. Will Allen, current leader of the organization, got started in Europe with Belgian composters. During off-hours from his European basketball team, he traveled the countryside studying compost piles. Inspired, he started a small farm at the team's housing complex outside Antwerp and was soon raising chickens and vegetables. When he retired from basketball in 1977, Allen settled outside Milwaukee, where his wife's family owned farmland. Soon he and his children were growing food for the family and for farmers' markets. It was serious work, notes his daughter Erika, who now runs Growing Power's office in Chicago. She recalls it as "farm labor, not chores."[a]

In the early 1990s, a YMCA group asked Allen to help them make their small organic garden profitable. Allen soon joined that effort with Neighborhood House, a family service agency in Milwaukee, so that he could receive a salary for what was becoming his life's work. In 1995, Allen formed a new organization called Farm City Link in an abandoned nursery as a way to standardize his work teaching young people about urban gardening and agriculture, and then forged a partnership with Growing Power, a Madison-based nonprofit that also ran gardening projects for youth. Allen and Growing Power founder Hope Finkelstein became codirectors of the combined program. A small grant from Heifer Project International allowed Growing Power to develop programs modeled on Chicago's God's Gang Worm and Fish Project, and soon Growing Power included greenhouses, food waste compost, and vegetables and fruit trees.[b] The Growing Power Web site describes its diverse programs including the Rainbow Farmers Cooperative, a multicultural network of more than 300 small family farmers, that supplies a year-round food security program, the Farm-to-City Market Basket Program.

a. Royte (2009).

b. The God's Gang Web site describes the organization as "a grassroots, nonprofit organization dedicated to supporting and strengthening the residents of the community." The group was established by residents in the neighborhood of St. Mary's A.M.E. Church in Chicago in the 1970s. The original incentive for children was a nutritious Sunday morning breakfast. Some youngsters who were also in the choir began to call themselves God's Gang, and they started an "emergency feeding program they called Mother's Cupboard Food Pantry." God's Gang supports diverse youth and community development programs. See http://godsgang1.net/aboutus.aspx.

farmstand with sliding-scale pricing that charges higher prices to better-off buyers in the community. Thus, within the neighborhood, those who can afford to subsidize food access for the less affluent.

Village Bottoms Farms is located in the Oakland neighborhood that unsuccessfully fought the post–Loma Prieta earthquake relocation of the Cypress Freeway structure into its midst. It reflects founder Marcel Diallo's embrace of self-determination in food justice. An artist who launched the Black Dot Artist Collective and the Bottoms Cultural District with a successful café and community center, he then purchased a vacant lot and turned his hand to farming. Diallo is explicit about his

The Pennsylvania Horticultural Society

The name "Pennsylvania Horticultural Society" (PHS) may not tend the imagination toward a hotbed of justice advocacy, perhaps because the organization was founded in 1827 and staged its first flower show two years later. Nonetheless, the PHS story raises interesting questions about the role of white folks in food justice organizations. It also provides a window into urban gardening as a tool of urban revitalization and community building over the long haul.

In the 1940s, PHS began to invest the proceeds from its famed garden show in revitalizing decaying urban neighborhoods. This effort was reorganized in the 1980s as Philadelphia Green, the program that now manages many of the city's more than 400 community gardens. Philadelphia Green coordinates a full suite of gardener training and youth empowerment programs (including a prison garden and a training program for released prisoners) as well as a harvest that distributes fresh produce to more than 1,000 families. Much of this is possible, it has been argued, because Philadelphia has approximately 40,000 vacant land parcels—more than 40 square miles of abandoned lots that drag down the economy and social structure of surrounding neighborhoods.[c] Nonetheless, decades of effort raise questions about the potential for gardens to resolve the problem: Philadelphia now has about the same number of deteriorating parcels as when the city identified the problem in the mid-1990s and began, with great fanfare, to address it.[d]

c. Econsult Corporation and Penn Institute for Urban Research (2010).
d. Pennsylvania Horticultural Society Web Site, "About Us," http://www.pennsylvaniahorticulturalsociety.org/aboutus/index.html (accessed February 2012). Lawson (2005) notes that the problem is not a recent one. The Philadelphia Vacant Lots Cultivation Association launched a program in 1897 that lasted at least thirty years.

priorities and is not happy that the "green movement" seems intent on replacing "the leadership of struggles by Black and other oppressed people in this country with liberal do-gooder environmentalists." He expresses a healthy disrespect for "green organizations" that derail "Black 'for us, by us' organizations." However, he simultaneously reaches out "to all of you who understand and/or have a willingness to understand the power dynamics of White Privilege and its function within gentrifying neighborhoods, and don't mind getting down and dirty assisting us in bringing the Bottoms up!" He is studying with Will Allen, learning "to work and condition the land so that we don't have to depend on Whole Foods, Trader Joe's, Safeway, Lucky's or any of these 'green' organizations to feed ourselves."[74]

The City Slicker and Village Bottoms models of urban agriculture match imperfectly with other food justice advocates' focus on and aspirations for sustainable economic growth. Certainly urban farmers make a substantial contribution to city food consumption and waste management in growing numbers of cities around the world. In Hong Kong, for example, urban farmers produce vegetables, pigs, chickens, eggs, and farmed fish from 160,000 tons of recycled food waste.[75] Nevertheless, even urban farms that are "nationally recognized, such as Earthworks in Detroit [and] Growing Power in Milwaukee . . . are not run as profit-making businesses," and rely heavily on grants and volunteers.[76]

It is unreasonable to imply that City Slicker Farms should operate without subsidies when behemoths J. G. Boswell, Cargill, Tyson Farms, and other food enterprises do not. However, creating reliable jobs that pay living wages is also important. A Detroit analyst has noted that if an urban farm was a business that employed ten workers at the level of the median family income—about $45,000 a year—it would need to generate about $500,000 before allocating a dime to seeds, tractors, insurance, and "everything else a farm business requires."[77] Whether Oakland's community gardens could generate that kind of returns, and how doing so would affect their ethos of community, is not clear. Detroit is one of the few places where for-profit urban agriculture has been seriously considered. A local entrepreneur has proposed commercially farming 1,000 acres of Detroit's approximately four square miles of abandoned land. Detroit's community gardening organizations are in general opposed to this and are defending greater community control while fighting the prospect of what they regard as industrial agriculture in the city.[78] These questions have yet to come up in Oakland's relatively recent and smaller-scale gardens. But they may need addressing as the urban agriculture element of the food justice community gains traction and moves into a next generation.

Local Grocery Stores in Oakland

Similar issues of for-profit versus nonprofit are etched into the debates that surround the most prominent aspiration of Oakland's food justice advocates: to build one or more grocery stores in Oakland's food deserts. The last full-service grocery store in West Oakland closed its doors in 1993. Many years of effort by churches, city government, and community groups failed to convince a conventional retailer to locate in the area after the smaller chains had shut down. Nonprofits have worked to provide alternative retail outlets that emphasize fresh produce, but they have been thwarted by both lack of capital and lack of experience in food retail. More recently, as Walmart, Target, and Tesco develop more neighborhood-friendly, urban-scale formats, the role of nonprofits in food retail has become less clear.

Mandela MarketPlace

A nonprofit leadership incubator that supports minority farmers and healthy food, Mandela MarketPlace began in 2001 as a project of the Tides Center Environmental Justice Initiative. Under the leadership of white Muslim Dana Harvey, Mandela MarketPlace emerged as an independent organization in 2005 and has continued Mo' Better Food's efforts to connect black farmers to a distribution network that includes farmers' markets and the first nonprofit-based food retail operation in town, the Mandela Food Cooperative.

In 2001, a group of concerned residents, community-based organizations, and social service agencies formed the West Oakland Food Collaborative. With a planning grant from UC Davis, the group undertook a community-based process that identified five goals: a farmers' market, a cooperative marketplace, liquor store conversion, small-business development, and community green space. The California Endowment led a funding effort to implement the agenda, and the Mandela Farmers Market opened in 2003.[79] Although the market is the frequent focus of

The Philadelphia Food Trust

Philadelphia's experience with retail offers insights into Oakland's prospects for food access: its Food Trust has developed a model for attracting and retaining retail grocers in underserved communities. The organization started with founder Duane Perry's work at Philadelphia's Reading Terminal Farmers Market and grew into a broader effort to improve access to nutritious, affordable food. Philadelphia's earliest and most successful markets were in low-income neighborhoods and housing developments; the upscale models are more recent. The Food Trust's programs are less focused on issues of local and sustainable foods and small farmers than many in the Bay Area. Everything the Food Trust does, current executive director Yael Lehman emphasizes, marries food access to education. In spite of its reputation in food retail, half of the trust's staff work on education, workshops, and demonstrations.

It is the Food Trust's successful focus on grocery retail policy that has raised its national profile. In 2001 the organization published one of the defining studies of U.S. food access issues, a report that linked concentration of diet-related disease to lack of access to supermarkets and focused on a clear, uncomplicated solution: the need for more supermarkets in Philadelphia. In fact, the report does not use food desert terminology; there is plenty of food in the alleged deserts, Lehman argues, just not healthy food, and using the term *desert* may discourage investment.[e]

Unlike the advocates in Oakland to date, the Food Trust has worked effectively with state and local government. The resulting Fresh Food Financing Initiative has leveraged $30 million in state funding to help supermarkets open in underserved neighborhoods where "infrastructure costs and credit needs cannot be filled solely by conventional financial institutions."[f] In the five years after it was established in 2004, the initiative helped open eighty-three supermarket projects, creating approximately 1.6 million total square feet of food retail and about 5,000 jobs. The Food Trust has also worked successfully with partners, including Oakland's Mandela MarketPlace, on the Healthy Corner Store Initiative to improve offerings at the much-lamented convenience stores (known as c-stores in the trade) and bodegas. With help from the Robert Wood Johnson Foundation, the Food Trust is starting a national program.

e. Food Trust (2001).
f. Lehman (2010).

studies that conclude that farmers' markets are inaccessible to persons of color or otherwise "inscribed white," it is woven into a number of Oakland's garden programs' teaching and distribution efforts.[80] Mandela MarketPlace is also engaged with promoting fresh, healthy food in corner stores and has collaborated with the Food Trust in this work as a small part of its programs.

The Mandela Food Cooperative has become the flagship of the collaborative. Its small but long-anticipated retail operation opened in June 2009 with much fanfare, but the operation attracted fewer neighborhood shoppers than hoped, perhaps because it is not a full-service grocery. The co-op is worker owned and managed by community residents, early graduates of Mandela Food Cooperative's entrepreneurship training program. The store sells produce sourced from local farms that are generally too small for larger retailers to deal with. In its first year, the store brought 50,000 pounds of fresh produce to West Oakland.[81] Dana Harvey recently won a Robert Wood Johnson Foundation Community Health Leaders Award for her work in establishing the cooperative, and James Berk, one of the Mandela Food Cooperative worker-owners, recently won the Art of Activism award from Robert Redford.[82] Before the co-op existed, he noted, "if you lived in West Oakland and you didn't have a car, your options were Hungry Man Dinners and Hot Pockets."[83]

People's Grocery

Running on a similar track, the People's Grocery emerged in 2002 from the bosom of City Slicker Farms. Founders Brahm Ahmadi, Malaika Edwards, and Leander Sellers, all of whom came to the task from environmental education and community organizing, began by replicating City Slicker's garden programs. They soon expanded these to include social entrepreneurship and aspirations for a community-owned and -controlled grocery store. People's Grocery programs have shifted frequently as its leaders have gained perspective and experience in the food business. Healthy food and self-determination plus economic development in West Oakland remain at the heart of their program, but its many experiments in format underscore its changing approach to larger issues about for-profit and nonprofit approaches to these problems.

Ahmadi came to food justice from an environmental justice and nonprofit background, having organized against a toxic dump next to a park near his home in Los Angeles. The dump shut down, giving Ahmadi "a sense of what could be done." Nevertheless, after working with and supporting diverse EJ campaigns—including a Hunters Point community fight against Pacific Gas & Electric,[84] toxic struggles at the Naval Shipyards,[85] and the Red Star Yeast air pollution fights in Oakland,[86] and running summer camps for inner-city youth with Urban Habitat—Amahdi turned to social enterprises that would build new infrastructure and deal with job losses. Describing himself as "super geographically focused," Amahdi wants to "create a pipeline of resources from within and outside the community to support this work." He too looks to leaders like Will Allen for inspiration: Allen's multimillion-dollar business is 75 percent self-funded, Amahdi notes, and has been able to both use and grow social capital.

People's Grocery's signature program was not economically successful. The group outfitted a former mail truck as a grocery store and converted it to solar and biodiesel power. Painted bright orange and purple, the Mobile Market traveled a regular route through West Oakland. Blaring rap music, the vehicle sold healthy foods and vegetables sourced from local farmers and supportive distributors to members. The truck was designed to appeal and to educate youth and in many tellings is likened to ice cream trucks of yesteryear. Low prices were an attractant, which People's Grocery could offer members because of subsidies, discounts from distributors, and grants. The truck was one of the few unambiguously popular programs in the area. But after four years, only about two hundred members had joined the program. Costs overwhelmed the operation and forced People's Grocery to eliminate the Mobile Market.

People's Grocery has also encountered difficulty growing food for its programs. A brief effort to run a two-acre garden in Sunol Ag Park did not attract sufficient volunteers and was discontinued.[87] A second partnership, with Dig Deep Farms, is off to a stronger start, and People's Grocery continues to sell produce through Grub Boxes, its second attempt at adapting a community-supported agriculture model.[88] However, People's Grocery's much-discussed major goal of opening a grocery store in West Oakland has proven elusive. In 2008, the organization split into a nonprofit with a new executive director and a for-profit arm, People's Common Market, that pursues the grocery store vision under Ahmadi's leadership.

Under a new executive director, the nonprofit side of People's Grocery maintains its well-established garden and Grub Box programs. But it is focusing the organization's considerable caché on building Oakland's food justice community and a national food system reform movement. People's Grocery's executive director Nikki Henderson is well suited to the task of negotiating across complex divides. One of nine siblings and the only daughter, she was raised by parents who insisted on brown rice and vegetables. But as she watched her friends and cousins "eat crap—Red Vines, Cheetos, hot chips, *home-made* hot chips," she also saw that many were dying. When Slow Food USA began trying to shake its image as the highwater mark of snooty food preoccupations and turned to engage justice issues, Henderson joined up. She was an obvious candidate to run the nonprofit side when People's Grocery split into for- and non-profit organizations. Henderson is thus one of relatively few leaders in the Oakland food justice field currently who did not start the organization she is running.

Under her vision, People's Grocery has developed as a center for leadership training and movement building. It is active in redeveloping West Oakland's California Hotel, a property that was a beacon for black artists and travelers during the Jim Crow era but became run down and was recast as housing for 250 low-income West Oaklanders. People's Grocery is developing gardens on the site and has organized a series of discussion sessions, leadership development programs, and community events that build on its history. The California Hotel is intended to become an incubator of projects, organizations, and activities supporting the organization's focus on what it calls "allyship" (i.e., being an "ally" to the community's work). Although the organization is explicitly using "food to build a movement," the emphasis is on foundational programs that build the leadership skills, social fabric, and the lasting relationships needed for long-term "campaigns for systemic change." Through dialogue circles, social justice internships, and participation in Fierce Allies, a network intended to help historically hostile groups work together, People's Grocery aims to develop new leaders for the food justice movement.[89]

Henderson's strengths are suited to People's Grocery's allyship emphasis. They include an ability to reach across racial and class lines, and she is almost unique in her emphasis on bringing people from diverse food movement efforts to work together at local and national scales.[90] But Henderson also asserts the necessity of nurturing black leadership in black communities. Even as her stump speech acknowledges that the food justice movement stands on the shoulders of those who worked previously to make food quality and food access mainstream issues, she is unambiguous about the importance of self-determination in minority communities.

Oakland's grocery store aspirations remain more complex. Mandela MarketPlace and People's Grocery are but two among many groups that have focused on food retail in Oakland. During the time that both have labored to develop their plans and financing, several small, independent operators have come and gone, and a new one, opened without fanfare, seems to be enjoying some success.[91] Both nonprofits' efforts to locate in Mandela Gateway, a city-supported housing project across the street from the West Oakland BART station, indicate some of the issues the sector is facing. Oakland's Housing Authority site includes affordably priced rental units, landscaped public spaces, and 20,000 square feet of retail space. City government financing of the space requires businesses operating there to meet living-wage requirements. Accordingly, the city council denied a lease to Fresh & Easy, a discount grocery that wanted to open at the site, when its parent company would not commit to living wages.[92]

The travails of the original supermarket at the Mandela Gateway site suggest additional challenges for food retail in the area. When the Acorn Supermarket opened in the mid-1980s, West Oakland had been without a grocery store for more than two decades, and community activists had urged the city to attract investors. Finally, an independent investor with a complex history in Oakland anchored a new strip mall with Acorn Super. But the mall was built with little regard for pedestrian access despite its neighborhood setting where fewer than half of residents owned a car in 1993 (figure 8.2). High hopes for the store crashed almost immediately, and the store was ultimately abandoned.[93]

People's Grocery and Mandela MarketPlace both have had their eyes on the empty Acorn Super spot. It was promised to Harvey, but she was unable to raise enough capital to open. When a 99 Cents Only store tried to lease the space, the Oakland Housing Authority equivocated, hoping that Mandela Marketplace would somehow make itself "credit worthy."

Figure 8.2
The formidable expanse of asphalt that comprises the parking lot of the Acorn Super grocery store in West Oakland, c. 1985, made access difficult for pedestrians even in a compact neighborhood where few residents had cars. Photograph by Howard Erker. *Source:* The Oakland Tribune Collection, the Oakland Museum of California. Gift of ANG Newspapers.

Oakland's Housing Authority is not authorized or inclined to act as a business incubator; it needed a solvent tenant and could or would not offer the local operators assistance or guidance. The city council offered Harvey a small grant to aid fundraising while many frustrated potential shoppers in the retail-deprived area fumed at the delay.[94] Even those on the city council who wanted to support Mandela MarketPlace were constrained: the city does not have a policy that identifies expectations for industrial redevelopers beyond the minimum wage or provides a basis for preferring or supporting a local business. When 99 Cents Only eventually secured a lease, the city restricted the amount of space the store could allocate to produce, meat, and dairy. That did nothing for the proposed co-op and angered those looking for a convenient place to shop.

City support, and the lack thereof, for the Mandela MarketPlace project, has intensified the grocery debate among citizens and within the food justice community.[95] Food justice advocates, however, recognize that there does not have to be competition among the different retail options: A People's Grocery market analysis demonstrated need in Oakland for at least five neighborhood operations of the type that they are planning.

Nonetheless, the situation is skewed by the city's redevelopment orientation toward big box retail. Numerous industrial food retailers have played hard-to-get, but FoodCo, a division of Kroger's, has shopped for a site. When the landowner refused to sell the site Kroger's selected, the city council offered to condemn the land. That stirred rivalries plus memories of decades of condemnations for proposed civic improvements that either never occurred or did not benefit those evicted from their homes.[96]

Who Is Investing in Alternative Food?

Clearly the financing and financial management expertise available to alternative food activists and entrepreneurs are among the key factors influencing the shape and direction of the institutions they create. In previous chapters, we have seen this dynamic play out in co-ops that developed out of buying clubs and volunteer labor; Alice Waters, the Cowgirls, and Albert Straus all got their start by borrowing from friends or mortgaging their family homes; Chez Panisse flirted with insolvency until Alice's businessman father redesigned the restaurant's financial management; and Odessa Piper includes her health and first marriage in her list of the investments she made in her Wisconsin restaurant.

By the early 2000s, public agencies, nonprofits, foundations, wealthy individuals, and social entrepreneurs were jostling with private sector and corporate investors in efforts to organize financing for healthier food systems, particularly grocery stores. The pros and cons of having Kroger's or Whole Foods or Walmart in an area as opposed to more community-oriented enterprises have been and continue to be debated. But food

Growing Alternative Capital

Claire's, a community-supported restaurant in Hardwick, Vermont, opened in 2008 and has become a headquarters for the area's growing numbers of agricultural entrepreneurs. Financing for the restaurant came in part from residents who invested $1,000 each in return for subsequent dinner discounts. Claire's menus are built around local products, and its Web site notes that 90 percent of the restaurant's food purchases are within the region. It also points out that restaurants purchase more than food and that 64 percent of Claire's total expenditures since opening, including for personnel, professional services, and laundry, have contributed directly to the local economy.[g] By the time the *New York Times* took note of Hardwick's accomplishments, food businesses in the area had lent each other about $300,000 and had started the community-run Center for an Agricultural Economy. The center has invested in greenhouses to support a year-round farmers' market and the Food Venture Center business incubator, among other local programs.[h]

The La Montanita Co-op began in 1976 in Albuquerque, New Mexico, with 300 member-owner families.[i] A decade later, during the downward spiral that pushed the Berkeley Co-op and so many others out of business, La Montanita began to invest more resources in becoming what it calls a "food shed co-op."[j] At first it simply offered the store as an alternative distribution venue for farmers who had leftovers after selling at the farmers' market or were willing to deliver their product to the co-op midweek. La Montanita established direct purchasing relationships with about 300 local providers by the mid-1990s.

The co-op board supported an employee proposal that reimagined the co-op as a hub for strengthening the regional food and farm economy. The board agreed to invest $150,000 of co-op profits over three years to rent and operate a warehouse as a distribution center. Managing more of its supply and distribution network allowed the co-op to reduce both its own and the producers' costs through pooling transportation expenses and joint purchasing of products.

In addition to sourcing its own stores, La Montanita encourages its producer members to supply other retailers and restaurants throughout New Mexico and neighboring Colorado.[k] Its distribution network facilitates such sales, often providing transportation and aggregating services. The co-op provides the insurance required by larger retailers such as Whole Foods, which small producers generally cannot afford. It also invests to solve problems in the food shed. When a hailstorm marred apple and pomegranate crops close to harvest, for example, the co-op bought the damaged but still wholesome fruits at a reduced price so its producers realized at least some income that season. It then contracted with a district juice company to produce an apple-pomegranate juice that has proven popular with customers.

La Montanita staff also worked with ranchers to address the spring glut of fresh goat milk and cheese. First they persuaded an Albuquerque restaurant chain to feature local chevre in its nine locations, which had a positive effect but could not absorb the entire supply. Then the co-op loaned $5,000 to a goat farmer who was willing to build the first cheese-aging caves in the state. That farmer now buys all the goat milk that other farmers want to sell him and produces five types of aged goat cheese sold throughout the year at La Montanita and other outlets.

La Montanita now carries more than fifteen hundred local products ranging from fresh and processed meats, produce, and beverages to condiments, herbs, and body care products. Local products make up 20 percent of La Montanita's total revenues and expenditures, a percentage the co-op systematically works to increase. In 2010 the co-op created the La Montanita Fund, a grassroots investing and microlending co-op, and invited members to join it in financing loans to the food producers that sell at its stores or through its distribution center.[l]

g. Claire's Restaurant and Bar Web site: http://www.clairesvt.com/
h. Burros (2005).
i. This section draws heavily on Seydel (2011).
j. La Montanita defined its food shed as a roughly 300 mile radius around Albuquerque. See La Montanita Web site "Food Shed," http://www.lamontanita.coop/index.php?option=com_content&view=article&id=29&Itemid=49.
k. The co-op currently operates three stores in Albuquerque (including the Grab & Go shop on the University of New Mexico campus) and a store each in Santa Fe and Gallup. The Gallup store reflects the taste preferences and traditional products of its largely Navajo customer base.
l. Robin Seydel, "La Montanita Fund: New Grassroots Investing Program Growing the Local Food System, Strengthening the Local Economy," at http://www.lamontanita.coop/index.php?option=com_content&view=article&id=74<emid=99.

justice activists in Oakland can draw on the successes and failures in other communities' attempts to generate and deploy capital resources to finance and sustain development of more socially just and ecologically based regional food systems.[97]

The Oakland Food Policy Council

Very aware of the conventional food retailers now lining up to open shop in their city, observers and activists alike are looking to Oakland's Food

Food Policy Councils

Toronto and Hartford are usually mentioned as having invented the Food Policy Council (FPC) concept in the early 1990s. However, Knoxville, Tennessee, got there first in 1982. A research project at the University of Tennessee led Knoxville to establish a body for addressing food comprehensively as an urban problem.[m] Like many of the food councils that followed, the Knoxville version is formally affiliated with city government, allowing it to foster cooperation between diverse actors in different sectors of the system, influence and evaluate policy, and launch or support programs.[n]

Toronto was among the first cities to follow the Knoxville example. Its Food Policy Council Web site declares that "in the absence of federal and provincial leadership on food security, the City created the Toronto Food Policy Council (TFPC)." After a decade in operation, the TFPC summed its accomplishments in 2001 in *Ten Years of the TFPC,* which urged the city to note that 10 percent of the jobs in the city are in the food sector. The TFPC successfully worked with the city's Economic Development Division to create "a consolidated approval process for public health regulation of small food processing businesses," including construction of a 2,000 square foot incubator kitchen for small food processors, one of the earliest such ventures.

Like Toronto, Hartford started a Food Policy Commission in 1991, but its effort to develop a systems approach to food issues began even earlier. Mark Winne founded the Hartford Food System in 1978, a grassroots organization dedicated to fighting hunger and improving nutrition that was critical to spurring the city to create an official commission. Like the Philadelphia Food Trust, the Hartford Commission is less concerned with sustainability and social entrepreneurship than many Oakland activists. Ensuring people are fed is its priority. As one Connecticut activist observes, her "primary job is to figure out how to get healthy food into people."[o] Hartford appears not to have the advantage of being part of a functioning food district or even a recognizable region, however. Although Connecticut has long been famous for its tony suburbs (Greenwich, Darien, New Canaan), its cities are among the poorest in the United States.

Hartford Food System has organized everything from "a parents committee advocating for healthier school meals," to an inner-city retail grocery and is promoting urban agriculture and a Commission on Food Policy.[p] Winne emphasizes institutional overload as a reason for founding the commission, noting that two rather uncoordinated tiers were operating in the city, one consisting of federal, state, and local agencies and the other, an array of private organizations doing everything from distributing free food to running community gardens. The commission has worked to connect those two tiers but has not solved the problem of organizational density and overlap.[q]

m. Southern Sustainable Agriculture Working Group (2005).
n. Harper et al. (2009).
o. J. Martin (2010).
p. Grow Hartford manages a twenty-four-acre nonprofit farm and two additional sites that grow food for local families and social service agencies. In 2010, the farms produced over 7,000 pounds of fresh food for distribution.
q. Winne (2007).

Policy Council to help sort out the grocery store issue. The process leading up to the council began when a study initiated by then Oakland Mayor Jerry Brown's Office of Sustainability noted in 2006 that key organizations were not "communicating and collaborating with each other in ways that could help strengthen their efforts and further their goals." The report also noted that most of the groups had not learned how to "maximize relationships with the City" and recommended creation of a Food Policy Council.[98] It took almost three years to establish the council, which held its first official meeting in 2009.

Oakland's council is made up of representatives of the city's various working communities but is not an official branch of city government. A new round of appointments to the council was announced in 2011, creating optimism in all sectors. It has begun to confront the agricultural zoning issues that threaten to cripple local urban agriculture, focusing initially on legalizing the urban gardens that have formed the basis of many food justice groups. The well-publicized travails of Oakland's celebrity urban gardener Novella Carpenter suggest as well that efforts to address agricultural zoning in the city are long overdue.[99] Having built hundreds of backyard gardens and seven urban farms, City Slicker Farm sites are vulnerable to the fact that city zoning does not permit urban agriculture, which is currently illegal within the city.[100]

Although growing vegetables and raising animals for personal use on residential property in Oakland has long been legal, selling, bartering, or even giving away what you grow is not. Neither is gardening on a vacant lot, and processed foods fall into an odd limbo. Fees for conditional use, permits, and business licenses can be prohibitively expensive. Animals are especially difficult because they might smell or make noise, and because some city residents object to raising animals for meat. In contrast, as one observer noted, "complaints about vegetables are rare."[101] The FPC has recommended that the city go beyond zoning reform and develop a coordinated policy that addresses public land and water access, promotion of sustainable growing practices, incubation and coordination, and other issues to support and expand urban agriculture.

The grocery retail challenge is also high on the FPC agenda, under pressure from the change in food retail that is now underway. Walmart, Tesco, Safeway, and even Whole Foods are developing smaller, "urban-friendly" alternatives to the megastore model. This transition in conventional food retail may not be a blessing for food justice, however.[102] Relying on firms that have a richly documented reputation for undercutting local businesses and lowering wages is probably not the optimal solution to the food access problems.

Area nonprofits have reached out to successful alternative district retail operators. People's Community Market has partnered with Rainbow Grocery and has invited Monterey Market proprietor Bill Fujimoto to help design its anticipated produce section, for example. The enterprise seems on track for funding from FreshWorks Fund, the California Endowment's $200 million loan program patterned on the Food Trust model.[103] In the midst of a public health crisis that turns on food access, many are hopeful that the FPC and the city government can define and implement a food retail program that addresses Oakland's food deserts and its need for economic development.

Food Crafters and Food Justice

These for-profit-versus-nonprofit grocery debates are taking place during a renaissance in Oakland's old food processing precinct and underscore some of the differences between a rising generation of food crafters and the previous generation of chefs. The new food crafters take for granted the expectations that the previous generation pushed hard for: that good food must be healthy and sustainable. Kathryn Lukas of Farmhouse Culture urges customers to check the provenance of her ingredients: "Got a question about any of our ingredients? We know who's growing every head of cabbage we use. Open a jar of Farmhouse Culture sauerkraut, and you're supporting a whole network of sustainable small farms and artisan producers who are preserving the land and its ecology."[104]

Key parts of the buzz around the new enterprises were framed in Berkeley's Gourmet Ghetto and similar haunts during the 1980s. However, new expectations for quality are being formulated where social ventures intersect with food justice. The nonprofit group Forage Oakland crystallizes a new view of gleaning in the Oakland community.[105] Founder Asiya Wadud came to the district from Washington, D.C., as an AmeriCorps volunteer teaching at the Edible Schoolyard. She soon mapped the fruit gardens near her home and began foraging, but not to supply upscale restaurants:[106] Forage Oakland encourages neighbors with surplus crops to offer them to neighbors. Wadud "works to address how we eat everyday, and how everyone can benefit from viewing their neighborhood as a veritable edible map."[107] Defining her goals in a manifesto, Wadud asserts that foraging and sharing build community:

> The gleaning of unharvested fruits; the meeting of new neighbors; the joy of the season's first hachiya persimmon (straight from your neighbor's backyard, no less); the gathering and redistribution of fruits that would otherwise be wasted—can . . . create a new paradigm. . . . You have the courage to introduce yourself (despite the pervasiveness and acceptance of urban anomie) and they reward your neighborliness with a sample of Santa Rosa plums, for example. Later, when you find yourself with a surplus of Persian mulberries, you in turn deliver a small basket to said neighbor. With time and in this fashion, a community of people who care for and know one another is built, and rather than being the exception, this could be the norm. . . . This project is about viewing food as a shared pleasure and a shared resource, redistributing it to those who will enjoy it.[108]

Nevertheless, for-profits, albeit very small ones, are actually the norm in Oakland's justice-oriented food-crafting world. These new businesses are so small that they are often discussed as a subset of the do-it-yourself movement, but they are unmistakably reshaping the food scene in Oakland: "The D.I.Y. movement has never tasted (or looked) better," opined the *New York Times* in its coverage of the Oakland crafters.[109] Real innovations are enabling young entrepreneurs to launch new businesses on shoestrings and find a role for themselves and an outlet for their preoccupations in hard economic times. These are not Sue Conleys and Albert Strauses hoping that the Small Business Administration might take interest in their new enterprises. One *New York Times* article observes, ironically perhaps, that "all you need is a Mason jar, your friends and a very sharp knife."[110]

These new and differently inspired artisans have multiple roots. Some worked at Chez Panisse and found the economic downturn challenging. Others are immigrant women who have turned to their native foodways to support their families. These new food entrepreneurs have developed approaches to production and marketing that do not require an enormous investment and insist that the food be affordable and culturally appropriate to a young and ethnically diverse population. The Oakland pizzeria Pizzaiolo lists "our farmers" on its Web site but also emphasizes fairness to workers. Its Chez Panisse–trained owner lives above the restaurant, and his two children watch what happens every day there, which he says makes fairness key to his business: "The people I need to be accountable to are right here, in my face every day. . . . My children watch me work, they have relationships with my co-workers, and so I am forced, everyday, to live with a level of integrity and integration that is hard to come by in this day and age." Shakirah Simley's Slow Jams ties her ambition to create a "socially conscious company" to her own childhood experiences with "lack of access."[111] Slow Jams works to "ensure that values such as 'high-quality, local and organic' and 'culturally appropriate and accessible' are not mutually exclusive":

> By creating positive economic activity through the vehicle of a local food/food-justice enterprise, I hope to stimulate the local economy through urban-ag and green-job development, utilize the untapped market of urban farmers and producers, provide opportunities for local nonprofits with similar value systems that serve as potential producers, and create more viable urban micro-foodsheds.[112]

The notion that everybody does better when everybody does better is a driving force in the Oakland food crafting scene, and many of these small crafters are supported by a formal or informal business incubator. Studebaker Pickles is the project of Pizzaiolo barista Kate Hug, for example. Her boss allows her to use the restaurant's facilities to concoct diverse pickle products, including (in an interesting throwback to fine food as French) Haricots Verts d'Oakland. Hug offers her small batch products for sale at the Pop-Up General Store, occasionally located at the nearby Grace Street Catering facility, which is the brainchild of two other Chez Panisse alums.[113]

Slow Jams has been accepted into support programs offered at La Cocina, a San Francisco nonprofit that works much as the Grace Street operation does, evincing what the *New York Times* calls the "foodie do-gooder trimmings" of real food.[114] La Cocina's communal commercial kitchen elides some of the differences between markets and communities. In a food crafting arena that is largely white, the kitchen provides training primarily to low-income, female immigrants "who want to build a food business but wouldn't have the resources to do so on their own."[115] However, as much for Slow Jams as for the Cowgirls, access to lucrative markets is a key to success. La Cocina completes a circle, running a booth at the Ferry Plaza Farmers Market and a permanent shop in the Ferry Building as well, giving its start-ups access to the region's highest prices and upscale consumers.

Both the pop-up and the incubator depend on carefully orchestrated sharing of underused kitchens and similar resources that provide young entrepreneurs an opportunity to experiment "without the risk of bankruptcy."[116] Any food enterprise is not without risk, however. In spite of their business incubators' and mentors' best efforts, some crafters operate without full awareness of the regulations or techniques that prevent foodborne disease.[117]

Scaling Up in Oakland

In addition to farmers' markets, food trucks, and pop-ups, Oakland has become home to a new wave of midscale regional businesses. This in part reflects a new wave of urban planning—known generally and without much empirical basis as smart growth—and is spurred by enthusiasms that highlight good food as an element of refined urban living. Gentrification has not addressed food access in West Oakland, but it has transformed the area in East Oakland around Jack London Square into a showcase of high-end, shiny lofts and small and midsized food-based enterprises.[118]

The butchers to whom Bill Niman bequeathed Oakland have become the leading edge of a growing artisanal meat scene in Oakland. As we write this book, innovators like Marin Sun Farms have become the establishment and conspicuously upscale. In lieu of going to Trader Joe's, Evans opened a new retail outlet in Oakland's swank Rockridge Market Hall.[119] Joining Evans at the high end, Boccalone, a salume and sausage production facility, opened in 2007 in an old Oakland Portuguese sausage company that had operated since the early 1900s and was soon selling from a stall in the Ferry Plaza Building.[120] Biagio Artisan Meats has recently opened in nearby San Leandro, and the owner has a butcher retail operation at the Jack London Market Place.

Food activist and consultant Anya Fernald is moving in the same direction. Her company, Live Culture, works with artisanal food crafters and sustainable farmers to develop sustainable and socially just companies. The food sector in Oakland has recently begun to attract investment on a scale that was hard to envision when People's Grocery first imagined a nonprofit grocery store. Entrepreneurs and private capital can see the market for high-quality foods, and, therefore, in Fernald's vision, socially responsible for-profit businesses should be developing the alternative food processing and food distribution businesses in the East Bay city. To prove her point, Fernald has taken a leadership role in a new $40 million organic meat product company. The meats will be traceable to a 10,000-acre farm, Bel Campo, in Siskiyou County. Fernald reports that meat processing is a high-margin undertaking that can be a significant engine for local jobs and development. Her plans with Bel Campo include a slaughterhouse and commercial kitchen that will support eight butcher shops around the region.

Those facilities will be integuments of a for-profit business, but Fernald is also deeply involved in food justice and is dedicated to the education of new food entrepreneurs and alternative food consumers. She recognizes gentrification and white, middle- to high-income "hipsters" as a challenge in democratizing the current food processing field, especially in the context of justice advocates calling for self-determination. But she also asserts that if recruitment, training, low-cost capitalization, and leadership development can be built into a social venture, it might be possible to cultivate equity and community development as central aspects of the new food businesses.

Alongside her for-profit business, Fernald has launched the Food Craft Institute, a social venture that will teach food-making skills and partner with similarly aligned ventures, like Inner City Advisors, to teach business, financial, and marketing skills and help network its fellows into the Bay Area food district's web of relationships. The institute is in the service of her vision of relocalizing food. For Fernald, local is not only a question of geography. It can also be a proxy for traceability, allowing one to know who made a food product and how it moved from farm to table. Thus, a traceable food, with all the positive connotations *local* currently carries, could just as easily be sold at a farmers' market where customers meet the producer, or online through a virtual marketplace, as long as there is a specific link to a particular person or group of people. Nevertheless, physically relocalizing food does facilitate a connection with social justice: it creates access to jobs and healthy food for people in serious need of it.

The Unfinished Agenda: New Models for Food System Labor

This wave of small food processors is also a part of rethinking food system labor in the district. Cesar Chavez's unionizing efforts were the peak of a golden age for agricultural labor in the United States.[121] However, even after state law gave California fieldworkers the right to organize that is denied them under federal law and in most other jurisdictions, membership in the United Farm Workers peaked and then dropped precipitously. As the union model has continued to erode throughout the U.S. economy, new generations of workers and some of their employers are shifting that model to redress the injustices visited on workers throughout the food chain. Recent public attention focused on fieldworkers has centered in the southeastern United States rather than California, and food processing and service workers have a growing place in the conversation as well. Bay Area organizations are part of that national effort to alter the dynamics of worker-management interaction. Chronically abused workers are developing new formats to address the fact that ignoring and violating the law is standard workplace practice in much of the food system.

This section looks at a spectrum of workers' rights action. It begins with two familiar tools: a UFW contract and a small but growing effort that relies on labels and certification. But a new form of organization, worker centers, seems to be having more success in the district. Although struggling to deal with unfair labor conditions and unscrupulous employers, worker centers rely heavily on ethical consumers and firms. All are aimed at changing the basis of interaction between workers and firms, and although their actions probably can resemble boycotts and work stoppages, they are also creating discussions that can resemble multistakeholder collaborations.[122] At their most promising, the worker centers' aspirations might be understood as a permutation of the recognition that firms, their suppliers, and even their competitors succeed best when everybody is doing well. However, although a happy thought and a familiar phrase, "everybody" has rarely included workers. The new worker centers aim to change that.[123]

Making a Union Contract Work

Swanton Berry Farm is appropriately an icon of what a union contract could accomplish for employers and employees alike. Owner Jim Cochran describes his UFW contract as a low-cost alternative to having to maintain a human relations department for his small operation. Because the union organizes insurance and profit-sharing programs for his workers, he does not have to do so. Similarly, the contract provides guidelines so that he can act consistently without having to remember precisely what decisions he made in a similar previous case. Although it is atypical, Swanton Berry Farm does frame possibilities. Cochran is fond of relating that when he told his workers that there would be no profit sharing in a particularly wet year when he anticipated losing $100,000, the workers rearranged the planting and harvesting to keep the farm in the black. That example of recognizing and working with skilled employees rather than abusing them is unusual and instructive.

Labeling and Labor

The Agricultural Justice Project (AJP) uses a very different approach to go beyond organic to secure workers' rights. Since 1999 AJP has worked to develop a certification and label scheme. An adaptation of the nineteenth-century practice of using a label to identify a product as union made, the AFP's version is a response to debate within the organic community.

Some organic movement founders say that labor issues were never part of the organic agenda, while others insist they were, but that they were pushed off the table in a compromise when the federal program was adopted in 1990. AJP's founders take the latter position, but they no longer seek to redefine organic standards to include fairness to workers. Instead, they are collaborating with diverse partners, including growers, labor groups, retailers, and nonprofits, to develop criteria for a label that would give farmers who use it a competitive edge among consumers willing to pay for fairness. Aimed in part at retail customers, AJP also offers businesses making ethical claims about their products a method for documenting that they source from growers who adhere to specific, measurable practices. The justice labeling enterprise was pioneered in the context of internationally traded goods, and AFP is still conducting discussions of how fairness might look and how to achieve and measure it in the U.S. context. Its application to the domestic, organic market remains to be seen.[124]

Worker Centers

Worker centers share characteristics with diverse social, mutual interest, service, education, advocacy, and research organizations.[125] Because they are not unions and do not represent workers in contract negotiations with their employers, worker centers are not subject to labor law and government regulation. These centers have been broadly adapted throughout the food system. Some have confronted unethical employers and resemble unions; others have been more focused on research and political activities designed to improve working conditions.

The Coalition of Immokalee Workers (CIW) is among the most visible of these relatively new institutional forms.[126] The CIW can resemble a union, as when it organized work stoppages protesting working conditions and wages in Florida's tomato industry, where organizing unions of fieldworkers is not legal. The CIW has gained much of its national following by providing testimony in cases involving charges of what the prosecuting attorney called "slavery, plain and simple" in Florida's fruit and tomato fields.[127] In these cases, employers have been found guilty of "beating workers who were unwilling to work or who attempted to leave their employ picking tomatoes, holding their workers in debt and chaining and locking workers inside U-Haul trucks as punishment."[128]

Although the successful litigation raised the visibility of the CIW, few growers changed their labor practices in response. In 2001, CIW began its Campaign for Fair Food, aimed not at the growers but at large-scale purchasers.[129] An Oxfam America study, *Like Machines in the Field,* explains the CIW strategy: Growers are confronted by increasingly consolidated buyers who press them to lower their price. They "pass on the costs and risks imposed on them to those on the lowest rung of the supply chain: the farmworkers they employ."[130] The CIW organized its campaign to convince not the employers but the tomato buyers that such squeezing is not tolerable and that they should "do the right thing."

CIW first aimed at Taco Bell and enlisted the support of churches, students at colleges and universities where Taco Bell had retail outlets, and shareholders in the parent company, Yum! Brands. A four-year campaign convinced Yum! to reverse the downward pressure on wages by offering a penny more per pound for its tomatoes and to establish a code of conduct for how its suppliers should treat agricultural workers. By the end of 2010, nine leading food retailers—including four of the largest restaurant companies in the world (Taco Bell, McDonald's, Burger King, and Subway) had signed on to the penny per pound promise and the CIW's Code of Conduct. When Florida Tomato Grower's Exchange refused to pass on the extra penny and threatened to retaliate against members that did, Whole Foods, the only supermarket in the industry to join the CIW effort, pressured two of its suppliers to break the stalemate and support the agreement.

The Bon Appétit Management Company (BAMCO), headquartered south of San Francisco in Palo Alto, provides an interesting district perspective on the CIW campaign. BAMCO operates 400 university and corporate cafés in twenty-nine states and buys about 5 million pounds of tomatoes a year. It has put leadership in sustainability at the center of its business model and has taken the lead in developing purchasing programs that address issues like "local purchasing, the overuse of antibiotics, sustainable seafood, cage-free eggs, [and] . . . the connection between food and climate change."[131]

Although the organization recognized that it ought to engage more actively in farm labor struggles, it had done and achieved little in this realm. BAMCO officials admit to having been taken aback in 2009 to receive a registered letter from the CIW, in part because the company was unaware that it was involved in the dispute. They were also ruffled by CIW's insistence on negotiating; their recitation of events depicts them as both relieved to have guidance on appropriate practice and anxious to sign on to the principles and pay the extra penny per pound. After a number of heated and unrewarding phone conversations with CIW, BAMCO corporate officers went to Florida to discuss an agreement with the group. The company added two details to the communication they committed to send to their tomato suppliers. First, following their own inquiries, they added a provision requiring growers to provide an employee-controlled health and safety committee. Second, while recognizing that discrimination based on sexual preference was not a top concern among the pickers, they concluded that they could not go home to the Bay Area with an agreement that did not squarely address the issue and added sexual preference to the CIW's list of "zero tolerance issues."[132]

BAMCO's efforts elicited mixed reactions from CIW supporters and its own customers. When the Florida Grower's Exchange was refusing to distribute the extra penny, BAMCO threatened to forgo tomatoes entirely. The CIW was not pleased at this stance, concerned that a boycott might cost workers jobs. *Fast Food Nation* author Eric Schlosser took issue as well, asserting that a boycott "smacks of grandstanding" and characterizing BAMCO's decision as a public relations ploy.[133] Schlosser's response reflects assumptions about what is likely to be popular. In fact BAMCO got some pushback from its customers. The Student Union Blog at Washington University in St Louis, for example, reported a series of comments from highly offended students:

> This is an outrage. If Bon Appétit refuses to serve tomatoes because of a political agenda, Washington University needs to change food service companies. . . . To say the workers are not being paid a fair minimum wage is really not accurate. The workers have agreed to work for a given wage and are being paid accordingly. No one is forcing them to work this job; it was a choice.
>
> I just want my Lycopene.[134]

The St. Louis response attracted strong critical response from other campuses.[135] CIW continues, at this writing, wrestling with food retail giants, most notably Trader Joe's and Publix, to address terrible working conditions in their supply chains. But the CIW experience with BAMCO suggests that ethical aspirations are neither self-evident nor self-enforcing. Action requires ongoing education even among those aware of and willing to act on justice issues.

Food Service Workers and the Restaurant Opportunity Center

The Restaurant Opportunity Center (ROC) is pushing on the education element in a very different arena. ROC was founded by and for surviving workers of Windows on the World restaurant at the top of the World Trade Center following the 9/11 attacks.[136] It has since grown into a restaurant worker center with branches nationwide (ROC United) and an innovative organization intent on changing how the food service industry operates. Like CIW, ROC can look like a union when its members picket abusive restaurants. However, ROC has emphasized research and training—for food service workers, restaurant managers, and regulators—in the requirements for food and worker safety. In addition, it runs its own restaurants, with on-the-job training programs for back-of-the-house workers (dishwashers, line chefs, and others never seen by customers) who want to move up in the industry.

ROC's early picketing was directed against labor practices at some of New York City's most renowned restaurants and chefs. ROC accused the upscale steak house chain Smith and Wollensky and Iron Chef celebrity and restauranteur Bobbie Flay of skimming tips. Daniel Boulud settled out of court after ROC demonstrators picketed his posh French restaurant in New York. ROC charged Mario Batali, impresario of very high-end eateries in New York (Babbo, Bar Jamon, Casa Mono, Del Posto) and Las Vegas (B&B, Carnevino Italian Steakhouse), with skimming tips and other unfair practices.[137] ROC's early campaigns focused on front-of-the-house abuses (those involving tipped employees), but the organization found unique purpose when it began to insist that front and back-of-the-house workers address shared problems together. It began distinguishing between what it called "high-road" and "low-road" employers: high-road restaurants that respond promptly to address abusive practices are rewarded and promoted to consumers, while low-road employers are shamed and picketed.

ROC has gained further traction with its research and consulting efforts. It assisted in the development of a training guide now used by New York City restaurants and is developing a certification program for high-road restaurants. It is also a formidable research organization that has documented numerous abuses and problems. ROC's *Serving While Sick* report details the salary and health care of restaurant workers and suggests that it may not be in the best interests of American food consumers to bar 80 percent of food workers from health care and sick days.[138] The group also has enemies: the National Restaurant Association, a major lobbying organization based in Washington, D.C., has invested heavily to make sure that the tipped minimum wage stays at $2.13 an hour and has asked members to notify it of ROC operations in order to mobilize against them.[139]

ROC is a high-road employer itself: it runs worker-owned cooperative restaurants, both called Colors, in Manhattan and Detroit. During the day, the facilities are training centers for restaurant workers. At dinner, they operate as restaurants. The first Colors, in New York City, experienced difficulties early on with the cooperative format and with disputes between former Windows workers and newcomers. However, the project righted itself and has steadily improved after a rocky start. Like the labels on the products made by Oakland's start-up crafters, Colors Restaurant menus highlight local, sustainable ingredients and artisanal producers. Nevertheless, ROC activists also argue that "sustainability" needs to be redefined to include worker rights and justice at every point in the food chain.

In addition to its own programs, ROC plays a role in the evolving Food Chain Workers Alliance. In 2008, several groups "struggling with ways to integrate their work with the new national interest in food systems" followed the suggestion of supportive funders to organize together. ROC United, the Coalition of Immokalee Workers, El Comité de Apoyo a los Trabajadores Agrícolas/The Farmworker Support Committee, and the International Labor Rights Forum raised funds from the Ford Foundation and invited organizations to Chicago in July 2009 to start the organization. Two years later the alliance participated with the CIW in a Do the Right Thing tour in Atlanta, which demonstrated against Publix Markets to address the abusive practices it had documented.[140]

Young Workers United

In the Bay Area, Young Workers United occupies a spot similar to that of ROC in New York, Detroit, New Orleans, and elsewhere. It also is focused on low-wage workers, primarily immigrants in low-wage service sector jobs, including food service. Organizing restaurant workers into a center or a union is difficult. These workers do not aggregate by the hundreds at a single factory or on predictable shifts; rather they are spread in groups of ten to twenty, or fewer, in hundreds of establishments at wildly different times.

YWU has simplified the challenge of locating restaurant workers by developing close ties with local schools and community colleges because so many restaurant workers congregate around college, English as a Second Language, vocational education, and similar classes. More clearly than many other worker centers, YWU regards itself as a preunion, that is, as as laying groundwork for a subsequent unionization effort.[141] It works closely with the Hotel and Restaurant Workers, Local 2, which has focused on hotels.

While workers change frequently from one employer to another, some problematic practices, for example, wage theft, are common across the sector.[142] Health care benefits and paid sick days are extremely rare. YWU's challenge to low-road employers has unionlike aspects. Its long campaign against the Cheesecake Factory ended when the chain settled workers' claims from across California for $4.5 million, for example. Again like ROC, YWU identifies and promotes high-road employers and is creating a handbook, *Dining with Justice: A Guide to Guilt-Free Dining*, that identifies high-road restaurant employers and includes awards for "food establishments that follow labor laws and treat their employees with dignity and respect."[143]

YWU has worked more directly in electoral politics than either CIW or ROC. In 2003, it successfully drafted and enacted Proposition L, which established a San Francisco-wide minimum wage of $8.50 an hour. Reasoning that young workers would be most affected, YWU organized for the referendum among low wage workers at schools and colleges. Three years later, YWU wrote Proposition F, legislation that would guarantee sick days to all workers in San Francisco. Since there was no model legislation to draw on, YWU members wrote the law themselves. The legislation provides that even part-time workers in the city earn an hour of paid sick leave for every thirty hours worked. To address small business concerns, the rule caps the sick leave that small businesses must provide at forty hours, while those with over ten employees must provide seventy-two hours of sick leave.[144] The proposition passed with 61 percent of the vote.

The continuum of abusive behavior includes everything from routine unfair practices to less common but still appalling instances of slavery. CIW's experiences suggest that when it is exposed, slavery is unacceptable in most quarters and may indicate a larger market for fairness than is traditionally supposed. Yet reaction to BAMCO's small role in this continuing drama illustrates that insisting on fairness for workers is not uniformly regarded as positive: incorporating fair wages into the true cost of food is not always popular, and even ending slavery still has its holdouts.[145] Having said that, we also note that the successes of food system worker centers are notable in both an era of declining state and

Figure 8.3
Oakland Jammers and Foragers (from left) Asiya Wadud, Forage Oakland; Rachel Saunders, Blue Chair Fruit; Dafna Kory, INNA jam; and Shakirah Simley, Slow Jams posed as celebrities in the extensive *New York Times Sunday Magazine* coverage of the second annual Eat Real festival in Oakland (photograph by Lucas Foglia).

national government labor law enforcement and declining numbers of unionized workers, and they are sending important signals to food system innovators within and beyond our district.

Creating and Celebrating a New Community

Much of the recent change we have described is on display at Oakland's new, and likely annual, festival of affordable, carefully crafted food, the Eat Real Festival. A hundred thousand people, more or less, have flocked into Jack London Square each year for the combination block party, street fair, and business development project since it started in 2009. Founded in Oakland by Anya Fernald's Live Culture Productions, the festival brings three critical groups together: food crafters, sustainable producers, and a new generation of consumers.[146] The organizers require participating food trucks and vendors to sell their products at $5 or less and to include at least two sustainably produced products. Fernald's for-profit company's mission reflects the belief "that to make real food America's everyday food, we need to build the skills that will be necessary to feed the demand for better food." Contests, bands, entertainments, and general hubbub have underneath them a serious purpose: to "build public awareness of and respect for the craft of making good food and to encourage the growth of American food entrepreneurs."[147]

A Sunday *New York Times* series of photos featuring the festival has made celebrity jammers and foragers visible among the growing list of celebrity chefs, farmers and urban farmers, and taco trucks (figure 8.3). Those taking the long view might regard the festival as a reenactment, in a different time and place, of the Tasting of Summer Produce that celebrated the connection between alternative growers and high-end restaurants. The crowd is younger, hipper, more diverse, and more reflective of the entire population of the district. Hence, the festival has a legitimate claim to "a place not in just a niche of food geekery, but in the community at large."[148]

9

Conclusion: The District and the Future of Alternative Food

An alternative food district continues to evolve in the Bay Area—expanding, thickening, and incorporating new institutions and new understandings of food and food quality. Although it is anchored, both symbolically and concretely, in a regional base of protected land, the district is not just a particular geography. The land opened possibilities for innovative food production and processing in a place that has a long tradition of intense attention to the creation and enjoyment of diverse foods. Those interests and possibilities intersected over many decades of personal and market relationships and shared histories, weaving a community and a common enterprise. The district has also been attuned to workers' rights and labor relations—perhaps in response to the fact that the treatment of labor in California's agricultural industry has been so dismal—and has pioneered creative programs for feeding the hungry. Nevertheless, it is only in the past decade that a more comprehensive justice agenda, including a full spectrum of food system workers and access to healthy food for all, has begun to drive innovation in the district.

Who Built the District, How, and Why?

The first wave of innovative institution builders in the district inherited a legacy of civic activism and parkland protection and adjusted that conservation tradition to value working landscapes and protect agricultural lands inside and beyond parks. Back-to-the-landers and farm-to-restaurant foragers then demonstrated the commercial potential of sustainable agricultural production. New, small-scale distribution models opened links to urban consumers, first for fresh, beautiful vegetables, and then increasingly for more sustainably produced food. In the maturing district, a third wave of urban priorities related to consumer health and fair wages and working conditions for labor are now shaping the development of its current institutions. A new set of voices—urban minorities and others suffering the consequences of eating a "normal" American diet, and young urban workers making and serving our food—have joined the conversation and are pushing the district to clarify its priorities around food access, self-determination, and fairness for food chain workers.

The standard set of industrial district values is important in this story: common enterprise, sharing what you know and what you have, and a willingness to learn. Districts are built on webs of personal relationships that generate trust and provide networks for knowledge sharing. These relationships mean that things that seem like private goods (skills, reputation, networks) in practice can become quasi-public goods, an evolution we have seen at work in the Bay Area. But the district is also never far from its particular history. In this case, organizations and events like the Black Panthers, the free speech movement, and the Summer of Love shaped many of the early food values and institutions in the district and many of their founders, some of whom are still doing business there.

As the district matured, that combination of common enterprise and shared history has altered collective expectations about appropriate behavior, production practices, and values. We argue that the Bay Area district facilitated increasingly demanding expectations about what alternative food could become. Competition and cooperation have combined in the region to urge ever-higher standards on the common enterprise. In this story, the idea of alternative food grew over time to encompass deeper understandings of a number of qualities that we see as generally positive; organic, sustainable, fresh, local, artisanal, healthy, fair, accessible, affordable, and supportive of community economic development. Not all products or institutions in the district include all of these qualities, but there are increasing pressures to consider, support, and display them. Pressures to do so are both market based and social. For the most part, new institutions in the district have not rejected earlier priorities. Instead, they have tended to begin from a new baseline, largely taking for granted the notion that alternative food is fresh, sustainable, and local, and driving forward on new priorities of access, fairness to workers, and community development.

The Bay Area's mixture of climate and geography, immigrant skills, cuisines, rootstocks, university support, and ready cash for entrepreneurial investment has been essential to the development of the district we describe, and these same factors also underwrote the early days of the conventional food system.[1] Several additional factors are critical to anchoring alternative food in the district: scale, proximity and district practices around the production of new knowledge.

Scale and proximity are intertwined in important ways. Although the growers in our story are not necessarily small compared to the average farm size in California or other states, they generally do not fit the conventional factory model that defines much of the nation's food system. District farmers are usually not simply investors; they are generally doing their own work—most often on their own land and with the help of farm laborers—in communities where they live their everyday lives and sell their products.

The district began when a new generation of growers rediscovered and adapted sustainable methods of growing food and then developed solid connections to alternative restaurants. It continues to emphasize, and expand through, new distribution networks: community-supported agriculture, farmers' markets, and Internet sales, but also farm-to-restaurant, farm-to-school, and farm-to-hospital links. These connections have helped to erode distinctions between urban and rural, bringing, for example, urban expectations about minimum wage to farms and rural concern for the preservation of working landscapes and processing infrastructure to cities.

Knowledge production and knowledge sharing in the district have not been simply accidental or lucky. A group of key players, those we have termed mavens, have engaged explicitly in producing and sharing new models of what food production, distribution, and consumption can look like in the region. The mavens saw both the need and an opportunity to make food better, and they were willing to invest their lives, careers, and values in showing the rest of the region (and often beyond) how to do things differently. Each of them expressed a preference for community before (but not to the exclusion of) profit and demonstrated a willingness to start small and invest returns and reputation in building a sustainable regional community rather than a commercial empire.

What Has Been Accomplished in the District?

The experiences and experiments of the district clearly have not rendered the conventional food system obsolete, but important changes have been made nonetheless. Least obvious, but nevertheless important, the district has helped make food a critical topic for personal and public attention in the district and well beyond. Americans are now generally aware that the U.S. food system is not working well and that significant changes are required to improve it. In the district and across the nation, people are talking and thinking about food—what they eat and feed their children—working long hours, and investing considerable financial and other resources to create alternatives to conventional food.

More specifically, the Bay Area community has articulated and consistently raised expectations about food quality for several decades. The district's alternative producers and restaurants created and distributed products that few people knew about, much less ate. Incrementally and subtly, products that improve diets and teach palates that vegetables come in flavors beyond potatoes, iceberg lettuce, and catsup spread all over the nation. District farmers and ranchers also demonstrated the possibilities for commercially viable ecological farming methods and adapted them to crops that are appropriate to the regional climate and markets, while pioneering the distribution of small amounts of noncommodity products on a regional scale.

Over time, a web of government, private, and nonprofit institutions has grown that supports continued learning and the evolution of new models of regional supply and demand. We believe this demonstration of the potential for regional-scale cultivation, processing, and consumption is likely to be a critical contribution to the future of the food system in California and beyond. Eventually, rising costs of inputs, especially fertilizers, due to peak oil and climate change, are likely to make much of today's standardized, commodified, global food trade prohibitively expensive or simply impossible. Transporting most foods over massive distances will become costly as well, and we expect calls for accountability, transparency, and traceability in the food system to grow more insistent as the public health consequences of the current conventional system become more apparent.

We suspect that while very local agriculture, in the form of urban gardens and farms, is an important tool of education and a source of some important vegetables and fruits, it is unlikely to offer sufficient efficiencies of scale to ensure food access for most people in the district, provide adequate supply for large markets or institutional buyers, or offer compelling enough product variation for consumers accustomed to purchasing any food in any season. Thus, we do not expect national and international trade in all food commodities to halt, but we do anticipate substantial growth in more regional, what some have called "appropriate scale," agriculture.[2]

Challenges in the District

Our emphasis on what has been accomplished should not conceal the many challenges along the way or suggest that all barriers have been or will be overcome. Internally, the district has had to make incomplete transitions and uncomfortable compromises between competing priorities, including park protection and working landscapes, rural producers and urban consumers, and aesthetics and access, many of which center on uneven distribution of entitlements. It has stumbled forward on questions of farm labor, supporting United Farm Workers boycotts but somnambulant on ranch worker housing. Nevertheless, it has raised the bar on what constitutes food, often in the face of those who seem to presume that fine cuisine and justice are oppositional and that better foods for the rich necessarily mean worse foods for the poor. While we are not surprised that well-heeled people have succeeded in making food better for themselves, we call attention to the degree to which district firms, activists, and civil society more broadly are focusing on making healthy food and decent jobs available to those who need them most.

Progress toward alternative food can be limited by those willing to settle for food indulgences—high-priced specialty items that will never be available at a scale sufficient to improve either diets or jobs in the food system. However, those who reflexively dismiss as elitist efforts to raise expectations regarding food quality can also slow progress. The unavoidable fact is that current public policy subsidizes the conventional food system, and many items produced outside that system are likely to be more expensive, at least in the short run.[3] Moreover, as we have repeatedly seen, paying a fair price for healthy food is a critical part of paying food system workers a living wage. It is therefore important to distinguish between "bourgeois piggery" and a faux egalitarianism that challenges higher food quality standards in ways that support the status quo.[4] In terms of food system reform, we see no need to limit the food choices of the rich or the zealous unless they actively inhibit access to healthy, sustainable food for others.[5]

The progression of events in the Bay Area district suggests, moreover, that voting with your fork is not just a tool of the privileged. Moving money and holding onto it are both important to change. Shifting money from urban to rural areas to pay "the real cost of food," including, but not only, fine food, has been critical to improving wages and working conditions in the district's alternative food chain. Furthermore, social entrepreneurship and economic self-determination are important themes among justice advocates in the district. Efforts to build a grocery store that sources in the community, employs from the community, and returns the profits to the community are variants on voting with your fork. Residents of West Oakland may be poorer than those of Marin, but as justice advocates and others have documented, they also spend a great deal of money on food.[6]

Opposition to scaling up may fall into that *actively inhibit* category.[7] Small-scale alternative food innovations have to grow and expand in order to create access to good food, fair jobs in the food system, and self-determination in underserved communities. If a food improvement is not scalable and cannot eventually become generally affordable and available, it is not going to play a role in creating alternative food systems. Therefore, characterizing scaling up as inherently bad can be destructive. The differences between even a very large Niman Ranch and McDonald's are important. Until Americans decide not to eat meat, raising it without antibiotics will be a critical public health priority, and projects that do so beyond the boutique level are enormously important. Keeping expectations about food quality low simply protects the worst elements of the conventional food system. Similarly, ensuring that poor people have access to healthy, nutritious food is essential, and worrying about how they will pay for it is central to that. But simply insisting on cheap food while calling anything else "elitist" is not constructive. Supporting innovations that hold promise of better food, employment opportunities, or financial returns in communities that need to build an economic base is critical. Achieving economies of scale, creating jobs, and providing better food for more people is the goal and cannot be regarded as a falling away from grace.

Throughout district history, food activists of every emphasis and persuasion have been drawn or forced into interactions with one another, with productive results for developing alternatives and innovations. However, just because access to safe food has moved to the center of food quality conversations in the district does not make it easy to produce healthy food or pay fair wages. The district's story underscores that even under the best circumstances, improving the food system will be very hard. There is a David and Goliath factor built into judging the success of the district against either the global food system or prevailing political reluctance to enforce health and safety standards, worker safeguards, labeling requirements, and chemical policies. And David is a memorable hero because Goliath usually wins.

The efforts we have highlighted in the district focus more on defining and demonstrating the commercial possibilities of an alternative approach to food systems than a rapid overthrow of the conventional one. Of course, even that smaller goal has been challenging, given the weight of the opposition, and it has become standard for the conventional system to adopt the language of fresh and local, sustainable and healthy, while moving in the opposite direction. Thus, even in a district with a lengthy growing season and relative abundance of resources, demonstrating the possibilities for a better regional system has been slow and not always steady. Still, we see reasons to be hopeful as a new generation of activists, food crafters, and investors continues pushing and innovating toward food democracy.

Markets, Politics, and Collective Action

While we do not offer the district as a model of what can or should be done elsewhere, we believe there are useful lessons in the narrative for those working to ensure the broadest possible access to good and healthy food. In particular, tracing multiple decades of alternative food action in the Bay Area suggests that sharp distinctions between markets and politics—and between voting with your fork versus changing policy—are unfounded and counterproductive.

Several decades ago, the market was recognized as an underexploited force for social change. That led to a raft of businesses and associated brands and marketing campaigns that could all be classified as doing good by doing well. That upsurge invited critics and analysts to focus on the inevitable shortcomings of such approaches, which include greenwashing (making overstated or false claims that a product is "green"—ecologically sustainable—when it is not) and other types of fraudulent social marketing. The net result has led some to undervalue consumers and entrepreneurial change. Clearly there are domains where collective action is required. For example, if you do not want meat in your diet, you can buy (or grow) something else. But even an ovolactarian, who abjures milk, eggs, and cheese, as well as meat, cannot avoid the antibiotic-resistant superbugs that spread from conventional animal feedlots. Political action to address this issue is as essential as it has been slow in coming. But we have found no basis for alleging that (to use the formulation in Leonard's *The Story of Stuff*) [8] the citizen muscle has atrophied in favor of the consumer one: efforts in the district combine tactics, tools, actors, and institutions from all across the board.

Exercising both muscles is possible, but scale matters. Healthy, sustainable foods have thus far done better politically at the state level, and even better at the county and city levels, than in national (and international) debates. There is a difference between current national policy, which is explicitly designed to advance a global, vertically integrated, high-input food system, and Bay Area county governments' priorities, which do not automatically tend in that direction. Indeed, most of the entrepreneurs in the narrative have worked side by side and often hand in hand, with both government and nonprofit agencies. The conventional food system is sustained by a combination of powerful corporations, government subsidies, and supportive regulatory and cultural environments. Changing it requires the same kinds of collaboration.

Entrepreneurs have also worked with ethical consumers to demand real changes in the district, a proof of concept that consumption can influence the food system. The United Farm Workers union was an early adapter, teaching consumers to vote with their forks for labor justice and to work with producers, retailers, and nonprofit activists to shape the food system by informing and directing consumer choices. And for all its insights, the literature on how labels do not work or how they help to replicate the neoliberal economic system generally misses two basic points: that boycotts are collective action and that both organizing boycotts and consumer education campaigns are political acts. Well-designed and supported boycotts can be effective. While we recognize that the profusion of alternative labels may overwhelm some consumers, they also educate others on some basic issues. If labels did not direct at least a meaningful subset of consumers toward ethical choices, conventional food purveyors would probably not fight so hard to prevent alternative producers from making intelligible information available—or to use words like *natural* or make dubious health claims on their own labels.

The important work that alternative food entrepreneurs do, creating and demonstrating food possibilities, is often a precursor to meaningful policy change. Those new products and alternatives constitute the basis for conversations about food options that simply did not exist before. Inescapable discussions about what a just, sustainable, healthy food system could look like, what and whose values it should embody and promote, and who will make money from it and how are a necessary, if not sufficient, part of redirecting the current conventional food system to operate on fairer and healthier norms and replacing it with something better.

Priorities for the Next Generation

We conclude with several observations about the next wave of innovation. Almost inevitably discussions of what needs doing in the food system focus on the Farm Bill, particularly on redefining and redirecting federal subsidies. We share that concern. We are particularly concerned about the durable link between encouraging intensification and overproduction and then shoving the inevitable surplus off on the poor, ostensibly to help them, and almost inevitably in the form of highly processed, calorie-dense, nutrition-light foods. But the Farm Bill is just one face of a many-headed dragon that must be addressed. We have discussed the Pure Food and Drug Act (as amended), federal pesticide regulation, and labor and antitrust laws that are not aligned with healthy food, and urge that federal policy barriers to a sustainable, fair food system do not begin or end with the Farm Bill.

Other important changes do not require Washington's cooperation. First, if the conventional food system tends toward deskilling consumers and producers, building pressure for change requires explicit focus on education and reskilling about food. Learning about growing healthy food, and doing it, is important, but building a population of home cooks and eaters who are able and willing to choose and prepare healthy, affordable food is critical. While education is challenging and incremental, and can be viewed as hectoring or condescending, stepping up education around food is probably the most important priority. Education must overcome both school meals that miseducate young Americans about food and health and the billions of dollars that the food processing industry invests in pushing its products and creating confusion about their health consequences. Engaging the Farm Bill on those issues is both critical and daunting. However, it is not merely faintness of heart to believe that the most effective education probably begins in communities and families and that supporting such efforts at the local and regional levels is vital. Educating people about healthy eating is something in which district innovators have some experience to build on and to share.

Second, the district community must identify a set of ground rules for the relationships between public priorities and private companies that prioritize public health and community development. Oakland, for example, could take a lot of the angst out of the retail food industry's new, ostensibly urban-friendly business footprint if it established minimum expectations—regarding fair wages and career opportunities, community development, nutritional education, and sustainable production and processing practices—for any firm expecting government support for entering the food business in its precincts.

Third, producers, retailers, and consumers in the district must focus more explicitly on recognizing and remunerating special skills, particularly in what are traditionally low-wage jobs, and on training and rewarding the district's food systems workforce. District entrepreneurs who are moving in that direction need support not just from ethical consumers, but also expectations and policies that reduce the incentives to exploit workers. Merely enforcing the requirements now on the books would produce enormous change in standard operating procedures. Similarly, the for-profit and nonprofit community should aggressively explore a range of creative financing options for developing small food businesses to a scale at which they can support community self-determination and economic stability.

Finally, the district must continue to protect land for farming and food production. It must add a missing link to ensure access to productive land for people who cannot afford to pay California's high land prices. Supporting urban agriculture with appropriate zoning and marketing regulations is a good place to start. Similarly, health regulations for food processors and vendors—which are essential—must be scale appropriate and designed to address the risks that occur in small-scale operations. District innovators and policymakers need to ensure that appropriate regulation of new food crafters does not drift toward overburdening small operators with inappropriate requirements.[9]

We recognize that alternative food requires efforts at multiple points in the system. Too many changes are needed at global, federal, state, and regional levels, for it to be otherwise. However, based on the results of the past half-century of small businesses, local governments, and civil society innovating to create a viable alternative to the conventional system in the district, we see greater hope for progress at that level. Our best hopes for creating a truly sustainable, healthy, and just food system lie in continuing to build that community.

Notes

Chapter 1

1. The redevelopment of the Ferry Building and the founding of the Center for Urban Education about Sustainable Agriculture and the Ferry Plaza Farmers Market are discussed in chapter 6.

2. Gilje came to PAN from Minnesota's Centro Campesino, founded in 1998 "to improve the lives of migrant workers and rural Latina/os and to create a strong southern Minnesota Latino/a voice." See the Centro Campesino Web site at http://www.centrocampesino.net/ (accessed February 2012).

3. Mission Pie is a Bay Area nonprofit focused on young people, sustainable agriculture education, and entrepreneurship, discussed more in chapter 7.

4. Simon (2006). We use the term *conventional* where others might say *industrial* to underscore that we have become accustomed to it. Conventional food is in no sense traditional or "natural," and it is not, as many have argued, clearly food. Michael Pollan (2008) was probably the first to use the now ubiquitous term "edible food-like substances."

5. Petrini's (2007) "good, clean, fair" formulation is similar, but the chronology of our narrative leads us to environment, food, justice.

6. Many different kinds of discussions of food system flaws begin with corn and federally subsidized grain commodities, and for good reason. See Poppendieck (1985), Pollan (2002), Nestle (2002), Simon (2006), Imhoff (2010), and Hesterman (2011).

7. Imhoff (2010).

8. As we shall underscore in chapter 8, access includes questions of both geography and cost.

9. Munroe and Schurman (2008).

10. Hesterman (2011).

11. The scholarly literature describes many kinds of social movements. In our understanding, two main theoretical models have dominated: a "political process" model, primarily understanding social movements as seeking legislative remedies and reform, and a "resource mobilization" model, which explores

efforts to increase the power and resources of those who are disenfranchised using tactics of direct action and mass mobilization. Scholars using those models—see McAdam and Snow (1997) and Johnston and Klandermans (1995)—would probably not regard our district as a "movement." Loosely affiliated groups organized around shifting goals and tactics are more recently described as "new social movements" (Buechler 1995). Because we view movements in the earlier model, we rely on the industrial district model, which seems to frame our observations more clearly. Whatever the terminology, we do not see a movement.

12. Belasco (1989).

Chapter 2

1. The singularity of M.F.K. Fisher is perhaps the exception that proves the rule. See Fisher (1937, 1941, 1942, 1943, 1946).

2. Two 1939 studies are classic anchors for an extensive literature: Carey McWilliams's much cited *Factories in the Field*, and less familiar, photographer Dorothea Lange's work with her husband, Paul S. Taylor, on *An American Exodus: A Record of Human Erosion*.

3. This disciplinary hegemony is not a recent or isolated phenomenon. We find a similar dynamic in chapter 4, when we locate the origins of sustainable agriculture in early-twentieth-century studies of composting in Asia rather than in efforts to understand the recent famines that had killed 12 to 29 million people on the Indian subcontinent between 1876 and 1902. Patel (2007) provides a starting point and references for probing the famines. See also Taubes (2011) for a history of the debate on fat and sugar.

4. W. Friedland (1991, 12).

5. White House (1967).

6. We consider the term *green revolution* a misnomer because there is nothing green about it, as we now understand the term. Researchers developed new crop varieties that were extremely responsive to water, fertilizer, and chemical inputs. The topic is discussed in chapter 3.

7. Lappé (1971, 5).

8. The subtitle of the volume is informative: *A Report of the Agribusiness Accountability Project on the Failure of America's Land Grant College Project.* The project was funded by the Field Foundation, which works as a "strategic enabler" of "innovative programs and organizations, with a primary emphasis on underserved and disadvantaged individuals and populations." See The Field Foundation of Illinois, "History," http://www.fieldfoundation.org/history.html (accessed February 2012). The same project produced Hightower's less-well-known but equally critical *Eat Your Heart Out* (1975).

9. Lappé's daughter, Anna, has recently expanded the family business, addressing issues of food justice and global climate change in *Diet for a Hot Planet* (2010).

Lappé also encouraged publication of Lerza and Jacobson's prescient 1975 book, *Food for People, Not for Profit: A Sourcebook on the Food Crisis.*

10. The crisis was related to the Carter-era Arab oil boycotts and the apparently permanent shift in the U.S. trade balance. The same crises of the 1970s are also generally associated with the rise of neoliberalism.

11. Sen was awarded the 1998 Nobel Prize in economics.

12. A selection, in order of appearance: Clark (1965), oysters; McPhee (1975), oranges; Mintz (1985), sugar; Zuckerman (1999), the potato; Pendergrast (2000), coffee; A. Smith (2001), the tomato; Dupuis (2002), milk; Kurlansky (2003), salt and (2006), oysters; Chapman (2009), bananas; and Niman (2009), pork.

13. *Charlotte Observer* (2008), Griffith (1993), Linder (1995), Boyd and Watts (1997), Boyd, (2001), Weinburg (2003), and Striffler (2007). The last reflects the famous 2002 Yale conference that considered the "Biological, Social, Cultural, and Industrial History" of the chicken, "from Neolithic Middens to McNuggets."

14. In 2010, the U.S. Department of Agriculture and the Department of Justice held hearings on anticompetitive behavior in agriculture, focused in part on the poultry industry. Proposed new rules for the Packers and Stockyards Act included long-overdue protections for contract poultry producers, but under pressure from industry and Congress, the USDA stripped most meaningful changes. See Gwin (2006, 2009). The "Final Rule" is available here: http://www.gipsa.usda.gov/Farmbill/final%20rule%20FR%20preview%202011-31618_PI.pdf (accessed February 2012).

15. Davis (2001) analyzes the famines in late nineteenth-century India in the context of British colonialism and expanding capitalism. Sen's analysis was consistent with and influential in the developing field of political ecology, which explores precisely this relationship between the distribution of the costs and benefits of environmental change and their relationships to social and economic inequalities. See, for example, the discussion of soil degradation in Blaikie and Brookfield (1985).

16. Zuckerman (1999), Davis (2001, Donnelly (2001), and Nally (2011).

17. We discuss the green revolution briefly in chapter 3. The Gates and Rockefeller foundations are currently in the midst of an effort to increase agricultural productivity in Africa. Their project is built on the premise of recreating the green revolution by using technology to address agricultural yield and resilience. Although their initiative does reference environmental and economic stability, many observers have noted that the pesticide- and fertilizer-intensive monocropping promoted by the original green revolution has had many deleterious effects on world food systems. Chemical and processed food companies are commonly viewed as financial beneficiaries of the first green revolution. The Gates Foundation purchased approximately 500,000 shares of Monsanto in 2010. McDonald's and Coca Cola are among the foundation's top five holdings. "Gates Foundation Buys Ecolab Inc., Goldman Sachs, Monsanto Company, Exxon Mobil Corp., Sells M&T Bank," *gurufocus*, August 17, 2010 (accessed September 2010).

18. The controversial term refers to areas in which access to fruits and vegetables is limited and food retail is largely confined to liquor stores, bodegas, and fast food outlets. We address this topic in chapter 8.

19. For an excellent introduction to connections between old colonial relationships and modern food scares, see Freidberg (2004) and French and Phillips (2000).

20. The term *food desert* appears to have first been used by a resident of a public sector housing scheme in the west of Scotland in the early 1990s (Cummins 2002). Most early research on the subject occurred in the United Kingdom, where the issue was both urban, produced by an exodus from city centers to suburbs, and rural, where a growing car culture and the dismantling of village centers and shops left those without access to transportation similarly without access to a variety of healthy food. The term is controversial in the food justice context, as we discuss in chapter 8.

21. J. Richardson (2009).

22. Schlosser was not the first to criticize McDonald's. See Boas and Chain (1976) and the surprisingly reflective volume by McDonald's impresario Kroc (1977). More recently, the company's own 2012 Twitter campaign (#mcdstories) backfired as users of the hashtag shared fast food horror stories instead of what McDonald's had apparently hoped would be personal tales of positive experiences with the restaurant chain.

23. Nestle's book was titled *Pet Food Politics: The Chihuahua in the Coal Mine* (2008).

24. Pollan's other contributions are listed, in part, in the Bibliography.

25. See also Nestle (2000).

26. "An international movement which coordinates peasant organizations of small- and middle-scale producers, agricultural workers, rural women, and indigenous communities from Asia, Africa, America, and Europe." They are a coalition of 148 organizations in 69 countries, advocating family farm–based sustainable agriculture and were the group that first coined the term *food sovereignty.* See Wittman, Desmarais, and Wiebe (2010) and http://viacampesina.org/en/ (accessed February 2012).

27. Gottlieb and Joshi (2010, 6).

28. Hassanein (2003). "Food Democracy," http://www.panna.org/issues/food-agriculture/food-democracy (accessed February 2012).

29. McWilliams (2009).

30. Battisti and Naylor (2009).

31. International Assessment of Agricultural Knowledge, Science and Technology for Development (2009).

32. Intergovernmental Panel on Climate Change (2007).

33. The International Assessment of Agriculture Knowledge, Science and Technology for Development was undertaken by a multistakeholder group of organizations, including the Food and Agriculture Organization (FAO), the United

Nations Development Program (UNDP), the United Nations Environment Program (UNEP), the United Nations Educational, Scientific and Cultural Organization (UNESCO), the World Bank, and the World Health Organization (WHO).

34. The first post–World War II Italy was characterized by large-scale industrial development, the second by a lack of development. The "Third Italy" is associated with the emergence of industrial districts as described by Marshall ([1890] 1920).

35. Allen and Guthman (2006) and Flanagan (2010).

36. Harvey (1982).

37. Arce and Marsden (1993) and Marsden and Arce (1995).

38. DuPuis and Gillon (2009, 43), Lyson (2004), Lyson and Geisler (1992), Bell (2004), Kloppenburg et al. (2000), and Hassanein (1999).

39. Thrupp (1995).

40. Striffler (2005, 163–164).

41. Moss (2009). As we note in chapter 7, an estimated 15 percent of hamburger sold in the United States contains "fatty process trimmings" treated with ammonia.

42. Moss (2009).

43. Marshall ([1890] 1920).

44. Porter (1998) lists several theoretical approaches from economic geography to agglomeration economies, but our categories are more reliant on Storper's (1997) presentation of the three main styles of industrial district.

45. Commons are usually associated with shared natural resources, but the term has expanded in both scale and subject matter such that both the ozone and digital information may be considered commons. We follow Hess and Ostrom (2007), who explore knowledge as a commons. As they point out, self-organized commons require collective action, self-governance, and social capital. These three aspects are also necessary and equally complicated in industrial districts, in which conventions, culture, and knowledge may function as a sort of knowledge commons. See Hess and Ostrom (2007) and Ostrom (1990).

46. *Thickening* is from Amin and Thrift (1994). See also MacLeod (1997).

47. Marshall ([1890] 1920), book IV chapter X section 3, available at http://www.econlib.org/library/Marshall/marP24.html#Bk.IV,Ch.X (accessed February 2012).

48. Glasmeier (2000) and Rinaldi (2002) discuss the collapse of iconic industrial districts.

49. Stevenson's framework (2008) associates three types of strategic change activity with three different categories of actor: warrior, builder, and weaver. Warriors resist the modern agrifood system politically and through analysis and boycotts. Builders work through markets and entrepreneurship. Weavers work to build a movement and connect the other two to civil society.

50. Gladwell (2000) familiarized the term, emphasizing the role of mavens in collecting and sharing information. While that part is central, we also value the emphasis in Wikipedia on the Yiddish origins of the term that invoke for us the *shadken,* or matchmaker, skills essential to the role. http://en.wikipedia.org/wiki/Maven (accessed February 2012).

51. Our analysis is long compared to most of the literature but a flash in the pan compared to Glasmeier's (2000) two-century-plus review of the rise and fall of the Swiss watch industry.

52. Goldschmidt's study has been attacked often but never seriously refuted. It has also been revisited and replicated, and its analyses extended in place and time. See Lobao (1990).

Chapter 3

1. A good place to start is with McWilliams (1939) and Hundley (2010). Our favorite brief introduction is actually in a cookbook (Lappé and Terry 2006). See also Gottlieb and Joshi (2010) for a very different treatment emphasizing justice issues.

2. For an exception, see Price (1939), discussed in chapter 4. See also Kurlansky (2009).

3. McWilliams (1939, 104).

4. Starrs and Goin (2010).

5. Real estate firms and railroad agents have long ballyhooed California's bounty, and the tradition thrives. This and similar passages appear frequently on contemporary real estate Web sites. For example: http://www.real-estate-2000.com/california.htm (accessed February 2012). On the long and colorful history of agriculture-related boosterism in the state, start with Sackman (2007) and Orsi (2005).

6. Hundley's discussion of the aboriginal waterscape of California observes that even the relatively extensive Paiute canals and ditches of the first millennium are "today hardly discernible" to the trained observer. Hundley (2001, 24).

7. Giving rise to still paramount pueblo rights. See Hutchins (1959–1960).

8. The literature on water in California is enormous. Kahrl, (1979), Pisani (1984), Hundley (2001), and Fradkin (1996) provide a good start.

9. For an introduction to that landmark in American environmental history, see Richter (2006).

10. The literature is extensive. Gutiérrez and Orsi (1998) is extremely useful, particularly the pieces by Preston and by Anderson, Barbour, and Whitworth. We have also delved into Engelhardt (1897, 1908), Jackson and Castillo (1995), Hackel (2005), and Lightfoot (2006).

11. Starrs and Goin (2010). Similarly, San Francisco Bay was plucked clean of shellfish in the first decade after 1849, perhaps explaining the enthusiasm for the

famous oyster dish, hangtown fry, among newly wealthy miners who wanted the most expensive dishes they could buy.

12. No one has attempted an accurate accounting of the regional flows, but the owners of wheat and cattle ranches at the time suggest California mining fortunes were in the thick of this agricultural investment (Walker 2004a). George Hearst, father of William Randolph, owned a mining empire that stretched from Nevada to Peru, and he invested heavily in ranchland, as well as newspapers. James Graham Fair and Billy Ralston became rich in the Comstock Lode, as did Charles Reed in Yolo County and Martin Murphy in Santa Clara County. All were among the larger purchasers of agricultural land.

13. Extensive landholdings gave Miller and Lux control over water as well. Litigation to loosen the grip of early riparian holders commenced in the 1880s. See Igler (2001) and Hundley (2001).

14. Liebman (1983).

15. See Orsi (2005), Walker (2004a), and Stoll (1998).

16. Guthman (2004). Periurban agriculture is thus caught in a bind: expensive land made more expensive by its development value. We discuss this in detail in chapter 5.

17. California Department of Parks and Recreation (1988) referencing the *Daily California Alta* (1862).

18. P. Martin (2003).

19. Similar statutes followed, and the Chinese exclusion program did not end until 1943, when the United States allied with China against Japan in World War II. By that time, immigration policy had capped overall immigration to the United States, allocating small numbers of permitted immigrants from mostly European countries while eliminating Asian immigrants entirely. See "The Chinese Exclusion Act (1882)," Harvard University Open Collections Program, "Immigration to the United States 1789–1930," available at http://ocp.hul.harvard.edu/immigration/ (accessed February 2012).

20. See P. Martin (2003). He notes, also ironically, that when the anti-Chinese homesteaders did try to homestead and farm their own land, they generally learned that family farmers made about as much money as Chinese laborers—not much—and moved on to other pursuits.

21. On the Meiji restoration and Japanese immigration, Jansen (2002) is a good source. The Japanese were part of a diverse wave of Asian immigrants at that time. See Takaki (1998).

22. Unlike the Chinese government, which viewed those leaving China as deserters, the Japanese emperor regarded the emigrants as his personal representatives and Japanese diplomats demanded that they be treated well in the United States. Having defeated Russia in a major war in 1904, Japan was such a major military power that the U.S. government could not entirely ignore the demand. That led to the 1907 "gentlemen's agreement" between President Roosevelt and the Japanese government: San Franciscans would end school

discrimination, and the Japanese government would stop issuing passports for Japanese to come to the United States. The agreement had almost no impact, but the issue of whether the federal government could use its treaty power to force a local school district to spend extra funds in order to provide schooling for offspring of noncitizens was debated at length in Congress and the courts. If that has a familiar ring to it, notice that Californians adopted Proposition 187, the Save Our State Initiative, in 1994 by a margin of 59.8 percent. The measure attempted to bar undocumented workers from using social services, including health care and public education in California. A federal court declared the measure unconstitutional. In 2009, President Obama promised that no illegal immigrant would benefit from proposed health care reform. In the 1920s, Japanese immigrants were excluded as the Chinese had been, but with a major difference: the Japanese immigrants' children were native-born citizens who, unlike their parents or the Chinese who preceded them, could own land. That led to the World War II internment of Japanese citizens during which they lost approximately 40 percent of their property, not including forced sales that preceded many relocations.

23. Quoted in Loftis (1998, 191).

24. *Free soil* refers to places where slavery was prohibited. *Free land* means land available to settlers at no cost. Although slavery was the most durable issue, Zahler (1941) reminds us that immigrant workers in East Coast factories began agitating for free land in the early nineteenth century.

25. Gates (1968).

26. Olmstead and Rhode (2008).

27. For a particularly enthusiastic period piece, see Davenport (1912). But note Walker's detailed discussion of how the California diverse soil types delayed the final triumph of fertilizer-dependent monocropping until after World War II (2004).

28. See Henke (2008). The Hatch Act of 1887 authorized federal extension efforts, and the Smith-Lever Act of 1914 brought the states into the program, requiring the land grant university in each state to "extend" the benefits of research to farmers on the ground. In California, the organization is formally known as the University of California Cooperative Extension (UCCE). Giroux's (2007) warnings about the "military-industrial-academic complex" resonate for us: we have worked together on a campus that manages the nation's nuclear weapons research and has accepted huge grants from Novartis and British Petroleum. See Dalton (2007), Press and Washburn (2000), and Heilbron and Seidel (1989). In this book, we focus in chapters 7 and 8 primarily on a far more creative and scale appropriate role for UCCE in the food district.

29. Redirecting some of those benefits to small-scale sustainable farming and farm workers has been a major priority of the sustainable agriculture movement.

30. It is probably not a coincidence that water development did not become important in California until the mid-twentieth century. See Walker (2004a).

31. The legendary plant breeder Luther Burbank and the University of California, Berkeley's first eminent agricultural scientist, Eugene W. Hilgard, both arrived in the Bay Area in 1875, "just in time to catch the fruit planting wave in California." Less familiar than Burbank, Hilgard made his reputation as a soil scientist. His deep involvement with early soil mapping and John Wesley Powell's surveys is reflected in Powell's *Report on the Arid Regions of the Pacific Coast* (1887). As expected, Hilgard served the state's booming agribusiness as well; hence his client-serving *Reports of Experiments on Methods of Fermentation and Related Subjects* (1888).

32. The collapse of California wheat was also caused in part by the development of railroads in Russia that made high-quality Siberian products readily available in European markets. See Starrs and Goin (2010).

33. Stoll (1998, 25).

34. Vaught (1999), Stoll (1998), and Walker (2004a).

35. The USDA Economic Research Service statistics for 2006–2008 show, for example, the average farm size in California ranging from 313 to 346 acres, but more than 70 percent of the state's farms in that same period were less than 99 acres.

36. Liebman (1983). Although alternative producers present smallness as a virtue, scale frequently has more to do with crop-associated technologies, market prospects, and relevant federal, state, or local subsidies and regulations. See Guthey, Gwin, and Fairfax (2003).

37. Benedict (1953).

38. Hays (1959), Wiebe (1967), and Woeste (1998).

39. Friedberg (2009).

40. Woeste (1998).

41. Walker (2004a, x).

42. Woeste (1998).

43. Walker (2004, 212–213), Vaught (1999), and Stoll (1998).

44. Muraoka (1978).

45. Woeste (1998). Baarda (1989) points out that agriculture co-ops are not completely exempt from antitrust regulations. The few instances when they are not exempt might confuse a co-op board member, but in this context, the simpler version is useful and generally accurate.

46. Woeste (1998).

47. The literature is endless. The discussion in this section draws from the structure of, and in part paraphrases, Paul Roberts (2008, 117–123). It is informed as well by Gardner (2002, especially chap. 7), and Potter (1998). Murray Benedict's classic *Farm Policies of the United States, 1870–1950* (1953) is for the truly determined reader.

48. Although California growers were not deeply involved in many of the grain crops that dominated federal commodity programs, federal marketing orders

based on preexisting state systems became important to California dairymen and specialty growers. Federal programs putting the unemployed to work were more important in California: they initiated major elements of the state's extensive water distribution system—the federal Central Valley Project and later the State Water Project. However, as we have noted, the water projects did not reach into the rather unproductive soils of our district. While these factors set the state partially aside, cotton, which has a small role in twenty-first-century California, was an important sector in the state during the Depression and in the federal policies that began at that time. Strikes in California's cotton fields precipitated major violence in the 1930s. See Vollman (2008) for the ups and downs of cotton in Imperial County.

49. Philpott (2008). For a slightly longer short version, one that gets into the marvelous language of the "ever normal granary," see Imhoff (2007, 39).

50. The collection is lengthy and is frequently shortened and confused by use of abbreviations. The one that follows is probably partial but indicates that a large bureaucracy of hunger is spread over numerous government departments: the National School Lunch Program, the Summer Food Services Program, the School Breakfast Program, the Child and Adult Care Food Program, the Nutrition Services Incentive Program, the Emergency Food Assistance Program, the Child Nutrition Commodity Supported Programs, the Food Assistance for Disaster Situations, the Food Stamp Act (programs now called SNAP), and the Food Distribution Program on Indian Reservations. We return to emergency food aid and school lunch programs in chapters 6 and 7.

51. Poppendieck (1985).

52. Matthew 25:32–46.

53. Poynter (1969).

54. Nestle (1999) notes that achieving that balance has proved difficult, citing Abbott (1940), de Grazia and Gurr (1961), and Klebaner (1976).

55. Poppendieck (1986) is the classic.

56. Poppendieck (1986) and Allen (1993).

57. Food Research and Action Center (2009).

58. Stein (1976) and Gregory (1991). This was, Martin (2003) points out, also a time when many believed that California's unusual system of large landholdings would be broken up.

59. Daniel (1981).

60. The issue has recently been raised again during Stephen Colbert's partnership with the United Farm Workers on a farmworker "Take Our Jobs!" campaign. The program encouraged unemployed Americans to sign up for the thousands of agricultural jobs available as harvest began in July 2010. There were few takers.

61. Daniel (1981). Martin (2003) describes a more complex intellectual environment in which some reformers, notably Paul Taylor, argued for breaking up the landholdings to return to the agrarian dream, while Cary McWilliams and others

argued that the fields were factories and the workers should have the rights of workers to unionize. Neither happened.

62. On the bracero program, start with Galarza (1964, 1977) and Martin (2003).

63. The transcript of Seeger's appearance before the House Un-American Activities Committee is available through the "Pete Seeger Appreciation Page," at http://www.peteseeger.net/HUAC.htm. Seeger was sentenced to a year in jail for contempt of Congress, and his appeal, which ultimately was successful, lasted seven years.

64. These lyrics are from Guthrey's 1948 song "Deportee (Plane Wreck at Los Gatos)."

65. Martin (2003, 31).

66. Dyl (2006) and Deutsch (2010).

67. Cronon (1991).

68. See Beckman and Nolan (1938) and Tedlow (1990). The Kroger Web site gives an idea of what an integrated retail enterprise entails, noting that at its 125th anniversary in 2008 the company serves "customers in 2,476 supermarkets and multi-department stores in 31 states under two dozen local brand names, including Kroger, Ralphs, Fred Meyer, Food 4 Less, Fry's, King Soopers, Smith's, Dillons, QFC and City Market. Kroger associates also serve customers in 779 convenience stores, 393 fine jewelry stores and 737 supermarket fuel centers the Company operates. The Company also operates 41 food processing plants in the U.S." See "About Us" on the Kroger Web site: http://www.kroger.com/company_information/careers/Pages/about_us.aspx (accessed February 2012).

69. Skaggs (1986) and Gwin (2009).

70. Longstreth (1999).

71. See Walker (2004a), also "Safeway History" at Groceteria.com: http://www.groceteria.com/store/national-chains/safeway/safeway-history/ (accessed February 2012).

72. S. Hamilton (2003a).

73. Lawrence and Burch (2007).

74. J. Walsh (1993), Humphrey (1998), and Deutsch (2010).

75. The consumers' movement of the 1930s is complex. Increasingly militant women sometimes allied with labor unions in efforts to increase purchasing power. They were sufficiently radical in that permutation to be regarded as Communists and investigated by Martin Dies of the House Un-American Activities Committee (M. Jacobs 2005, 171). In other circumstances consumers are presented as bargain hunters lining up for savings. Cohen (2003, 1–3) draws fairly standard distinctions between "entitled" or "citizen" consumers and "purchasing consumers," with the latter less threatening to big business. But she also makes clear that for women to be an object of enormous policy concern during the New Deal was empowering—and particularly important to black women, who had enjoyed precious little political attention in previous decades.

76. Childs (1986), Fitzgerald (2003), Berger (1980), and Rhul (2000).

77. S. Hamilton (2008).

78. M. Rose (1990) and Gutfreund (2004).

79. Rose (1990) and Hamilton (2008).

80. Witzel (2002).

81. Schlosser (2001).

82. Schlosser (2001, 23). Innovation continued, of course: for example, computerization and, later, bar codes allowed more complex coordination between distributors and retailers (Walsh 1993; Konefal et al. 2007), allowing the supermarket chains to grow steadily larger, more centralized, and more powerful relative to growers and food processors.

83. Lowe and Gereffi (2009).

84. Dean (2006) and Cochrane and Ryan (1976).

85. Brannan would have added to the list of basic commodities (wheat, corn, cotton, rice, soy, tobacco) such products as whole milk, eggs, farm chickens, hogs, beef cattle and lambs. See Dean (2006). Note that neither list includes fruits or vegetables. Of the $260 billion that taxpayers spent to subsidize agriculture between 1995 and 2010, $16.9 billion went to subsidize four "common food additives, corn syrup, high fructose corn syrup, corn starch and soy oil" and $262 million subsidized apples, "which is the only significant federal subsidy of fresh fruits or vegetables" (Russo 2011, 1).

86. Gottlieb (2002).

87. Norman Borlaug, who is frequently styled as the Father of the Green Revolution, received a Nobel Peace Prize for his work. New varieties of wheat introduced in India and Pakistan were extremely responsive to fertilizer, irrigation, and pesticides. However, the system had been promoted as scale neutral: all farmers, it was alleged, would benefit from the new technologies. That turned out to be inaccurate: new technologies destabilized traditional agriculture as small farmers were forced off their land by growers increasingly producing for export.

88. Although discussion has focused on peak oil, input-dependent agriculture may stub its toe first on phosphorus, for which there is no known synthetic alternative. Experts estimate that the globe may hit "peak phosphorus" in about thirty years (Lewis, 2008).

89. Insects rapidly develop resistance to any pesticides. First more, and then new, chemicals are required to keep up. See Wier and Shapiro (1981).

90. Olmstead and Rhode (2008) and Lyman (2001).

91. It is critical that *efficiency* was defined without regard to the social and environmental costs of industrial agriculture.

92. Lauck (2000), Dean (2006), and Clarke (1994). Least cost to whom, while externalizing what costs to others, was and remains a central issue of course.

93. Swampbuster provisions of the 1985 and 1990 farm bills were, in theory, designed to reduce Butz-era incentives to convert wetlands, but the rule of exemptions applied. For example, Swampbuster did not prohibit conversion of wetlands, and it applied only when converted wetlands were planted. Golstein (1996).

94. See Ruhl (2000), Looney (1993), Grossman (1994), Chen (1995), and Runge (1997). The picture can be complex. Water pollution control put on Bay Area ranchers occurred in spite of exemptions from environmental regulation, in part from state regulation of the oyster industry and in part from protection for the endangered Coho salmon.

95. Ruhl (2000).

96. Niman (2009).

97. Imhoff (2010).

98. Olmstead and Rhode (2008), Lauck (2000), Clarke (1994).

99. Walker (2004a).

Chapter 4

1. Fighting the use of lead arsenate as a pesticide continued for thirty years after it was demonstrated that washing did not remove residues. The pesticide did not lose favor until 1948, when the codling moth had developed resistance to the lead arsenate and DDT proved to be more effective (Peryea 1998).

2. This topic was discussed in chapter 3.

3. King (1911) describes another front in the soil wars: more than thirty years of "prolonged war" was required to convince soil chemists that organisms in the roots of legumes were "largely responsible for the maintenance of soil nitrogen."

4. Howard's work was undertaken in Indore, India, and in 1931 he wrote (with Y. D. Wad) *The Waste Products of Agriculture*, describing the ancient practice of protecting soil fertility with agricultural waste products. His major work, *An Agricultural Testament* (1943), anchored early efforts to evaluate so-called scientific farming methods that relied on chemical fertilizers.

5. Selg (2010).

6. Balfour was an organizer of the Soil Association, which remains the leading organic certifier in the United Kingdom, and an advocate of sustainable agriculture policy and education.

7. The Rodale operation has yet to find its historian, but its home page has an interesting brief history. See http://www.rodale.com/rodale-story.

8. Wargo (1996). See also Van Den Bosch (1978). See "Talk: Rachel Carson," in Wikipedia for insight into the ongoing vilification: http://en.wikipedia.org/wiki/Talk%3ARachel_Carson (accessed August 2010).

9. The Farm, now known as the Center for Agroecology and Sustainable Food Systems, was way ahead of the curve. Most of today's numerous collegiate food and agriculture programs began forming in the 1990s. See "A Directory of Student Farms" at the Rodale Institute Web site: http://newfarm.rodaleinstitute.org/features/0104/studentfarms/directory.shtml (accessed February 2012).

10. Fukuoka's *One Straw Revolution* (1978) provided new inspiration with a "do-nothing" or no-till method of farming.

11. See Schrader (n.d.). California law predated the 1990 federal law.

12. Garrett (2000).

13. Garrett (2000).

14. Shapiro (1986). Alice Fay Morgan, an early leading nutritionist, was educated as a chemist but entered the field of food science in order to get a job. Morgan organized the Home Economics Department at the University of California, Berkeley, and coordinated early analyses of the nutritional status of U.S. citizens. Ironically, chemists studying farm animals made many important advances in human nutrition. For example, chemist Elmer McCollum first identified vitamins A and B while researching improved cattle feed.

15. Harvard Medical School evinces the generally condescending regard for the nutrition field, as well as a change of direction in medicine generally: a century after the first nutrition professor began work at the University of California, Berkeley, Harvard announced that its "Medical School developed the Division of Nutrition in 1996 in order to establish nutrition as a discipline." Harvard Medical School, Division of Nutrition, Welcome Page: http://nutrition.med.harvard.edu/ (accessed February 2012).

16. Nestle (2001). For her history of the way food industry interests shaped, for example, USDA recommendations for what Americans should eat, see the discussion of the evolution of the "food pyramid," pp. 51–66.

17. Samuel Hopkins Adams's articles on patent medicines, collectively entitled "The Great American Fraud," appeared in *Collier's Magazine* in 1905.

18. The 1906 act forbade manufacturing and interstate trade in "adulterated" or "misbranded" food and drugs.

19. Kallet and Schlink (1933).

20. This paragraph draws from U.S. Food and Drug Administration, *History of the FDA: The 1938 Food, Drug and Cosmetic Act,* available at http://www.fda.gov/AboutFDA/WhatWeDo/History/ProductRegulation/ucm132818.htm (accessed February 2011). Amendments in 1959 added the Delaney clause, now repealed, which provided that "no additive shall be deemed safe if it is found to induce cancer when ingested by man or animal." The clause was unique because it set a zero-risk standard. Balancing risk and benefit, or a "negligible risk" standard, is more common and preferred by the pesticide industry, which fought for decades to remove the clause. See Cornell University Cooperative Extension, "Pesticide Management Education Program. The Delaney Paradox and Negligible Risk" at http://pmep.cce.cornell.edu (accessed September 2011).

21. Laskawy (2011a).

22. Nancy Langston, personal communication (August 15, 2011). Price (1939) remains a cult classic. Sally Fallon, founder of the Weston A. Price Foundation, wrote *The Cookbook That Challenges Politically Correct Nutrition and Diet Dictocrats* (1999).

23. Hawley (1966, 261).

24. Quoted in Beckman and Nolan (1938). See Spector (2005).

25. M. Jacobs (2005).

26. Food Marketing Institute ([2005]).

27. Fulda (1951). The term *housewife* appears to have had several connotations, including both a salt-of-the-earth gloss and an opposite dithering, easily dismissed overlay.

28. Luck (1946). University of Chicago economics professor (and later Illinois senator) Paul H. Douglas provides a summary of the early co-op movement's critique of conventional capitalism and, by implication, how to differentiate the co-op movement from the early consumers' movement. Douglas raised the basic question about the vote-with-your-fork strategy: "In saying that capitalism can be vanquished more easily on the conventional [market] field than on the political, does not the author overlook the fact that it would be difficult, if not impossible, for consumers' cooperation to win such a battle?" Douglas (1924, 385).

29. Sinclair invested the proceeds from *The Jungle* in a cooperative community that was destroyed by a mysterious fire. Despite the uproar about Sinclair, the federal government soon encouraged cooperative markets in sites as diverse as Japanese internment centers and camps for migratory farm labor. The national organizations that strengthened Sinclair's proposals "seem relatively tame in retrospect, . . . cautious in comparison to the bold ambitions of the Civilian Conservation Corps (CCC), which by 1935 had more than half a million unemployed men working . . . under military supervision." Starr (1997, 132).

30. The Berkeley Co-op's book, *The Berkeley Co-op Food Book: Eat Better and Spend Less* (Black 1980), "climax[ed] more than twenty-five years of daily, dedicated labor by . . . a member-owned, member controlled consumers' cooperative that operates twelve supermarkets and several nonfood facilities in the San Francisco Bay Area, California."

31. The Consumer's Cooperative of Berkeley (CCB) formed in 1937 and opened a retail food store. A year later, Berkeley's substantial Finnish community formed a buyer's cooperative as well, including a gas station (one of the first to offer unleaded gas) and a hardware store. Oakland minister Roy Wilson established the New Day Cooperative Club seeking a "Christian alternative" to the dominant food system. These organizations supported each other, met together, and ultimately consolidated in 1947 under the CCB name. The CCB included several pharmacies, "a wilderness supply outlet, bottle shops, a garden nursery, and one of the first natural foods stores using a supermarket format. . . . At its peak, CCB ran twelve supermarket stores [and] . . . had the highest sales per square foot of any grocery store west of the Mississippi" (Zimbelman 1992).

32. Northern and southern California co-ops began meeting to coordinate efforts in 1939 at "Camp Sierra," an annual summer practice that lasted thirty years (Neptune 1971).

33. Neptune (1971).

34. Fullerton (1992, 93).

35. Fullerton (1992, 94).

36. In chapter 8, we discuss those policies in the context of our district.

37. Eisenhauer (2001).

38. Mack (2006, 94, 112).

39. Mack (2006, 119).

40. Mack (2006).

41. Ralph Nader revitalized the muckraking genre and brought consumer advocacy to a peak. His 1965 volume, *Unsafe at Any Speed,* focused on the "designed-in" dangers of automobiles, but he also went after the meat processing industry with two 1967 articles in the *New Republic:* "We're Still in the Jungle," and "Watch That Hamburger." Congress soon passed the 1967 Wholesome Meat Act (amending the 1906 act) that gave primary authority over meat processing to the federal government. Nader and Upton Sinclair, then eighty-nine years old, attended the signing ceremony for the bill (Marcello 2004).

42. Petersen was also the first director of the President's Commission on the Status of Women and, as assistant secretary of labor under President Kennedy, the highest-ranking woman in government at that time. See the AFL-CIO Web site, "Esther Eggertsen Peterson (1906–1997)," at http://www.aflcio.org/aboutus/history/history/peterson.cfm (accessed February 2012).

43. Mowbray (1967) and Cross (1976).

44. U.S. Congress, House Committee on Government Operations (1968), U.S. Department of Agriculture (1968), U.S. Federal Trade Commission (1969).

45. Cross (1976).

46. Cross (1976).

47. Eisenhauer (2001).

48. This discussion is inappropriately brief given the importance of the AWOC and UFW, but it is extensively treated elsewhere. Start with P. Martin (2003).

49. P. Martin (2003).

50. Gordon (1999) discusses the mainstream environmental groups' surprising lack of interest in banning DDT.

51. California Agrarian Action Project, Inc. v. The Regents of the University of California, 210 Cal. App. 3d 1245 (1989). The War on Poverty also precipitated key groups in Oakland, which we discuss in chapter 8.

52. By the early 1980s, CAAP began a gradual journey to a different programmatic emphasis, working on behalf of family farmers and to support organic and sustainable farming. That led the organization, reframed as the Community

Alliance with Family Farmers (CAFF), to take a very different position when the short-handled hoe issue returned in 2003. See chapter 7.

53. Bennett and Reynoso (1972).

54. Bennett and Reynoso (1972).

55. Diana v. State Board of Education (CA 70 RFT, N.D. Cal. 1970).

56. For a brief discussion of charges and countercharges regarding board bias, see P. Martin (2003, 5).

57. Murray (1982).

58. Murray (1982, 30, citing Perelman 1977, 5).

59. Brown and Pilz (1969, 74).

60. Steinhart and Steinhart (1974, 184).

Chapter 5

1. A Bay Area was not discussed until the early 1940s—perhaps first imagined in a proposal for a regional planning organization, the Bay Area Council, which was formed in 1944 (California State Planning Board, 1941). "Marvelous Marin" is an attitude in county civic life and a booster club organized in the 1920s.

2. Griffin (1998).

3. Gerhart (2002).

4. See Walker (2008), Gerhart (2002), and Dyble (2007).

5. FDR's interior secretary also had his eye on Point Reyes (Mackintosh 198).

6. Dyble (2007).

7. Dyble (2007, 57).

8. Dyble (2007).

9. Scott (1985). Self (2003) argues that the focus of civil rights historians on the South omits important urban and western issues. As we discuss in chapter 8, land use planning in Oakland was its own kind of Freedom Summer.

10. Dyble (2007).

11. Nobel (2010, 12). The history of land protection in Marin inclines us to question the applicability of that familiar line in coastal Marin, where the ranchers opposed both development and federal intervention.

12. Rancher Boyd Stewart, broadly recognized as "the" rancher-conservationist spokesperson of the early Point Reyes National Seashore years, married a Marin Conservation League activist from suburban Marin, Joseffa Conrad Stewart, and shared her deep involvement in the conservation movement.

13. See Drakes Bay Land Company v. U.S., 424 F.2d 574 (1970).

14. Thomas (2008, 2010).

15. Clawson and Knetsch (1966). The early National Park Service (NPS) had tried to pass mildly disturbed lands on to state and regional parks, maintaining

its focus on the jewels. The agency soon changed course to connect with congressional representatives of eastern and more urban districts. Cape Cod National Seashore, established a year before PRNS, included a similar straddle, but without the agriculture (Fairfax et al. 2005).

16. See Lage and Duddleson (1993, 142) and Hart (1992). Peter Behr, coordinator of the Save Our Seashore movement in the early 1970s, recalls both 21,000 and 26,000. Harold Ickes's Department of the Interior invented the seashore notion among other less than "crown jewel" categories as part of its competition with the U.S. Forest Service for land: "park purism" limited the agency's reach, requiring less restrictive designations (Fairfax et al. 2005).

17. Fairfax et al. (2005).

18. The NPS was not, of course, Wilkins's only problem: the creamery to which the Wilkinses sold milk was bought out by a Texas company that canceled all the milk contracts. Their proposal to subdivide their land fell through, leaving the family selling firewood (Livingston 1995). See also Guthey, Gwin, and Fairfax (2003).

19. Livingston (1995).

20. The $14 million authorization was based on the presumption that land in the pastoral zone would not be acquired. It also reflects a standard legislative strategy: the sum was a starter to accompany a controversial designation; it was to be supplemented at a later time. Discussed in Lage and Duddleson (1993). The NPS expected to acquire some land through an exchange with the Bureau of Land Management in Oregon, but that hope died when Oregon officials objected. There seems little doubt, however, that the inflated land prices that caused the NPS such grief in the 1970s were at least in part their own creation.

21. Lage and Duddleson (1993). We have found frequent reference to the proposed subdivision but never the perspective drawings that NPS director George Hartzog presented at the May 13, 1969, hearings.

22. A 1962 Sierra Club coffee table book, *Island in Time*, ignored virtually all human habitation. It mentioned the ranchers in connection with the 1906 earthquake, and Native Americans were treated only in relation to the intensely promoted possibility that Sir Francis Drake landed at what is now called Drake's Beach. Ironically, perhaps, Interior Secretary Stuart Udall made the only reference in the book to modern ranchers, noting in the Preface that Point Reyes "owes its wide uncluttered dimensions to ranchers—Western husbandmen—who have maintained its land for a century." Gilliam (1962, 8) with photographs by Philip Hyde.

23. Watt (2001); Hart (1992); Walker (2008).

24. Medvitz, Sokolow, and Lemp (1999).

25. Dyble (2007). Figure 3.1, showing the number of California farms over time, suggests that the Williamson Act had some role in slowing the decline in number of farms.

26. Tryphonopoulos (1967), Medvitz et al. (1990), and Guthey (2004).

27. Rolland (1997, 7) and Hart (1991, 63-64).

28. Hart (1991, 87).

29. Fairfax et al. (2005) and Merenlender et al. (2004).

30. People for Open Space Farmlands Conservation Project (1979, 1980a, 1980b, 1980c, 1980d).

31. Nan McEvoy is forthright about starting her olive oil business as a cover for a Tuscan-style villa in which to entertain her grandchildren. She also kindled a renaissance in the state's artisanal olive oil business (Mitchell 1996).

32. Peter Warshall, formerly a *Whole Earth Catalog* editor and currently working on Dream New Mexico, an ecological and social restoration project in Santa Fe, reminisced at the thirty-first EcoFarm conference about his early days in Bolinas working for and with the young Niman and Weber. He alleged that most of the funding for the early innovators and most of their knowledge about seeds came from close study of cannabis culture. "Can You Imagine a Better Food System? It's Easy if You Try" (Thursday Plenary, January 27, 2011, Asilomar, CA). Available on CD from EcoFarm or the authors.

33. Fourth-generation rancher Amanda Stewart Wisby currently raises beef cattle on the family ranch.

34. Weber has served as vice president of MALT, president of California Certified Organic Farmers, and president of the Board of Marin Organic, and he was a founder of the Organic Farming Research Foundation and a founding participant and board member of the Ferry Plaza Farmers Market.

35. Warshall (2011).

36. Straus Family Creamery (n.d.).

37. Morse (2008).

38. Including the Environmental Action Committee, the Marin Conservation League, the Greenbelt Alliance (originally People for Open Space), the Tomales Bay Advisory Committee, and the Associated Dairymen of California. She also took a turn on the Marin Community Foundation board and worked in the founding of the West Marin Growers Group, which morphed into Marin Organic.

39. Morse (2010, 6).

40. See Straus Family Creamery (2006). The Straus operation is tagged "the first organic dairy west of the Mississippi." Mark Lipson of Organic Farming Research Foundation suggests that Engelbert Farm in Nichols, New York, led the way.

Chapter 6

1. Levenstein (1988).

2. For a broader context, see Kurlansky (2005).

3. Rosen and McGrane (2000).

4. Berkeley students, many of whom had spent time in Mississippi before and during Freedom Summer (1964), came back to a surprisingly restricted campus where the administration had only recently lifted a ban that barred Communists

from speaking. It still barred political activity on the central Sproul Plaza. When a fundraiser set up an illicit leafleting table, police tried to haul him away. Students spontaneously gathered around the police car and sat down. Jack Weinberg remained in the car for thirty-two hours and is still known as "the guy in the car." Mario Savio, himself recently returned from Mississippi, was among those who climbed atop the car to address the crowd, by then numbering in the tens of thousands.

5. See Digger Archives (1968), T. Miller (1999), Kisseloff (2006), and Berg and Goldhaft (1982).

6. When Black Panthers learned that it was legal to carry guns into the California State legislature, they shocked and outraged the nation by showing up there heavily armed in May 1967. It is informative to contemplate the Black Panthers and gun advocates in today's Tea Party organization as aligned on this issue.

7. Black Panther Party Platform (1966).

8. The Black Panthers are discussed more thoroughly and in their Oakland context in chapter 8.

9. FBI assaults on the Black Panther Party and its members are inventoried in the 1976 "Church Committee Report," the result of an investigation of the Senate Select Committee to Study Governmental Operations with Respect to Intelligence Activities, chaired by Sen. Frank Church of Idaho. The FBI also targeted church leaders who offered space for the breakfast and censured its own officers who reported that the party was not dangerous but primarily engaged in feeding school children. The literature is enormous. Start with Murch (2010). The Church Committee Report is available at http://www.aarclibrary.org/publib/contents/church/contents_church_reports.htm (accessed February 2012).

10. Although the Diggers were not aligned with the Panther's Black Power and Black Nationalist agenda, they shared common turf and "faced common enemies." They worked together when Huey Newton sought the Diggers' advice on how to produce a newspaper. See Coyote (1998).

11. Curl (2009, 197).

12. Red Star Cheese in Sonoma did the dairy; Peoples' Bakery in San Francisco provided baked goods, and Veritable Vegetable handled the produce.

13. See Drew (1998) and Curl (2009). See also Jesse Drew, "People's Food System" at FoundSF.org: http://foundsf.org/index.php?title=People's_Food_System (accessed February 2012).

14. Drew (1998, 328).

15. Older (1982, 39–40). Similarities with subsequent farmers' market critiques are striking. See Alkon (2008).

16. Cox (1994) tells a similar story about the co-ops in the Twin Cities.

17. Edgar (2010, 10).

18. See Edgar (2010). Because cheese is made from rennet, a natural product of a harvested cow's stomach, cheese has presented some issues for Rainbow, which nevertheless sells a full range of quality product.

19. Rainbow's Web site hints at some of the challenges of being profitable and progressive in the early twenty-first century: "Rainbow's place in this new agri-business is at times uncomfortable and challenging. . . . Soymilk can now even be found at many local corner stores that typically used to only sell chips, beer and beef jerky. (The opposite is also now true, for unlike in the old days, you can find chips, beer and (vegi) jerky on our shelves.) With the advent of Genetically Modified foods and the lack of government requirements when it comes to testing and labeling, we face even more challenges. And occasionally, certain government organizations decide they want to change organic standards to include practices that we abhor. Despite these challenges there are enough people who are interested in organically grown and locally produced foods to keep our doors open. We continue to stay true to our mission and hope to inspire others in the realms of good food and cooperative living." From "Rainbow Grocery History" at http://www.rainbow.coop/history2/ (accessed February 2012).

20. Cooperative Whole Grain Education Association (1983, 10–11).

21. Volumes of the book circulated in 1967 and 1972. Part of the purpose of the books seemed to be to turn African Americans from slavery-era food that enjoyed a brief renaissance as soul food beginning in the 1960s. Elijah Muhammad urged that "peas, collard greens, turnip greens, sweet potatoes and white potatoes are very cheaply raised foods. The Southern slave masters used them to feed the slaves, and still advise the consumption of them. Most White people of the middle and upper class do not eat this lot of cheap food, which is unfit for human consumption." Elijah Muhammad was concerned with gas and starch and fat, which he described as "friends to diabetes" (1967, 4).

22. Hill and Richman (2007) and Kuruvila, Fagan, and Van Derbeken (2008).

23. Survivors of high school Latin know that Caesar announced, "The die is cast," when crossing the famous river in Italy to begin a civil war against Pompey the Great.

24. Grossner (2012).

25. See the Mission Pie Web site at http://missionpie.com/?p=213 (accessed August 2011).

26. Museum of the City of San Francisco, Virtual Museum, Gold Rush Chronology, http://www.sfmuseum.org/hist/chron1.html (accessed June 2006).

27. The name is probably a mispronunciation of Le Poulet D'Or (The Golden Chicken). The Poodle Dog persisted until the 1960s. Ethnic and fine restaurants have always been a feature of the region. The nation's oldest Italian restaurant, Fior D'Italia, is also in San Francisco but it is a parvenu, founded in 1889, forty years after Croatian immigrants founded the city's oldest restaurant, Tadich's Grill. Tadich's remains at the apex of San Francisco seafood establishments and was sixty-three years old when the equally popular (Zagat listed) Swan Oyster Depot (1912) opened. Chinese cooks and their food were major features of San Francisco dining until they became the focus of racist attacks during the Chinese Exclusion. The Chinese restaurant did not recover its former stature until after World War II. Coe (2009) provides an excellent introduction to Chinese food

in the region. See Wells (1939) for insight into San Francisco restaurant history more generally.

28. Conlin (1986).

29. Livingston (1995), Guthey, Gwin, and Fairfax (2003), and Sadin (2007).

30. Kann (1993).

31. Levenstein (2003).

32. See Fowler (2005, 203). Fowler argues that Jefferson's role in U.S. food is "lost between exaggeration and inattention." Some legends credit him with introducing ice cream, macaroni, tomatoes, vanilla, and French fries (Fowler, 2005, 1).

33. Reichl (2007).

34. Culinary Institute of America, "The Story of the World's Premier Culinary College—A History of Excellence, Professional Advancement, and Innovation," http://www.ciachef.edu/about/history.asp (accessed February 2012).

35. Beard and Child are among the early food celebrities featured in Kamp (2006).

36. The first of the three-volume set was written with Simone Beck and Louisette Bertholle. The latter dropped out of subsequent books.

37. This is not to suggest that regional cuisine died during the Depression. See Kurlansky (2009).

38. Schlosser (2001).

39. Zuckerman (1998).

40. Ogden et al. (2004). The "edible food-like substances" phrase probably originated with Michael Pollan.

41. Peet's Coffee and Tea, "The Original Coffee Revolution," http://www.peets.com/who_we_are/history_vine.asp (accessed March 2012).

42. Cheese Board Collective (2003).

43. Cheese Board Collective (2003).

44. This story has many versions and nuances. See McNamee (2007 especially chapter 3) and Waters and friends. (2011).

45. See Brenneis (2009).

46. Fullerton (1992).

47. Johnson (2007) coined the striking term.

48. David (1962, 4).

49. David (1962, 4–5).

50. McNamee (2007, 90).

51. Tower (2003, 111).

52. Kuh (2001, 146–147). The soup was in Charles Ranhofer (1894): *A Selection of Interesting Bills of Fare of Delmonico's from 1862 to 1894*. Ranhofer was chef at New York's most famous restaurant and invented such familiar dishes as

delmonico potatoes, lobster newberg, and probably eggs benedict and chicken à la king (Wirth 2009, 272).

53. Tower (2003), citing (Kuh 2000, 147). For Waters's take on David, see Waters and friends (2011).

54. McNamee (2007).

55. See Flammang (2009).

56. McNamee (2007, 90).

57. Hank Rubin's Berkeley-based restaurant, The Pot Luck, preceded many of Chez Panisse's priorities. It is also reputed to have been the first restaurant in the area to hire black waiters (see Glassner 2002).

58. Johnson (2008).

59. Dicum (2007).

60. And granddaughter of Harvey Mudd, founder of the Pitzer College that bears his name.

61. The demonstration and community garden have since disappeared under city management.

62. Tower (2003 263).

63. Discussed in chapter 5.

64. The Small Farm Center's fifteen-year history highlights the Tasting of Summer Produce as "one of the most important food events in the country" (Stumbos 1990).

65. Kraus (1983).

66. Kraus (1983).

67. We have heard diverse stories about how "yuppie lettuce" was born. This one resonates with us.

68. Kraus (1983, 12).

69. Kraus (1983).

70. Alice Waters, interview on *Fresh Air* with Terry Gross, August 22, 2011.

71. Kraus (1983, 15).

72. The contract is appended to Kraus's 1983 report. Note that the exuberance ran counter to national trends. A year later, Willie Nelson's first Farm Aid concert, a gesture of relief for farmers who tried and failed to get big in the Butz mode, responded to suicides among farmers.

73. Sims (1988).

74. See Calvin Trillin's Foreword in Waters and friends (2011, 2).

75. Cheese Board (2003, viii). See also McNamee (2007).

76. Dyble (2008).

77. Farmers' markets vary enormously in their ability to attract poor and minority customers. Fisher reports that San Francisco's Heart of the City market began in 1981 as a Quaker project to bring produce to the Tenderloin, one of the city's

poorest and "most heavily immigrant neighborhoods." In that, it has succeeded. The market has "rock bottom prices," and the white-collar workers who shop at the market provide a stabilizing cash flow. Growers bring surplus and cosmetically imperfect produce that they would not take to Ferry Plaza, for example.

78. The California Marketing Act of 1937 gives the director of the Department of Food and Agriculture authority to issue state marketing orders that have the weight of law (Muraoka 1978).

79. See DuPuis and Gillon (2009).

80. These are Certified Farmers Markets, tracked by the California Department of Food and Agriculture. Available at the California Department of Agriculture Certified Farmers' Market Program website: http://www.cdfa.ca.gov/is/i_&_c/cfm.html (accessed February 19, 2012). Noncertified markets are different and are viewed by many of the earlier markets as problematic because they saturate the field.

81. Payne (2002), Brown (2001), and O'Hara (2011).

82. California direct marketers are also much younger as a group than farmers in general, reflecting the interest of beginning farmers who don't have the resources to establish large-scale operations but are willing to experiment and innovate (Kambara and Shelley 2002).

83. Abel, Thomson, and Maretzki (1999).

Chapter 7

1. rBGH is a synthetic growth hormone used to increase milk production. In addition to the hazards of the additional hormones in waste streams, the Monsanto product increases mastitis in cows and use of antibiotics in feedlots. rBGH is not permitted in Canada or Europe. Using it at all is somewhat curious given that overproduction is the constant crisis in the dairy industry.

2. Fromartz (2006) coined the term. See also Guthman (2004).

3. Knuhtsen (1993).

4. J. Richardson (2010).

5. See the Code of Federal Regulations, Title 7: Agriculture PART 205—National Organic Program, http://ecfr.gpoaccess.gov/cgi/t/text/text-idx?c=ecfr&tpl=/ecfrbrowse/Title07/7cfr205_main_02.tpl (accessed February 2012).

6. The champion may be Aurora Dairy, which the USDA charged with fourteen "willful violations" in 2007. Cornucopia Institute (2010).

7. That was among the factors that motivated Straus to oppose pasture requirements designed to curtail CAFO organic by requiring year-round pasture access.

8. Powell (2009).

9. Seventy percent of the antibiotics in the United States are administered to healthy livestock in CAFOs—cattle, swine, and poultry primarily—because they are at risk of illness due to stress from crowding. Antibiotics also boost feed

efficiency (growth per unit of feed). Some observe that using antibiotics to care for an occasional sick animal is quite different and should not be precluded by organic rules. Caring for sick animals is both humane and economically reasonable and does not create a risk to human health. Others fear that "some" rather than "none" creates a slippery slope that invites abuse (Cashulin 2011). Overuse of antibiotics creates antibiotic resistance in microorganisms that cause diseases in humans, in the process compromising the efficacy of antibiotics used against those diseases. The resistant organisms are transported in both the waste stream and the food we eat. See the Union of Concerned Scientists, "The Mounting Scientific Case Against Animal Use of Antibiotics," http://www.ucsusa.org/food_and_agriculture/science_and_impacts/impacts_industrial_agriculture/the-mounting-scientific-case.html (accessed September 2011).

10. McCue (1999).

11. Morse (2008, 10).

12. Morse (2008).

13. Just four reuses balance the energy needed to manufacture, wash, and transport a reusable glass bottle relative to a plastic bottle. The other expected two to four uses are clear gain. Some also believe that the glass improves the taste of milk.

14. Morse (2008, 18).

15. Straus (2008). Another district yogurt maker, Saint Benoît, has already moved into that niche, offering yogurt in small crocks and glass bottles.

16. The Cowgirls have deployed their connections to advance their vision for land conservation in the district. Thus, it is not mere coincidence that R. W. Apple, the legendary *New York Times* writer, came to visit a number of start-up young cheesemakers. In the counties north of the Bay, Apple (2001) found "a New World counterpart to Lombardy and Normandy." "A Cowgirl's Sweet Dream," he called it.

17. Smith presents the restaurant as a fine graduate program: her colleagues "tasted everything, constantly read and collected cookbooks, sought out new skills and knowledge, and had a reverence for and a scholarly approach to food. There was always a sense of discovery." Compare McNamee (2007) and Kamp (2006).

18. Founded in 1936, a decade before the Culinary Institute of America, the two-year program has a diverse constituency. It maintains close ties to San Francisco's powerful Local 2 of the Hotel Employees & Restaurant Employees Union.

19. Under *servis compris,* the diner's bill includes a set fee that will be divided among all the employees, not just the front-of-the-house workers. Establishing health care programs and enforcing minimum wage requirements would be better, but until the required social policies are in place in the United States, *servis compris* is a useful, but infrequently deployed, stop-gap.

20. Conley's best-known producer was Ig Vella, whose operation was an earlier effort to help ranchers: he started in a brewery abandoned during prohibition at

the request of dairymen seeking an outlet for their milk. Conley's line also included Jose Matos, who made cheese from his own milk for the large Portuguese community around Petaluma. Matos was not interested in haute food or Conley's services. But Conley needed the product and distributed it without charge. Cindy Callahan retired from nursing to establish Bellwether Farms to raise lamb. Several trips to Italy inspired her to make aged sheep cheeses. Jennifer Bice started milking goats as a 4-H project. When she took over the family farm in the 1970s, she began selling raw goat milk at natural food stores. Soon she too was making cheese at Redwood Hill Farm and Creamery.

21. Buford (2006) and Bourdain (2000) and even Disney-Pixar's *Ratatouille* (2007) paint a grim picture of the professional kitchen.

22. Dubbink (1984) summarizes the tensions between the ranching and more urban locals—who sometimes try to protect rural nature of a place while eliminating the smelly farms.

23. O'Connell et al. (2001, 3–4).

24. A report from the third gathering, in 1992, noted a pending "marina, restaurant, retail stores, ten unit inn, ten cottages and nine houses. A short distance to the south a similar development has been considered. Across the bay from Inverness and adjacent to Millerton Creek, a greatly expanded quarry operation is proposed, raising concerns about siltation. And at the head of Tomales Bay at Tomasini Creek, a sevenfold expansion of the West Marin Sanitary Landfill is being planned. It would accept industrial and other waste from the San Francisco Bay Area." Also noted, without irony, was that recreationists' suntan oil was a pollutant. Third Biennial State of Tomales Bay Conference 1992, Agenda, October 24, 1992, available at cemarin.ucdavis.edu/files/30430.doc (accessed February 2012).

25. Note that in the 1960s some considered manure in the water a boon, believing that it helped the oysters grow faster (Gerhart 2004).

26. Blood (1998).

27. Tomales Bay Association (2000).

28. Gerhart (2004, 23).

29. An architect who "trained" in creamery design with the Cowgirls' creameries was also a major resource for the Lafranchis.

30. Among the lessors was Bill Barboni, whose parents had sold the easement to MALT. A local veterinarian, Barboni uses land leased from Barinaga to anchor his start-up alternative beef operation. He supplied Marin Sun Farms until he received a better offer from Whole Foods and also has his own direct sales customers.

31. Marin County's agriculture commissioner, Stacy Carlsen, was also concerned about the idea of Marin being as associated with cheese as Napa and Sonoma counties are with wine: "I wouldn't want to see us as just a cheese region, but as a region that creates all the products associated with grass-based agriculture: premier organic vegetables, grassfed beef and cheese" (Rogers 2010).

32. Schell (1984) updated Sinclair's 1906 classic. He left the firm in 1996 to become dean of the School of Journalism at UC Berkeley.

33. The Federal Meat Inspection Act allows "custom-exempt" processing if the meat will be used by the owner, his or her family, or nonpaying guests. Small producers have used this exemption by preselling animals to new owners, who are then responsible for the processing. Such processors are not unregulated; they must follow federal and state sanitation regulations and are typically inspected annually by a state agency. However, they are not inspected daily and do not have to operate under the Hazard Analysis and Critical Control Point system. The typical custom-exempt butcher provides mobile, on-farm slaughter. California requirements that the animal be taken off the seller's farm for slaughter were not enforced until the sales became common.

34. He sold 49 percent of Niman Ranch to Mike McConnell. Rob Hurlbut, operations and marketing manager, became a co-owner in 1997 (W. Brown 2000).

35. When TJ's wanted to use its own label on Niman's product, Niman withdrew the hot dogs; Niman's branded bacon continues to be too popular with TJ customers for the firm to let it go.

36. It is raised without hormones or antibiotics. Neither grassfed nor grain-finished beef is a recent innovation. Early U.S. colonists fed cattle on grain, and even conventional CAFO-raised beef eat grass for part of their lives, typically six to eight months, before they are sent to the feedlot.

37. Grassfed cattle, like buffalo and elk (which are also becoming available in markets), are lower in fat, calories, and cholesterol.

38. Salatin (1996). Evans is one among several district alternative beef producers who raise eggs and chickens as part of their diversification and pasture pest management. The supply is sufficient that some local farmers' markets now limit egg sales to those that are pasture-raised.

39. Only a year later, a major bovine spongiform encephalopathy (BSE, or "mad cow disease") outbreak reiterated Pollan's basic message. BSE is a progressive neurological disorder in cattle that appears related to a human variant of Creutzfeldt-Jakob disease, a fatal brain disorder with no known cure. The disease was spread in the British cattle industry by feeding rendered, bovine meat-and-bone meal to young calves and peaked in 1993 at almost 1,000 new infected cattle per week. By 2008, more than 184,500 cases of BSE had been confirmed in more than 35,000 herds. The time lapse between exposure to contaminated meat and infection in humans is approximately ten years. See Centers for Disease Control and Prevention, "BSE (Bovine Spongiform Encephalopathy, or Mad Cow Disease)" March 17, 2011, available at http://www.cdc.gov/ncidod/dvrd/bse/ (accessed July 2010).

40. California ended its program in the 1970s.

41. Rancho's days appear numbered. Housing developers are interested in the site, and it has been a victim of fires for which the Animal Liberation Front claimed responsibility. Closing Rancho would impose significant hurdles

for the district's alternative beef producers and for dairies that rely on its cull service.

42. A "retail exemption" in Federal Meat Inspection Act allows a retailer that is not USDA inspected to cut up and sell meat from animals slaughtered under inspection, up to a certain annual dollar amount and percentage of total sales.

43. *Stockman Grass Farmer* (2009).

44. Dairymen, Straus notes, will produce more because the price is low and then do the same thing because the price is high. The government is no smarter. The feds have tried to control supply with cow kill programs that are temporary fixes in the milk market and flood the beef market.

45. Harrell (2006).

46. Burros (2005).

47. "Ayesha, Altitude and the Mile High Creamery," *Cowgirl Creamery Newsletter* April 2009, http://www.cowgirlcreamery.com/nl_april2009.asp (accessed February 2012).

48. Vishal (2010) and MALT (2009).

49. Thiboumery and Lorentz (n.d.).

50. Casey (1990), Dowie (1990).

51. Russell (1992).

52. Reese (1995).

53. CRLA played a small role. See Agustin Gonzalez et al. v. Spaletta Ranch Corp. discussed on the Talamantes Villegas Carrera, LLP Web site: http://www.talamantes.org/work.html (accessed September 2011).

54. Cashulin (2011). Note as well that unlike other workers in California, agricultural workers do not start earning overtime wages until they have worked sixty hours per week rather than the usual forty. In 2010 the state legislature passed a bill, SB 1121, that would have ended the exemption from overtime for agricultural workers, but it was vetoed by Governor Schwarzenegger. Ironically, when the state did move to enforce agricultural labor laws vigorously, it was to fine small organic farmers who worked with interns and volunteers.

55. The new CAFF is a reorganized version of the earlier group, CAAP, that led the fight against university extension programs that support agribusiness at the expense of workers and communities. In 1993 CAAP merged with the California Association of Family Farmers (CAFF), taking a third and only slightly different name, Community Alliance with Family Farmers. CAFF II has remained at the forefront of food systems work in the state ever since.

56. Allen et al. (2003).

57. Getz, Brown, and Shreck (2008, 500).

58. Hendricks's history (1993) and Bernstein's novel (2008) are good places to start on Kaiser's history. The opposition to prevention and health maintenance from the medical profession and the AMA was impressive.

59. We focus on K–12 programs. But we note the California Student Sustainability Coalition, a key participant in the Real Food Challenge, a national project of the Boston-based Food Project. The challenge focuses on students' supporting the sourcing of "real food" at colleges and universities. The goal is to shift 20 percent of higher education's annual food expenditure of $4 billion toward real food by 2020. http://www.sustainabilitycoalition.org/projects/west-coast-real-food-challenge (accessed October 2011).

60. Birth and Gussow (1970), Poppendieck (1985, 1998, 2010), Cooper (2007) and Levine (2008).

61. Hunter (1904).

62. Spargo (1906).

63. Spargo (1906, 117). The two academic volumes did not have the same impact as Sinclair's almost simultaneous novelistic revelations about meatpackers. Politically, addressing hidden risks in meat that almost everybody eats, though not a simple task, turned out to be an easier one than ensuring access to healthy food for poor people or safe conditions for factory workers. Sinclair lamented later in life that he had failed with his packinghouse novel: aiming for Americans' hearts, he hit their stomachs instead.

64. Levine (2008) and Woodward (1938).

65. Levine (2008,).

66. Levine (2008), Levenstein (2003).

67. Levine (2008).

68. Levine (2008). Another irony is that sustainable agriculture advocates note that because the hunger lobby is convinced that the survival of the lunch program requires continued support from conventional agriculture, it is not an ally in many efforts to redirect farm subsidies to a more sustainable path.

69. Cooper and Holmes (2006).

70. Poppendieck (2010).

71. Poppendieck (2010). If you are not familiar, start with the Hot Pockets Web site: http://www.hotpockets.com/products/allproducts/index.aspx (accessed October 2011).

72. Urban gardening and agriculture play such an important role in the food justice movement that we discuss them more fully in chapter 8.

73. Flanagan (2010, 1).

74. Miller (1904), Greene (1910), Dewey (1917), Waters (2008).

75. See Chez Panisse Foundation's evaluation of Edible Schoolyard undertaken by nutritionists from UC Berkeley and UCLA (Rauzon et al. 2010). The report is discussed in J. Black (2010). For those suspicious of self-evaluations see Yost and Chawla, (2009), Blair (2009), Robinson-O'Brien, Story, and Heim (2009), and McAleese and Rankin (2007).

76. Wong (2008, 8).

77. *Forbes Magazine* identified Wolfgang Puck as the "patriarch of celebrity chefdom." We agree with the designation of patriarch, but numerous others got there earlier. Note that celebrity seems tied to empire building and does not require haute cuisine. The 2008 top earner, Rachael Ray ($18 million a year), introduces olive oil and endorses Dunkin' Donuts (Vorasarun 2008).

78. Guerrero (2011).

79. Waters was excoriated for suggesting that the cost to feed all school children a decent meal (about $5 each, or about $29 billion) was worth it (Waters and Heron 2009). The child nutrition bill, introduced in 2010 limited junk food in schools and increased funding for school lunches for the first time in thirty years by about 6 cents per lunch (about $2.6 billion) total. The money was deducted from food stamps.

80. "Pink slime" is a frozen mash of "fatty processing trimmings" treated with ammonia to reduce danger from *E. coli* and constitutes about 15 percent of all hamburger sold in the United States (Moss 2009). The leadership of Berkeley's less entitled neighbor, Oakland Unified School District, welcomes the raised expectations. It has, for example, prepared an "ideal meal," that included "organic, local, and environmentally-friendly food items." And along with school districts in San Diego, Portland (Oregon), and Denver, it has served grassfed beef during National School Lunch Week "in an attempt to show farmers and the USDA that there is a viable market for grassfed beef in schools." The ideal meal demonstrated "what we could do if we could have a funding increase," noted one school official. "This is what our kids are asking for, this is what our parents are asking for, and this is what administrators are asking for" Chin and Pennington 2010).

81. Health Care Without Harm, "Issues: Toxic Materials" available at http://www.noharm.org/us_canada/issues/toxins/ (accessed February 2012).

82. This support for regional agriculture, sometimes called a food hub, differs markedly from the Walmart regional scaling model (see Flaccavento 2010). The Healthcare Without Harm Food Pledge committed signers to work with their suppliers to source food that is healthy—meaning "produced without synthetic pesticides and hormones or antibiotics given to animals in the absence of diagnosed disease." It is available at http://www.noharm.org/us_canada/issues/food/pledge.php (accessed February 2012).

83. Note that institutional thickening can create confusion. This is the same CAFF that opposed legislation restricting use of the short-handled hoe.

84. *San Francisco Bay Crossings*, December 11, 2003.

85. The EPA awarded the farm a Stratospheric Ozone Protection Award as the "pioneer . . . in developing the technology of farming strawberries . . . without relying on the soil fumigant methyl bromide." Methyl bromide was standard in conventional strawberry production until it was phased out, only to be replaced by the even more toxic methyl iodide, another serious ozone depleter.

86. The comparison is based on the lowest price available on the date of the research and does not reflect freshness, quality, flavor, or environmental benefits

from fewer food miles or, CUESA points out, the pleasure of "knowing where your food comes from." CUESA Executive Director Dave Stockdale observes that earnings vary widely based on farm size, products, staffing, and other characteristics. Some smaller farms earn as much as 75 to 100 percent of revenues at FPFM, but a few larger ones earn 10 to 20 percent and use FPFM as an access point for restaurants, community supported agriculture (CSA), and other sales strategies.

87. *Gleaning*, or gathering after the harvest, is an age-old term now applied more broadly than to crop fields. USDA, "A Citizen's Guide to Food Recovery. Online," http://www.usda.gov/news/pubs/gleaning/content.htm (accessed September 2011).

88. Cuellar and Webber (2010).

89. Reports on Timothy Jones' findings: Lance Gay, "Americans are Tossing $100 Billion of Food a Year," Scripps Howard New Service, August 10, 2005, on the Organic Consumers Association Web site: www.organicconsumers.org/Politics/foodwaste081005.cfm; Jeff Harrison, "Nation Wastes Nearly Half its Food," *UA News*, November 18, 2004, http://uanews.org/node/10448 (accessed September 2011).

90. Parson (2010).

91. The most comprehensive outsider analysis is Parson (2010). See also Butler and McHenry (1992), McHenry and Butler (2005) and the Food Not Bombs Web site.

Chapter 8

1. Boyle et al. (2010) and Kessler (2009).

2. Summarized in Taubes (2011). See Nestle (2007) and Kessler (2009).

3. Best described in Gottlieb and Joshi (2010).

4. Bada, Fox, and Selee (2006) and Liu and Apollon (2011).

5. We use the term *food justice* when discussing food access–related issues, and refer to *labor* or *workers' issues* when that is our main topic. We avoid glossing the two together as food justice but use the term *food democracy* to include both.

6. Fine (2006).

7. Brechin (2006).

8. See Wilkerson (2010) on the black diaspora.

9. See generally McClintock (2008), Self (2003), Walker (2004), and Scott (1985).

10. Bullard and Johnson (1997, 2000). The classic case challenging the "run-a-road-through-an-African-American-community-to-save-on-land-acquisition-costs" strategy, *Citizens to Preserve Overton Park* v. *Volpe,* 401 U.S. 402 (1971), put some constraints on locating highways, but that precedent came too late for West Oakland.

11. See Self (2006). In spite of the Panthers' tolerance of criminality, Self argues that Black Power was in fact "an extraordinarily plastic concept. It meant everything from revolution to electing African American school board members to wearing a dashiki" (218).

12. Despite their national organization, Oakland "was still home" for the Panthers (Self 2006, 299). When the Cypress structure collapsed in the 1989 Loma Prieta earthquake, the reconstruction process reanimated these earlier struggles. The NAACP legal defense fund filed suit on behalf of the Church of the Living God Faith Tabernacle and the Clean Air Alternative Coalition to protect the neighborhood from both the air pollution and other health impacts of the highway and the disruptions resulting from the proposed location, just fifty feet from the church. At its best, the case was less a blow for justice or preventing environmental assaults in black neighborhoods than a conflict among potential losers scrambling to avoid further loss. No matter what route was selected, a poor black neighborhood would suffer. The out-of-court settlement (*Clear Air Alternative Coalition* v. *U.S. Department of Transportation,* ND Cal. C-93-O72LVRW) changed the course of the freeway slightly but left residents of the Lower Bottoms neighborhood, rather than their equally affected neighbors slightly to the east, holding the short end of the stick yet again. The issues are discussed briefly on the Federal Highway Administration Web site. See "Environmental Justice Case Studies: Cypress Freeway Replacement Project," http://www.fhwa.dot.gov/environment/environmental_justice/case_studies/case5.cfm (accessed February 2012).

13. Details about the port are distilled from Cannon (2008). Bay Area enthusiasts frequently insist that the port's container hoists were the models for the AT-AT walkers in *Star Wars*. Whether this is true or not, they are a certainly a major feature of the Oakland skyline. Established in 1927, the port today occupies nineteen miles of waterfront, about one-quarter devoted to maritime activities and the rest to Oakland International Airport. In 2006, Oakland handled more than 99 percent of the containerized goods moving through northern California. The port is a big source of funds for the city: in 2005, it paid the City of Oakland $25.3 million from its revenue and generated 28,522 direct, induced, and indirect jobs; $2.0 billion of total personal income and consumption expenditures; and $1.8 billion in direct operating revenue for businesses supporting the port. The port also presents problems for the people of West Oakland, leading to the formation of the West Oakland Toxic Reduction Collaborative and the creation of the Oil Free Oakland Initiative among other responses.

14. Ravallion (2011) underscores how unusual the period was. He searched three centuries of scanned books in the Google library looking for evidence of a belief that poverty "could and should be eliminated." He found two "poverty enlightenments" since 1700: the first at the end of the eighteenth century and a "striking resurgence" of interest in the 1960s.

15. The Kerner commission, appointed to investigate the causes of a number of urban riots around the country, famously concluded that the nation was "moving toward two societies, one black, one white—separate and unequal." The report

mentioned food once, in the context of providing food during emergencies, and health once, noting standard life expectancy and similar data. It did, however, encourage income support for the needy. For an introduction to the civil rights era more generally, see Self (2006), Branch (1988, 1998, 2006), and Arsenault (2006).

16. Maney (1989) has argued that the People's Campaign did not help hunger advocates: "Johnson did not want to be criticized by members of Congress (whose help he needed on other legislative matters) for giving in to the demands of street demonstrators" (109). Head Start, originally a summer educational enrichment program for low-income preschoolers, and the companion Job Corps for older children, both continue. For a small window into the Job Corps program in Oakland, see Seale (1970). Some OEO programs that worked too well from some perspectives—like Legal Services, which played an important role in agricultural labor disputes in California discussed in chapter 6—were attacked and gradually diminished or eliminated.

17. Self (2003, 200). A 1966 Ford Foundation report commented on the racist local politics in Oakland, noting that a large part of Oakland's race relations problem lies in the "power structure, the attitudes of which are fully as backward as many claim. . . . And there's little doubt that Oakland is a much worse city in which to live because of the predominance of that elite" (Gordon 1966, quoted in Self 2003, 249).

18. Self (2003, 218–219).

19. Renamed the Department of Health and Human Services in 1980 after the education responsibilities were shifted to a separate department.

20. CCAP was a collaboration of labor unions, business groups, and religious groups founded in 1964. It merged with the Center for Community Change in 1969. Walter Reuther, head of the United Auto Workers, chaired CCAP during the five years in which it spearheaded training and education programs in Watts and Delano, California; the Mississippi Delta; and several northern cities.

21. Nixon (1969).

22. The measure was related to Nixon's ideas for a "New American Majority" that would move blue-collar workers into the Republican Party. Both liberals, who opposed ending the Aid for Dependent Children program, and Republicans, who invoked the standard fears of the unworthy poor, opposed it. The Family Assistance Program did incubate the Supplemental Security Income program that provides a guaranteed income to elderly and disabled people. The literature on the Family Assistance Program is enormous. Start with Moynihan (1972) and Zigler, Marsland, and Lord (2009, especially chap. 2).

23. Maney (1989). Harrington (1962, 2) observed presciently, "If these people are not starving, they are hungry, and sometimes fat with hunger, for that is what cheap foods do."

24. The USDA partisans prevailed in spite of evidence that in the Mississippi Delta, "county officials were refusing to certify blacks for participation in the food distribution program if they had registered to vote." USDA officials were

accustomed to "having the department's programs adjust to local political conditions" (Maney 1989, 42).

25. Purchasing imported foods with stamps was not allowed. Unfortunately, a House prohibition on purchase of soft drinks did not make it into the final bill (Landers 2007).

26. Although their position complicates efforts to reform the Farm Bill (Imhoff 2010), a suboptimal bedfellow is arguably preferable to no bedfellow (or bed) at all.

27. Martin Luther King Jr. lived long enough to observe that in Vietnam, "we spend $322,000 for each enemy we kill," while the War on Poverty spends $53 for each poor person (Self 2003, 202).

28. The first major legislation of the Reagan era, the Omnibus Budget Reconciliation Act (OBRA, 1981), cut food assistance as well as income support, Medicaid, and funding for housing and the unemployed. In the wake of further cuts, and apparently embarrassed by the protests that resulted, Reagan also approved USDA's distribution of 30 million pounds of surplus cheese in chaotic mob scenes. Describing the action as "A Mess However It's Sliced," *Time* magazine lamented the intersection of farm problems caused by subsidized overproduction and social problems revealed by crowds of poor people surging to capture bricks of moldy cheese (Time 1982). A task force appointed by Reagan to study hunger in America was unable to find any. See President's Task Force on Food Assistance (1984). The administration of George W. Bush went further: in 2006 the USDA removed the term *hunger* from its policy statements (Himmelgreen and Romero-Daza 2010). Although "very low food security," the substitute term, may be more precise, it probably communicates less urgency to the average reader.

29. Poppendieck (1998). The first food bank in California (and the second in the United States) opened in Santa Cruz in 1972. The Alameda Food Bank started in 1977 and initially operated from a church closet. By the mid-1980s, however, a sharp recession intersected with OBRA cuts to create a "sudden, dramatic expansion" in charitable food distribution. As homelessness became a visible and divisive issue in urban areas, food banks multiplied all over the country, from a few dozen in 1980 to more than a hundred by 1985. Programs started to meet emergencies first became permanent and then totally inadequate to meet the growing need. Today food banks are inarguably essential to many people who rely on them to eat regularly and feed their families. But they are problematic as a primary policy tool for addressing hunger: the public–private collaborations benefit industrial food producers while allowing beneficiaries and the broader public little direct voice. Poppendieck also notes that food banks, the largest of which is now called Feeding America, have "proven extraordinarily useful" to the food industry. The government uses taxpayers' money to subsidize conventional farmers to grow more food than can be sold. To deal with the overages, companies can avoid dumping fees and associated hauling costs by donating them to food banks. Taxpayers pay again with tax deductions for the corporations.

30. Patel (2010c) notes that the McDonald's Happy Meal can reasonably be viewed as a poverty program in an economy where real wages have stagnated for almost forty years.

31. Senator Gaylord Nelson (Democrat, Wisconsin) is usually credited with conceiving of Earth Day in response to a major oil spill off the Santa Barbara, California, coast in 1969. Although this claim is not uncontested, Gaylord's own account, and those of others, makes clear that he modeled the national event on the then-widespread teach-ins against the Vietnam War.

32. Nixon signed critical air and water legislation and created the Environmental Protection Agency after the first Earth Day, but the point remains valid. The membership may vary with the list maker, but the Big Ten usually includes large national and multinational environmental groups that have budgets in the tens of millions of dollars, large staffs of lobbyists in Washington, D.C., and frequently international offices as well. It includes some or all of the following: Defenders of Wildlife, Environmental Defense (or Environmental Defense Fund, depending on when the list was made), National Audubon Society, National Wildlife Federation, Greenpeace, Natural Resources Defense Council, the Nature Conservancy, Sierra Club, Conservation International, and the World Wildlife Fund. Some of these groups have missions that either ignore humans or regard them, regardless of race or class, as the problem. See Dowie (2005).

33. Bullard (1998). Big Ten groups were also generally unsupportive of growing numbers of regional and toxics-oriented environmental groups that worked in alliance with minority communities at the time. See, for example, Maura Dolan, "Race, Poverty Issues Grow Among Environmentalists," *Los Angeles Times* April 21, 1990. http://articles.latimes.com/1990-04-21/newa/mn-1114_1 (accessed February 2012).

34. 42 USC. 4321, et seq.

35. Beginning in the 1960s and accelerating over the next two decades, plastics and other industrial manufacturers began locating along the Mississippi River between Baton Rouge and New Orleans. The region now known as Cancer Alley rapidly became infamous for an enormous diversity of diseases. In 2004 Margie Eugene-Richard won a Goldman Prize for her work in founding Concerned Citizens of Norco (LA) and forcing Shell Oil Company to address problems caused by its release of 2 million pounds of chemical toxicants a year. Bullard (1990).

36. Bullard (1998). African Americans began protesting hazardous waste dumping in 1982. Further demonstrations followed the selection of predominantly African American Warren County in North Carolina as the official site for storing soils contaminated with PCBs (polychlorinated biphenyls). Previously these soils had been disposed of by dumping along roadsides.

37. Communities at the edge of polluting industrial and military facilities.

38. B. Taylor (2000), Gottlieb (1996), and Cole (1993). Compare Nordhaus and Shellenberger (2007).

39. Bean v. Southwestern Waste Management Corp., 482 F. Supp. 673 (S.D. Tex. 1979).

40. First National People of Color Environmental Leadership Summit (1991).

41. See *El Pueblo para el Aire y Agua Limpio v. County of Kings,* 22 Envtl. L. Rep. (Envtl. L. Inst.) 20,357 (Sup. Ct. Dec. 30, 1991), discussed in Cole (1993, 530) and *Angelita C., et al. v. California Department of Pesticide Regulations* (CDPR) Title VI Complaint 16R-99-R9.

42. See generally Cole (1993) and Chitewere (2010).

43. "Environmental Justice calls for universal protection from nuclear testing, extraction, production and disposal of toxic/hazardous wastes and poisons... that threaten the fundamental right to clean air, land, water, and food." First National People of Color Environmental Leadership Summit (1991).

44. See, for example, a preconference paper that addressed "toxic food" in the form of foodborne diseases rather than malnutrition. Environmental Justice Resource Center (2001, 4).

45. Kessler (2009). For Flegal's comments see "Katherine Flegal Discusses Prevalence of Obesity in the US" *sciencewatch.com* www.sciencewatch.com/ana/st/obesity2/10sepObes2Fleg/ (accessed September 2011).

46. The results were striking enough that Flegal worried that she and her colleagues had made an error; they rechecked the data multiple times for errors that could explain the difference but found none (Kessler 2009, 3-5).

47. The term is familiar so we stick with it, but it is contested. To some, *food desert* suggests that there is no food. That is incorrect: there is plenty of food, but it is both expensive and unhealthy. To others, the term suggests that the goal is to build a normal grocery store. However, while many want one of those, others prioritize local economic development and regard conventional food retail with justifiable reservations.

48. Quoted in Kane (1984, 7). Kane linked the trend to federal policy that "explicitly encouraged both racial segregation and 'redlining' practices" and limited "investment in racially homogenous neighborhoods," contributing to a "vicious cycle of neighborhood decline" among minority neighborhoods.

49. Sen (1981).

50. Kaufman et al. (1997, 1): "Low-income households may face higher food prices for three reasons: (1) on average, low-income households may spend less in supermarkets—which typically offer the lowest prices and greatest range of brands, package sizes, and quality choices; (2) low-income households are less likely to live in suburban locations where food prices are typically lower; and (3) supermarkets in low-income neighborhoods may charge higher prices than those in nearby higher income neighborhoods. Despite the prevailing higher prices, surveys of household food expenditures show that low-income households typically spend less than other households, on a per unit basis, for the foods that they buy. Low-income households may realize lower costs by selecting more economical foods and lower quality items. In areas where food choices are limited due to the kinds and locations of food stores, households may have sharply higher food costs."

51. The grandmother of this work is Food First, which is located in Oakland.

52. Shaw (2007) and Beaulac, Kristjansson, and Cummins (2009), cited in Gottlieb and Joshi (2010, 24).

53. Luttrell (1973). Briefly, in the summer of 1972, the United States sold Russia about 440 million bushels of wheat for about $700 million. That absorbed much of the U.S.'s surplus but some have argued that the farmers did not benefit.

54. Food and Agriculture Organization of the United Nations (1996). The U.S. government has consistently rejected any notion of a right to food.

55. Food and Agriculture Organization of the United Nations (2003, 29).

56. Note that charity is not security. Beginning in the Reagan administration, Poppendieck argues, and particularly following Clinton's mid-1990s welfare reform, the federal government redefined hunger alleviation from its brief appearance as a publicly supported entitlement to food back to a privately supported, and therefore voluntary, increasingly overstretched, and unreliable charity (Poppendieck 1998).

57. Rossett (2005, 2).

58. An international movement that "coordinates peasant organizations of small and middle-scale producers, agricultural workers, rural women, and indigenous communities from Asia, Africa, America, and Europe." The coalition includes more than 100 organizations, all advocating family-farm-based sustainable agriculture. Rossett (2005, 2).

59. The definition on the Via Campesina Web site has varied slightly over the years. This version was the one shown when we accessed the site in 2009, http://viacampesina.org/en/ (accessed February 2012).

60. Schanbacher (2010) underscores that there is a "fundamental antagonism" between the way hunger and malnutrition are conceived within food security compared to within food sovereignty: "Ultimately the food security model is founded on, and reinforces, a model of globalization that reduces human relationships to their economic value. Alternatively, the food sovereignty model considers human relations in terms of mutual dependence, cultural diversity, and respect for the environment" (ix). At best, food security will be achieved within the WTO/World Bank paradigm of trade liberalization by "the reduction of trade-distorting subsidies" (13). But the requisite phasing out of or substantial reductions in global North government subsidies to agriculture have not happened. The United States and others have circumvented WTO regulations by "manipulating the language that describes different types of supports or subsidies" (11). Schanbacher also notes problems associated with the "fulfillment of human rights" (xv).

61. The Community Food Security Coalition is a "nonprofit 501(c)(3), North American organization dedicated to building strong, sustainable, local and regional food systems that ensure access to affordable, nutritious, and culturally appropriate food for all people at all times. We seek to develop self-reliance among all communities in obtaining their food and to create a system of growing, manufacturing, processing, making available, and selling food that is regionally

based and grounded in the principles of justice, democracy, and sustainability." See the Food Security Coalition's Web site: http://www.foodsecurity.org/.

62. Food Security Coalition, "What is Community Food Security?" online at http://www.foodsecurity.org/views_cfs_faq.html (accessed February 25, 2012).

63. Ibid.

64. Gottlieb and Joshi (2010, 6).

65. Harper et al. (2009).

66. While recognizing that the boundaries of the neighborhoods and their sub-communities are a matter of some meaningful dispute (Smoothe 2008).

67. Bell and Burlin (1993), but compare Kaufman et al. (1997).

68. Booth (2008).

69. Analyzed in a Youth Radio series. See Densie Tejada, et al: "Trafficked Part I," November 2, 2010, http://www.youthradio.org/news/part-one-trafficked; "Trafficked Part II," November 12, 2010, http://www.youthradio.org/news/part-two-trafficked; "Trafficked: Oakland Parents Organize to Protect Girls from Sex Trafficking," September 19, 2011, http://www.youthradio.org/category/bureau/yr-bay-area (all accessed February 2011).

70. Fong-Torres (1994, 93).

71. Alameda County Building Blocks Collaborative, "Healthy Babies, Healthy Families, Healthy Communities (blog)," http://buildingblocksalamedacounty.wordpress.com/ (accessed October 2011).

72. Mo' Better Foods has preserved its story on its Web site: http://www.familyhoodconnection.org/. See also Smart (2007) and Roach (2006).

73. The best overview is McClintock (2008). Oakland's garden activism is not the first in communities of color. Earlier interest was shown just after the Civil War when Booker T. Washington and George Washington Carver collaborated on efforts to bring agricultural education to recently freed slaves. They enlisted Thomas Munroe Campbell in their project, and his efforts to improve the skill levels and resources available to sharecroppers led him to a position with the USDA. Campbell is frequently described as the nation's first cooperative extension specialist. His very early innovation, the mobile classroom, is described in Campbell (1936). The major funding for the extension program came from the Hatch Act of 1887 and the Smith-Lever Act of 1914. Campbell's mule-drawn portable classroom demonstrated scientific methods and was known as the Jessup Wagon to honor the original donor of funds (Jones 1953; Lawson 2005). In the 1970s, San Francisco used federal OEO funds to support garden-based employment and worker training programs. When that money dried up, the Parks Department deployed the San Francisco League of Urban Gardeners, a private group, to maintain many of the gardens (Peirce 1994). In the 1980s Catherine Sneed's Garden Project began to provide support and transition aid for inmates at some of the Bay Area's more famous prisons, working along the lines of the Rubicon Bakery. The similar Alemany Youth Farm began to evolve in 1994 with the same goal: to employ youth and residents from the surrounding low-income communities in the construction and maintenance of an urban farm (Lawson 2005). The goal was to

combine "job training and education to make urban agriculture an avenue of economic and educational opportunity" for project residents. By the end of the decade, the Alemany Farm was the site of a burgeoning business, Urban Herbals, which sold herbs in about fifty retail outlets and natural food stores in the area (Feenstra, McGrew, and Campbell 1999; Cook 1994; Peirce 1994).

74. Minister of Information JR (2009).

75. Borkowski (2010).

76. Gallagher (2010, 60).

77. Gallagher (2010, 60).

78. Gallagher (2010, 63–69).

79. The Ecology Center, which runs farmers' markets in neighboring Berkeley, helped the Mandela market set up to accept an electronic benefit transfer system. The system allows purchasers to use food stamps in farmers' markets and was intended to overcome a common barrier to poor people's participation. This was and remains more complicated than one might predict. The Ecology Center in Berkeley also developed tools to allow WIC (Women, Infants, and Children), food stamp, and other food aids at farmers' markets. See Ecology Center (2008). The Ecology Center has also worked through much iteration of formats designed to bring underserved populations into community supported agriculture and farmers' markets' modes of produce distribution. Its most recent iteration, Farm Fresh Choice, is a community supported agriculture variant that focuses on school sites where parents come to pick up their children. At those convenient locations, they can also obtain low-cost produce, broken out of the traditional box format and available for purchase at subsidized prices. See the Ecology Center Web site, "Farm Fresh Choice," http://www.ecologycenter.org/ffc/. For a list of Farm Fresh Choice producers see http://www.ecologycenter.org/ffc/ffc-farmers.html.

80. Alkon (2003) and Guthman, Morris, and Allen (2006).

81. Mandela Foods Cooperative Web site, "Programs," http://www.mandelamarketplace.org/5.html (accessed February 2012).

82. Robert Wood Johnson Foundation, "Winners of the 2010 *Community Health Leaders Award*," at http://www.communityhealthleaders.org/news_features/pr/62569 (accessed February 2012)

83. Donohue (2010).

84. PG&E was forced to close an archaic plant. Compare http://www.pge.com/about/environment/pge/goodneighbor/ and *San Francisco Chronicle* reports: Leslie Fulbright, "As PG&E closes its old, smoky power plant, the neighborhood breathes a sigh of relief," http://www.greenaction.org/hunterspoint/press/sfgate051506.shtml. (both accessed May 2010).

85. Compare the community Web site, "San Franciscans for Our City's Health," http://www.hunterspointnavalshipyard.com/, and the redevelopers', http://www.hunterspointcommunity.com/ (both accessed May 2010).

86. Greenaction led a fight against air pollution from an underregulated yeast factory that eventually closed. See Cecily Burt, "Regulators go too easy on yeast

factory, neighbors say," *Oakland Tribune* December 18, 2002, available at http://www.greenaction.org/westoakland/press/oaktrib121802.shtml (accessed February 2012).

87. Sibella Kraus manages the agricultural park at Sunol under the umbrella of her new nonprofit organization, Sustainable Agriculture Education. She continues to focus on sustainable agriculture education and protecting periurban farms as central to food system reform.

88. Dig Deep Farms is itself a complex nonprofit partnership. Kaiser Permanente and the San Francisco Foundation support the program, which is run by the Alameda County Deputy Sheriffs' Activities League.

89. Discussed at the People's Grocery Web site, "Urban Ag and Food Justice Allyship," at http://www.peoplesgrocery.org/article.php?story=FJI (accessed September 2011).

90. For example, Henderson taught a UC Berkeley food course with Michael Pollan that was funded by the Chez Panisse Foundation in connection with the restaurant's fortieth anniversary in 2011.

91. Produce Pro opened on San Pablo Avenue almost in the center of West Oakland in a spot that People's Common Market had discussed as a grocery site with the owner.

92. If the city makes a substantial investment in a site, businesses occupying the site are bound by an ordinance that requires employers to pay workers a specified hourly wage if they offer benefits and slightly more if they do not. This is a popular requirement. In 2002, by a 72 percent majority, Oakland voters approved extending the requirements to low-income workers in the Port of Oakland. However, it burdens those who operate on city-owned sites while not affecting those who operated in fully private spaces.

93. Fagan (1995a, 1995b). Customers complained about "old produce, lack of selection, and rotten meat," and a survey of groceries in four Oakland neighborhoods revealed that Acorn had by far the highest prices: 11 percent higher than the next most expensive market and nearly 30 percent higher than the least expensive market (Bell and Burlin 1993). The investor, Sang Hahn, conceded failure, but complained that "of course some of my food sits on the shelf longer and some of my prices have to be higher. But what can I do?" (Fagan 1995a). The lease expired in February 1996, leaving a gutted strip mall in the middle of the beleaguered neighborhood. Hahn and his partners actually fared well when the mall was condemned a few years later: the city paid $3 million in federal redevelopment funds, well over its assessed value. The Hahn partnership has been criticized for manipulating eminent domain in locations throughout the city (Bair 2011). In 1999, the city transferred the site to the nonprofit Bridge Corporation, which refurbished the area using federal and city loans and grants (DeFao 1999). The Mandela Gateway Village opened in 2004. The mixed-use project included retail space and housing. This time, advocates promoted a nonprofit approach to grocery retail (Fagan 1995a). People's Grocery and Mandela Market both eyed the spot, and Dana Harvey began to raise funds.

94. *East Bay Express* provides some insight into the Co-op's financing: "With their eyes set on the village's prime 11,000-square-foot retail space, it was agreed that if Harvey and company could raise $500,000 to guarantee their first three years of rent, [the project] . . . would lease them the space and even contribute $250,000 to the store's start-up costs. Harvey and company set about raising their start-up money and building public awareness. First, the West Oakland Project Area Committee approved a $200,000 grant for the project and agreed to guarantee an additional year's rent. Then Oakland councilwoman Nancy Nadel contributed $100,000 of city funding to assist them. Harvey announced that the store would be open by that summer. However, as the months passed, this goal became less and less certain, and by December 2006 Mandela Foods was still unable to raise the remaining funds needed to open. Bridge eventually rented the space to the Southern California-based chain, 99 Cents Only. Still determined to make the store a reality, in 2007 the project received another $220,000 in grant money from the California Endowment and had a New Year's Eve fund-raiser that brought in $10,000. By September, Mandela Foods had signed a lease with Bridge for a smaller retail space, but further delays held up the project until June of 2009" (Taylor 2010).

95. Smoothe (2008).

96. While not directly opposed to Kroger's, Henderson argues that West Oakland needs grocery stores that "provide jobs, ownership opportunities and quality economic opportunities for the community, that create wealth in the community rather than pulling wealth out, and that provide healthy, organic local produce and items that nourish our bodies and minds." Nikki's Blog, October 14, 2010, http://revolutionandevolution.blogspot.com/.

97. Start with Tasch (2008) and Pons and Long (2011). The nonprofit Tasch founded Slowmoney, http://www.slowmoney.org (accessed February 2012).

98. Unger and Wooten (2006, 90).

99. Carpenter (2009) and Carpenter and Rosenthal (2011) are a good place to begin. For a start on urban agricultural zoning, see Grown in the City, "Interactive Urban Ag Zoning Map," at http://growninthecity.com/interactive-urban-ag-zoning-map/ (accessed September 2011).

100. Urban agriculture was legalized in San Francisco as recently as April 2011. See Antonio Roman-Alcalá, "San Francisco Passes Progressive Urban Agriculture Policy," April 14, 2011, at "Civil Eats (blog)," http://civileats.com/2011/04/14/san-francisco-passes-most-progressive-urban-agriculture-policy-in-u-s/.

101. Kuruvilla (2011); Matthai Kuruvilla, "Oakland urban farming prompts plan to redo rules," *San Francisco Chronicle,* May 9, 2011, at "Civil Eats (blog)," http://www.sfgate.com/cgi-bin/article.cgi?f=/c/a/2011/05/08/BA7O1J74O5.DTL#ixzz1mgH9sjX7 (accessed February 2012).

102. Teicholz (2005) introduces Walmart's food retail and how it might play out.

103. Henry (2011).

104. Farmhouse Culture Web site: http://www.farmhouseculture.com/ (accessed March 2011). Note that this is sauerkraut, not chevre.

105. Foraging for high-end restaurants is also returning to fashion, albeit not without some concerns for food safety. See Strand and DiStefano (2010).

106. Wadud also had a day job bartending at Chez Panisse (Hopper 2008).

107. Forage Oakland, "What is Good is Given Back (blog)," at http://forageoakland.blogspot.com/ (accessed September 2011).

108. Ibid.

109. *New York Times Magazine* photo essay "Food Groups" at http://www.nytimes.com/slideshow/2010/10/10/magazine/food-groups.html (accessed February 2012).

110. "Portfolio: Food Groups: Brewers Bitters-Makers," *New York Times Magazine* October 10, 2010, available at http://query.nytimes.com/gst/fullpage.html?res=9804E7D7143AF933A25753C1A9669D8B63&scp=2&sq=eat%20real%20mason%20jar&st=cse (accessed February 2012).

111. Laskawy (2010).

112. Ibid.

113. The occasional food market "pops up" every few weeks or so in Grace Street Catering facilities. Founded in 1987, Grace Street caters private functions and runs cafés in major venues: the Oakland Museum and Lawrence Hall of Science, for example. With her own operations stabilized, founder Erin McKinney now shares her space with local start-ups and the pop-up General Store. Store founders Samin Nosrat and Chris Lee trained at Chez Panisse and were displaced when Eccolo, a posh Berkeley eatery, closed. In a tight economy, Nosrat sought a way to "make and share the foods we so love to cook without carrying the burden of a restaurant." Drawing on Bay Area colleagues and "our heritage as the students of Alice Waters," the producers continue the tradition of supporting local farmers and ranchers and "teaching others about eating seasonally and locally" (Ness, 2010). The pop-up raises interesting issues. Some have noted that it sells products that most Oaklanders cannot afford. In Birdsall (2007) diverse bloggers wonder whether the General Store and its boudin blanc at $14 per pound are "hopelessly out of touch" with the economic realities of its surroundings. The pork sausage made without the blood is a traditional Cajun dish, and we are not upset by the possibility that sausage fanatics can occasionally pay an artisan a high price to get a very good one. Pop-up founder Nosrat has a slightly different perspective: noting that the $14 per pound price tag barely covers the cost of the ingredients and a wage for the workers, she suggests that consumers may be out of touch with the actual cost of high-quality, labor-intense food.

114. R. Walker (2010).

115. Ibid.

116. Dicum (2010).

117. San Francisco has, for example, begun to develop a permitting process for overseeing the rising tide of food carts. The new wave "has given talented, spirited, but underfunded cooks the chance to investigate the demand for their wares,

and . . . it has given drunk people easily accessible food at times and locations where a licensed seller would fear to tread. In its disfavor, there is this: It's totally in violation of California health code." The cart operators view the move as an effort by local restaurants "to bar the gate against interlopers" (H. Smith 2010, but see Hirsch 2011). Similarly, the USDA Federal Safety and Inspection Service's small plant program is assembling resources to assist small and very small plants in complying with safety regulations. See http://www.fsis.usda.gov/Science/Small_Very_Small_Plant_Outreach/index.asp (accessed February 2012).

118. Booth (2008).

119. Lucchesi (2010). MSF is beginning to connect to its new community. Business manager Daniel Kramer has joined the Oakland Food Policy Council. Kramer has worked at both Swanton Berry Farm in northern Santa Cruz County and at Frog Hollow Farm in Brentwood. As a congressional staffer Kramer has also worked on agricultural, environmental, and energy policy.

120. Boccalone Web site: http://www.boccalone.com/pages/About-Boccalone.html (accessed February 2012).

121. P. Martin (2011). Subsequent legislation has made organizing labor increasingly difficult, and the percentage of workers in unions has declined dramatically since its peak in the mid-1940s. Most scholarship on unions focuses on the possibility of a revival (see Rose and Chaison 2001, for example) or on the flagrant violation of labor laws that has become routine in many fields, none more dramatically than the food chain. See Bernhardt, McGrath, and De Filippis (2007), Bauer and Ramirez (2010), and Bon Appétit Management Company Foundation and United Farm Works (2011).

122. The most explicit and comprehensive multistakeholder discussions that we are aware of have been convened by the New York chapter of the Restaurant Opportunity Center (ROC-NY), discussed below, but small roots and branches are apparent throughout the United States. See, for starters, Sen and Mamdouh (2008).

123. Exceptions discussed above are an important part of district values. Saxenian (1994) has noted that her "protean place" in Silicon Valley valued vertical bonding between top executives and those lower on the organizational chart but was notably exploitative of low-wage, unskilled labor and generally female laborers.

124. Getz and Shreck (2006).

125. Fine (2006, 5). For a brief history, see Fine's discussion in chapter 1.

126. Bowe (2003).

127. Gillespie (2008).

128. Grimm (2007).

129. Start with the Coalition of Immokalee Workers Web site: http://www.ciw-online.org/101.html (accessed February 2012).

130. Oxfam (2004, 36).

131. See the Bon Appétit Management Company Web site: http://www.bamco.com/about-us (accessed February 2012).

132. The BAMCO code of conduct is available at http://www.bamco.com/uploads/documents/CODE%20OF%20CONDUCT%20FOR%20SUSTAINABLE%20TOMATO%20SUPPLIERS%20.pdf (accessed February 2012).

133. J. Black (2009).

134. Morgan DeBaun, "Tomatoes on Campus," Press release via Bon Appétit, November 17, 2009, at http://su.wustl.edu/blog/no-more-tomatoes.

135. "Editorial: Fruits of Labor," *The Daily Pennsylvanian*, December 8, 2009. Available at http://www.dailypennsylvanian.com/article/editorial-fruits-labor (accessed February 2012).

136. Sen and Mamdouh (2008). See also Madelein Brand," Window on the World Owner Seeks Fresh Start," NPR broadcast May 30, 2012. http://www.npr.org/programs/morning/features/2002/may/noche/index.html (accessed February 2012).

137. Severson and Ellick (2007) and Ellick (2007). Batali's appalling treatment of coworkers was romanticized in Buford (2006).

138. ROC United's studies include *The Great Service Divide* (2009); *Serving While Sick: High Risks and Low Benefits for the Nation's Restaurant Workforce* (2010); *Waiting on Equality* (2010); and *Behind the Kitchen Door* (2011). ROC's 2006 publication, *Dining Out, Dining Healthy*, first demonstrated that "the same restaurant owners that violate employment laws also create unhealthy conditions for consumers." See ROC-United's Web site: http://www.rocunited.org/what-we-do. See also Liu and Apollon (2011).

139. See the National Restaurant Association "Local Advisory Alert," at http://2.bp.blogspot.com/_uWnXm4WvKC0/TAfvEGDg_nI/AAAAAAAAHJc/wQfklNM0B8E/s1600/NRA+ROC+Advisory+Alert_Page_1.jpg (accessed June 2011). See also Young Workers United (2009).

140. See Food Chain Workers Alliance Web site, "March 2011 Workers Exchange," at http://foodchainworkers.org/?page_id=955 (accessed August 2011).

141. Fine (2006).

142. Wage theft includes undercounting hours worked, not paying overtime, paying workers late or with checks that bounce, not providing breaks during working hours, and skimming tips.

143. Young Workers United (2011). The notion that restaurant rating services could include information about restaurant labor practices in the information they make available to consumers is interesting. Based on the exchange between ROC cofounder and codirector Saru Jayaraman and restaurant reviewers Tim and Nina Zagat (regarding Zagats' response to workers' litigation against restaurant owners), however, it does not appear likely to happen soon. Attributing labor problems in the industry to a few unscrupulous bad apples and confusing regulations, Zagat and Zagat appear to ignore well-documented substandard wage and labor conditions in an industry where worker abuse and violation of labor laws are widely recognized as part of the culture and the business plan.

Compare Zagat and Zagat (2010) and Jayaraman (2010). Those who doubt that abuse is not just tolerated but admired in some quarters might start with Buford (2006) and Reed (2006).

144. Young Workers United (2009).

145. Martin and Midgley (2006) have demonstrated that a 40 percent increase in farmworkers' wages would result in an average increase of $9 per year in household food expenditures (in 2005 dollars) if all the additional labor costs were passed on to consumers. However, note that CIW has had only limited success with grocery stores. "We don't have any plans to sit down with the CIW," Publix's media and community relations manager, Dwaine Stevens, is reported to have said, citing the fact that "the company sells around 36,000 products in the stores and it cannot get involved with each product's labor issues. 'If there are some atrocities going on, it's not our business. Maybe it's something the government should get involved with'" (Golden 2010).

146. Fernald recently sold the enterprise to others who have expanded it while retaining the same mission.

147. Oakland Eat Real Festival Web site: http://www.eatrealfest.com/eb/11/ER_9_28_2011.html (accessed February 2012).

148. Ibid.

Conclusion: The District and the Future of Alternative Food

1. Alongside the University of California's role in subsidizing and advising conventional agribusiness, UC has also consistently attracted a steady stream of socially concerned critics who have worked to create alternatives to that system.

2. Of course, this appropriate scale will differ by geographies based on a variety of factors, including climate, precipitation, growing season, population, and distribution networks.

3. With public money or exemptions from the regulatory regimes around labor, antitrust, and the environment. The Environmental Working Group publishes a Farm Subsidy Data Base that traces the money to states, counties, and individual recipients: http://farm.ewg.org (accessed February 2012).

4. J. Johnson (2008).

5. We particularly object to the idea that it is elitist to enjoy a tasty peach or healthy, sustainable, well-prepared food of any kind.

6. People's Community Market CEO Brahm Ahmadi suggests about $58 million per year. See People's Community Market 2011 slideshow available at http://slowmoneynocal.org/wp-content/uploads/2011/07/PeoplesCommunityMkt_ShowcaseJune2011.pdf (accessed October 2011).

7. We understand, however, that Niman Ranch beef is no longer on the cutting edge of food quality and therefore not routinely featured by alternative trendsetters. To illustrate, Chez Panisse seems focused on expectations about food quality, imagining, financing, teaching about, and serving ever better food. Not restricted

only to what is available, its contribution has been to imagine something better and to make it happen.

8. Leonard (2010).

9. California has some history here that should be useful in this regard, including over fifty years of experience with large retail operations using government regulation to prevent direct sales from farmers to customers, as described in chapter 6.

Bibliography

This research project has lasted long enough to allow us to watch successive cohorts of innovators lay the groundwork for three generations of change in the district, with more clearly on the way. One of us produced the mandatory copies of her doctoral dissertation with sheets of carbon paper. But because we began working with the enormous brew of Web-based resources long before we understood it, we have not been able to recreate references for all that we accessed prior to figuring it out. Thus, our bibliography will look odd to the carbon paper cohort and inadequate to subsequent tech-savvier generations, for which we apologize. We also acknowledge Jennifer Kao's creativity and instruction in these arts. She joined our project early in her undergraduate career and has stuck with us well beyond graduation. Without her, there might not be any bibliography at all.

Abbott, Edith. *American Principles and Policies*. Vol. 1: *Public Assistance*. Chicago: University of Chicago Press, 1940.

Abel, Jennifer, Joan Thomson, and Audrey Maretzki. "Extension's Role with Farmers' Markets: Working with Farmers, Consumers, and Communities." *Journal of Extension* 37 (1999). Accessed March 13, 2012. http://www.joe.org/joe/1999october/a4.php

Adams, Samuel Hopkins. "The Great American Fraud." *Collier's,* October 7, 1905. Accessed June 28, 2011. http://www.museumofquackery.com/ephemera/oct7-01.htm.

Aglietta, Michael. *A Theory of Capitalist Regulation: The U.S. Experience*. London: Verso, 1979.

Agyeman, Julian. *Sustainable Communities and the Challenge of Environmental Justice*. New York: NYU Press, 2005.

Alexander, Jacquelyn, Mary-Wales North, and Deborah K. Hendren. "Master Gardener Classroom Garden Project: An Evaluation of the Benefits to Children." *Children's Environments* 12 (1995): 256–263.

Alkon, Alison. "Black, White and Green: A Study of Farmers Markets." Ph.D. diss., University of California, Davis, 2008.

Allen, Patricia, ed. *Food for the Future: Conditions and Contradictions of Sustainability*. New York: Wiley, 1993.

Allen, Patricia. "Mediating Entitlement and Entrepreneurship." *Agriculture and Human Values* 16 (1999): 117–129.

Allen, Patricia. *Together at the Table: Sustainability and Sustenance in the American Agrifood System*. University Park: Pennsylvania State University Press, 2004.

Allen, Patricia. "Realizing Justice in Local Food Systems." *Cambridge Journal of Regions, Economy and Society* 3 (2010): 295–308.

Allen, Patricia, David Goodman, Harriet Friedman, and Douglas Warner. "Shifting Plates in the Agrifood Landscape: The Tectonics of Alternative Agrifood Initiatives in California." *Journal of Rural Studies* 19 (2003): 61–75.

Allen, Patricia, and Julie Guthman. "From Old School to 'Farm-to-School' Neoliberalization from the Ground Up." *Agriculture and Human Values* 23 (2006): 401–423.

Allen, Patricia, and Carolyn Sachs. "Sustainable Agriculture in the United States: Engagement, Silences and Possibilities for Transformation." In *Food for the Future: Conditions and Contradictions of Sustainability*, ed. Patricia Allen, 139–168. New York: Wiley, 1993.

Amin, Ash, and Nigel Thrift. *Globalization, Institutions, and Regional Development in Europe*. New York: Oxford University Press, 1994.

Amundson, Ronald, and Dan H. Yaalon. "E. W. Hilgard and John Wesley Powell: Efforts for a Joint Agricultural and Geological Survey." *Soil Science Society of America Journal* 59 (1995): 4–13.

Anderson, M. Kat, Michael G. Barbour, and Valerie Whitworth. "A World of Balance and Plenty: Land, Plants, Animals, and Humans." *California History* 76 (1997): 12–47.

Apple, R. W. "A New Normandy North of the Golden Gate." *New York Times*, November 28, 2001. Accessed February 28, 2011. http://www.nytimes.com/2001/11/28/dining/a-new-normandy-north-of-the-golden-gate.html.

Arax, Mark, and Rick Wartzman. *The King of California: J. G. Boswell and the Making of a Secret American Empire*. New York: Public Affairs Press, 2005.

Arce, Alberto, and Terry Marsden. "The Social Construction of International Food: A New Research Agenda." *Economic Geography* 69 (1993): 291–311.

Arsenault, Raymond. *Freedom Riders: 1961 and the Struggle for Racial Justice*. New York: Oxford University Press, 2006.

Baarda, James R. "Antitrust Laws." In *Cooperatives in Agriculture*, ed. David W. Cobia, 397–417. Upper Saddle River, NJ.: Prentice Hall, 1989.

Babcock, Ernest Brown, Cyril Adelbert Stebbins, and Eugene Woldemar Hilgard. *Elementary School Agriculture: A Teacher's Manual to Accompany Hilgard and Osterhout's "Agriculture for Schools of the Pacific Slope."* New York: Macmillan, 1911.

Bada, Xochitl, Jonathan Fox, and Andrew Selee, eds. *Invisible No More: Mexican Migrant Civic Participation in the United States*. Washington, DC: Woodrow Wilson International Center for Scholars, 2006.

Bailey, Kenneth W. *Marketing and Pricing of Milk and Dairy Products in the United States*. Ames: University of Iowa Press, 1998.

Balfour, Lady Eve. *The Living Soil*. London: Faber and Faber, 1943.

Barron, Hal S. *Mixed Harvest: The Second Great Transformation in the Rural North, 1870–1930*. Chapel Hill: University of North Carolina Press, 1997.

Battisti, David S., and Rosamond L. Naylor. "Historical Warnings of Future Food Insecurity with Unprecedented Seasonal Heat." *Science* 323 (2009): 240–244.

Bauer, Mary, and Monica Ramirez. *Injustice on Our Plates: Immigrant Women in the U.S. Food Industry*. Montgomery, AL: Southern Poverty Law Center, 2010. Accessed May 25, 2011. http://www.splcenter.org/sites/default/files/downloads/publication/Injustice_on_Our_Plates.pdf.

Baumgartner, Stefan, and Martin Quaas. "What Is Sustainability Economics?" *Ecological Economics* 69 (2010): 445–450.

Beard, James. *Hors d'Oeuvres and Canapés*. New York: M. Barrows, 1940.

Beaulac, Julie, Elizabeth Kristjansson, and Steven Cummins. "A Systematic Review of Food Desserts, 1966–2007." *Preventing Chronic Disease* 6 (2009): A105.

Beckman, Theodore N., and Herman C. Nolan. *The Chain Store Problem*. New York: McGraw-Hill, 1938.

Belasco, Warren. *Appetite for Change: How the Counter Culture Took on the Food Industry*. Ithaca, NY: Cornell University Press, 1989.

Belasco, Warren. *Meals to Come: A History of the Future of Food*. Berkeley: University of California Press, 2006.

Belasco, Warren, and Roger Horowitz. *Food Chains: From Farmyard to Shopping Cart*. Philadelphia: University of Pennsylvania Press, 2009.

Belasco, Warren, and Philip Scranton, eds. *Food Nations: Selling Taste in Consumer Societies*. New York: Routledge, 2002.

Bell, Judith, and Bonnie Maria Burlin. "In Urban Areas: Many of the Poor Still Pay More for Food." *Journal of Public Policy and Marketing* 12 (1993): 268–270.

Bell, Michael M., ed. *Farming for Us All: Practical Agriculture and the Cultivation of Sustainability*. University Park: Pennsylvania State University Press, 2004.

Belzer, Michael H. *Sweatshops on Wheels: Winners and Losers in Trucking Deregulation*. New York: Oxford University Press, 2000.

Benedict, Murray. *Farm Policies of the United States, 1970–1950: A Study of Their Origins and Development*. New York: Twentieth Century Fund, 1953.

Bennett, Hugh Hammond, and William Ridgely Chapline. "Soil Erosion: A National Menace." U.S. Department of Agriculture Circular 33. Washington, DC, 1928.

Bennet, Michael, and Cruz Reynoso. "California Rural Legal Assistance (CRLA): Survival of a Poverty Law Practice." *Chicano Law Review* 1 (1972): 1–79.

Bently, Amy. *Eating for Victory: Food Rationing and the Politics of Domesticity*. Urbana: University of Illinois Press, 1998.

Berg, Peter, and Judy Goldhaft. Interview by Marty Lee and Eric Noble. San Francisco, CA, April 29, 1982. Accessed February 19, 2012. http://www.diggers.org/oralhistory/pb_jg_0482.htm.

Berger, Michael L. *The Devil Wagon in God's Country: The Automobile and Social Change in Rural America, 1893–1929*. Hamden, CT: Shoe String Press, 1980.

Bergin, Mary. "L'Etoile Plans to Move, Expand, Broaden Menu." *Wisconsin State Journal*, January 23, 2010. Accessed February 28, 2011. http://host.madison.com/wsj/business/article_ac50bcfc-07d4-11df-aae2-001cc4c03286.html.

Berlin, Linda. "Marshall to Get Organic Dairy." *Point Reyes Light*, June 25, 1992.

Bernhardt, Annette, Siobhan McGrath, and James DeFilippis. *Unregulated Work in the Global City: Employment and Labor Law Violations in New York City*. New York: Brennan Center for Justice, 2007. Accessed May 25, 2011. http://nelp.3cdn.net/cc4d61e5942f9cfdc5_d6m6bgaq4.pdf.

Berry, Wendell. *The Unsettling of America: Culture and Agriculture*. San Francisco: Sierra Club, 1976.

Besson, Jean-Marc, and Hartmut Vogtmann, eds. *Towards a Sustainable Agriculture*. Aarau, Switzerland: Wirz-Verlag, 1978.

Bhattacharjee, Riya. "The Cheese Board at 40 Is a Vibrant Collective." *Berkeley Daily Planet*, August 21, 2007. Accessed February 28, 2011. http://www.berkeleydailyplanet.com/issue/2007-08-21/article/27821?headline=the-cheese-board-at-40-is-a-vibrant-collective.

Birch, Herbert Gee, and Joan Dye Gussow. *Disadvantaged Children: Health, Nutrition and School Failure*. New York: Harcourt, Brace and World, 1970.

Birdsall, John. "Muddled No More: Thirty Years in the Making, Mudd's Has Become a Destination." *East Bay Express*, July 25, 2007. Accessed August 8, 2010. http://www.eastbayexpress.com/ebx/muddled-no-more/Content?oid=1083870.

Bittman, Mark. "The Master of Meat." *Wine Spectator*, November 15, 2000. Accessed February 28, 2011. http://www.winespectator.com/webfeature/show/id/The-Master-of-Meat-_4.

Bjerklie, Steve. "Niman's New Model." *Meat Processing*, February 25, 2004.

Black Panther Party Research Project. "History of the Black Panther Party, Black Panther Party Platform and Program: 'What We Want, What We Believe.'" *The Black Panther*, Nov. 23, 1963. Accessed February 23, 2011. http://www.stanford.edu/group/blackpanthers/history.shtml.

Black, Helen, ed. *The Berkeley Co-Op Food Book: Eat Better and Spend Less*. Palo Alto: Bull, 1980.

Black, Jane. "A Squeeze for Tomato Growers: Boycott vs. Higher Wages." *Washington Post*, April 29, 2009. Accessed February 28, 2011. http://www.washingtonpost.com/wp-dyn/content/article/2009/04/28/ar2009042800835.html.

Black, Jane. "Waters Gets Proof That Edible Schoolyard Works." *Washington Post*, September 23, 2010.

Blaikie, Piers M., and Harold Brookfield. *Land Degradation and Society*. New York: Methuen, 1985.

Blair, Dorothy. "The Child in the Garden: An Evaluative Review of the Benefits of School Gardening." *Journal of Environmental Education* 40 (2009): 15–38.

Blair, Madeleine. "Profiting from Eminent Domain." *East Bay Express*, January 5, 2011. Accessed September 21, 2011. http://www.eastbayexpress.com/ebx/profiting-from-eminent-domain/Content?oid=2335282.

Blood, Richard. "Results of the Ranking Process, East Shore of Tomales Bay." Draft, County of Marin Environmental Health Services Department, San Rafael, California, July 1998.

Boas, Maxwell, and Steve Chain. *Big Mac: The Unauthorized Story of McDonald's*. New York: E. P. Dutton, 1976.

Boisard, Pierre. "The Future of a Tradition: Two Ways of Making Camembert, the Foremost Cheese of France." *Food and` Foodways* 4 (1991): 173–207.

Bon Appétit Management Company Web Site. Accessed March 28, 2012. http://www.bamco.com.

Bon Appétit Management Company Foundation and United Farm Workers. *The Inventory of Farmworker Issues and Protections in the United States*. Palo Alto: Bon Appétit Management, 2011. Accessed February 17, 2012. http://bamco.com/sustainable-food-service/farmworker-inventory.

Bonacich, Edna, and Jake Wilson. *Getting the Goods: Ports, Labor, and the Logistics Revolution*. Ithaca, NY: Cornell University Press, 2008.

Booth, Kwan. "The Fine Art of Community Building." *OakBook*, March 19, 2008.

Borkowski, Liz. "Feeding the Hungry Cities: Backyard Chickens, Rooftop Gardens, and Vertical Farming." *ScienceBlogs* (blog). December 7, 2010. Accessed January 13, 2011. http://scienceblogs.com/thepumphandle/2010/12/feeding_the_hungry_cities.php.

Borron, Sarah Marie. *Food Policy Councils: Practice and Possibility. Community Report*. Eugene, OR: Congressional Hunger Center, 2003.

Bottle Shock. DVD. Directed by Randall Miller. Los Angeles: Freestyle Releasing and 20th Century Fox Home Entertainment, 2008.

Bourdain, Anthony. *Kitchen Confidential: Adventures in the Culinary Underbelly*. New York: Harper Perennial, 2000.

Bowe, John. "Nobodies: Does Slavery Exist in America?" *New Yorker* (April 21, 2003): 21.

Bowe, John. *Nobodies: Modern American Slave Labor and the Dark Side of the New Global Economy*. New York: Random House, 2007.

Boyd, William. "Making Meat: Science, Technology, and American Poultry Production." *Technology and Culture* 42 (2001): 631–664.

Boyd, William, and Michael Watts. "Agro-Industrial Just-in-Time: The Chicken Industry and Post-War American Capitalism." In *Globalising Food: Agrarian*

Questions and Global Restructuring, ed. David Goodman and Michael Watts, 139–165. Oxford: Basil Blackwell, 1997.

Boyle, James P., Theodore J. Thompson, Edward W. Gregg, Lawrence E. Barker, and David F. Williamson. "Projection of the Year 2050 Burden of Diabetes in the U.S. Adult Population: Dynamic Modeling of Incidence, Mortality, and Prediabetes Prevalence." *Population Health Metrics* 8 (2010): 1–12.

Branch, Taylor. *Parting the Waters: America in the King Years, 1954–63*. New York: Simon & Schuster, 1988.

Branch, Taylor. *Pillar of Fire: America in the King Years, 1963–65*. New York: Simon & Schuster, 1998.

Branch, Taylor. *At Canaan's Edge: America in the King Years, 1965–1968*. New York: Simon & Schuster, 2006.

Brand, Madeleine. "Windows on the World Owner Seeks Fresh Start." *National Public Radio,* May 30, 2002. Accessed February 1, 2011. http://www.npr.org/programs/morning/features/2002/may/noche/index.html.

Brechin, Gray A. *Imperial San Francisco: Urban Power, Earthly Ruin*. Berkeley: University of California Press, 2006.

Brown, Allison. "Counting Farmers Markets." *Geographical Review* 91 (2001): 655–674.

Brown, Stephen L., and Ulrich F. Pilz. *U.S. Agriculture: Potential Vulnerabilities (Work Unit 3525A) for the United States Office of Civil Defense, Office of the Secretary of the Army*. Menlo Park, CA: Stanford Research Institute, 1969.

Brown, Steven A. *Revolution at the Checkout Counter: The Explosion of the Bar Code*. Cambridge, MA: Harvard University Press, 1997.

Brown, Steven E. F. "Ahead of the Herd." *San Francisco Business Times*, February 15, 2002.

Brown, William J. "Niman Ranch—A Natural Meat Processor Case Study." *International Food and Agribusiness Management Review* 3 (2000): 403–421.

Brucato, John G. *The Farmer Goes to Town: The Story of San Francisco's Farmers' Market*. San Francisco: Burke, 1948.

Brusco, Sebastiano. "The Emilian Model: Productive Decentralisation and Social Integration." *Cambridge Journal of Economics* 6 (1982): 167–184.

Bruske, Ed. "New Report Challenges Dairy Campaign Promoting Chocolate Milk in Schools," *Chef Ann Cooper Blog* (blog), February 22, 2011. Accessed February 28, 2011. http://www.chefann.com/blog/archives/category/general.

Buck, Daniel, Christy Getz, and Julie Guthman. *Consolidating the Commodity Chain: Organic Farming and Agribusiness in Northern California*. Oakland, CA: Institute for Food and Development Policy, 1996.

Buck, Daniel, Christy Getz, and Julie Guthman. "From Farm to Table: The Organic Vegetable Commodity Chain of Northern California." *Sociologia Ruralis* 37 (1997): 3–20.

Buechler, Steven M. "New Social Movement Theories." *Sociological Quarterly* 36 (1995): 441–464.

Buechler, Steven M. *Social Movements in Advanced Capitalism*. New York: Oxford University Press, 2000.

Buford, Bill. *Heat: An Amateur's Adventures as Kitchen Slave, Line Cook, Pasta-Maker, and Apprentice to a Dante-Quoting Butcher in Tuscany*. New York: Knopf, 2006.

Bullard, Robert D. *Dumping in Dixie: Race, Class, and Environmental Quality*. Boulder, CO: Westview Press, 1990.

Bullard, Robert D., and Glenn S. Johnson. *Just Transportation: Dismantling Race and Class Barriers to Mobility*. Gabriola Island, BC: New Society, 1997.

Bullard, Robert D., and Glenn S. Johnson. "Environmental Justice: Grassroots Activism and Its Impact on Public Policy Decisions." *Journal of Social Issues* 56 (2000): 555–578.

Burch, David, and Geoffrey Lawrence, eds. *Supermarkets and Agri-Food Supply Chains*. Cheltenham, UK: Edward Elgar, 2007.

Burros, Marian. "Cowgirls Going Home." *New York Times*, October 5, 2005. Accessed February 28, 2011. http://www.nytimes.com/2005/10/05/dining/05chee.html.

Burt, Cecily. "Regulators Go Too Easy on Yeast Factory, Neighbors Say." *Oakland Tribune*, December 18, 2002. Accessed February 2011. http://www.greenaction.org/westoakland/press/oaktrib121802.shtml.

Burt, Justine. "Forage Oakland." *Greenwala* (blog), November 26, 2008. Accessed February 28, 2011. http://www.greenwala.com/channels/other/blog/596-Forage-Oakland.

Busch, Lawrence, and Carmen Bain. "New! Improved? The Transformation of the Global Agrifood System." *Rural Sociology* 69 (2004): 321–346.

Busch, Lawrence, and William B. Lacy, eds. *Food Security in the United States*. Boulder, CO: Westview Press, 1984.

Butler, C. T. Lawrence, and Keith McHenry. *Food Not Bombs: How to Feed the Hungry and Build Community*. Philadelphia: New Society, 1992.

Butler, L. J., and Christopher A. Wolf. "California Dairy Production: Unique Policies and Natural Advantages." *Research in Rural Sociology* 8 (2000): 141–161.

California Department of Parks and Recreation. *Office of Historic Preservation. Five Views: An Ethnic Historic Site Survey for California. A History of American Indians in California: 1849–1879*. Sacramento: Office of Historic Preservation, California Department of Parks and Recreation, 1998.

California Health Benefits Review Program. *2010 Analysis of Senate Bill 1104: Diabetes-Related Complications: A Report to the 2009–2010 California Legislature. Report to California State Legislature*. Oakland: California Health Benefits Program, 2010.

California Human Development Corporation. *Evaluation of the Need for Ranch Worker Housing in Marin County, California*. Santa Rosa: California Human Development Corporation, 2008.

California State Planning Board. "A Proposal for the Formation of a Temporary Regional Planning Organization for the San Francisco Bay Area." Sacramento: California State Planning Board, 1941.

Cammack, Carol, Tina M. Waliczek, and Jayne M. Zajicek. "The Green Brigade: The Educational Effects of a Community-Based Horticultural Program on the Horticultural Knowledge and Environmental Attitude of Juvenile Offenders." *HortTechnology* 12 (2002a): 77–81.

Cammack, Carol, Tina M. Waliczek, and Jayne M. Zajicek. "The Green Brigade: The Psychological Effects of a Community-Based Horticultural Program on the Self-Development Characteristics of Juvenile Offenders." *HortTechnology* 12 (2002b): 82–86.

Campbell, Thomas Monroe. *The Movable School Goes to the Negro Farmer*. Tuskegee, AL: Tuskegee Institute, 1936.

Canaris, Irene. "Growing Foods for Growing Minds: Integrating Gardening and Nutrition Education into the Total Curriculum." *Children's Environments* 12 (1995): 134–142.

Cannon, James S. *U.S. Container Ports and Air Pollution: A Perfect Storm. An Energy Futures, Inc. Study*. Boulder, CO: Energy Futures, 2008.

Carman, Hoy F. "Tax Loss Farming: A Perennial Problem." *California Agriculture* 32 (1978): 12–14.

Carpenter, Novella. *Farm City: The Education of an Urban Farmer*. New York: Penguin Press, 2009.

Carpenter, Novella, and Willow Rosenthal. *The Essential Urban Farmer*. New York: Penguin, 2011.

Carson, Rachel. *Silent Spring*. Boston: Houghton Mifflin, 1962.

Case, John. "The Ultimate Employee Buy-in." *Inc Magazine,* December 2005. Accessed July 28, 2011. http://www.inc.com/magazine/20051201/employee-stock_pagen_4.html.

Casey, Conor. "San Francisco's 'Butchertown' in the 1920s and 1930s: A Neighborhood Social History." *Argonaut* 18 (2007). Accessed February 28, 2011. http://foundsf.org/index.php?title=butchertown%27s_beginnings.

Casey, Dan. "Freitas Family Center in Olema: Urging Mexicans to Speak Up for Their Rights." *Point Reyes Light,* May 3, 1990.

Cashulin, Kyle. "Cost of Organic Livestock Purity." *Point Reyes Light,* February 24, 2011.

Castells, Manuel. *The Power of Identity*. 2nd ed. London: Blackwell, 2004.

Center for Urban Education about Sustainable Agriculture. *Does Ferry Plaza Farmers Market Produce Cost More?* San Francisco: Center for Urban Education about Sustainable Agriculture, 2009.

Centro Campesino, Inc. Web Site. Accessed March 28, 2012. http://www.centrocampesino.net.

Chan, Sucheng. *This Bittersweet Soil: The Chinese in California Agriculture, 1860–1910*. Berkeley: University of California Press, 1987.

Chapman, Peter. *Bananas! How the United Fruit Company Shaped the World*. New York: Canongate U.S., 2009.

Charlotte Observer. "Observer Special Report: The Cruelest Cuts." *Charlotte Observer*, February 10–15, 2008. Accessed February 28, 2011. http://www.charlotteobserver.com/poultry.

Cheese Board Collective Staff. *The Cheese Board: Collective Works: Bread, Pastry, Cheese, Pizza*. Berkeley: Ten Speed Press, 2003.

Chen, Jim. "Get Green or Get Out: Decoupling Environmental from Economic Objectives in Agricultural Regulation." *Oklahoma Law Review* 48 (1995): 333–352.

Child, Julia, Louisette Bertholle, and Simone Beck. *Mastering the Art of French Cooking*, Vol. 1. New York: Knopf, 1961.

Childs, Williams R. *Trucking and the Public Interest: The Emergence of Federal Regulation, 1914–1940*. Knoxville: University of Tennessee Press, 1986.

Chin, Teresa, and Whitney Pennington. "Oakland Students, Administrators Try Organic Lunches." *Oakland North*, October 15, 2010. Accessed September 17, 2011. http://oaklandnorth.net/2010/10/15/oakland-students-administrators-try-organic-school-lunches.

Chitewere, Tendai. "Equity in Sustainable Communities: Exploring Tools from Environmental Justice and Political Ecology." *Natural Resources Journal* 50 (2010): 315–340.

Clancy, Kate, and Nadine Lehrer. "A Priority Research Agenda for Agriculture of the Middle." *Agriculture of the Middle*, May 2010. Accessed January 1, 2011. http://www.agofthemiddle.org/pubs/AOTM_research.pdf.

Clark, Eleanor. *The Oysters of Locmariaquer*. New York: Pantheon Books, 1965.

Clarke, Sally. *Regulation and the Revolution in United States Farm Productivity*. Cambridge: Cambridge University Press, 1994.

Clawson, Marion, and Jack L. Knetsch. *The Economics of Outdoor Recreation*. Baltimore, MD: John Hopkins Press for Resources for the Future, 1966.

Coalition of Immokalee Workers Web Site. Accessed March 28, 2012. http://ciw-online.org.

Cobia, David W., ed. *Cooperatives in Agriculture*. Upper Saddle River, NJ: Prentice Hall, 1989.

Cochrane, Willard Wesley, and Mary Ellen Ryan. *American Farm Policy, 1948–1973*. Minneapolis: University of Minnesota Press, 1976.

Coe, Andrew. *Chop Suey: A Cultural History of Chinese Food in the United States*. New York: Oxford University Press, 2009.

Cohen, Lizabeth. *A Consumers' Republic: The Politics of Mass Consumption in Postwar America*. New York: Knopf, 2003.

Colborn, Teo, Dianne Dumanoski, and John Meyers. *Our Stolen Future: Are We Threatening Our Fertility, Intelligence, and Survival? A Scientific Detective Story*. New York: Plume, 1996.

Cole, Luke. "Environmental Justice Litigation: Another Stone in David's Sling." *Fordham Urban Law Journal* 21 (1993): 523–545.

Community Food Security Coalition. "Community Food Security Coalition." 2011a. Accessed February 28, 2011. http://www.foodsecurity.org.

Community Food Security Coalition. "Weaving the Food Web." 2011b. Accessed January 14, 2011. http://www.foodsecurity.org/CFSCguide-foodweb.pdf

Community Food Security Coalition. "What Is Community Food Security?" 2011c. Accessed January 17, 2011. http://www.foodsecurity.org/views_cfs_faq.html.

Conkin, Paul K. *A Revolution Down on the Farm: The Transformation of American Agriculture since 1929*. Lexington: University Press of Kentucky, 2009.

Conlin, Joseph. *Bacon, Beans, and Galantines: Food and Foodways on the Western Mining Frontier*. Reno: University of Nevada Press, 1986.

Coodley, Lauren, ed. *The Land of Orange Grooves and Jails: Upton Sinclair's California*. Berkeley, CA: Heyday Books, 2004.

Cook, Christopher. "The Green Link: Jobs and Education." *Slug Update*, Spring 1994. Accessed August 18, 2011. http://www.alemanyfarm.org/pdf/slug_update_spring_94.pdf.

Cooper, Ann, and Lisa Holmes. *Lunch Lessons: Changing the Way We Feed Our Children*. New York: HarperCollins, 2006.

Cooperative Whole Grain Educational Association. *Uprising: The Whole Grain Bakers Book*. North Carolina: Mother Earth News, 1983.

Cornell University Cooperative Extension. "Pesticide Management Education Program. The Delaney Paradox and Negligible Risk." 2008. Accessed September 7, 2010. http://pmep.cce.cornell.edu.

Cornucopia Institute. "New USDA Rules Establish Strong Organic Standards for Pasture and Livestock." Cornucopia Institute, February 12, 2010a. Accessed July 20, 2010. http://www.cornucopia.org/2010/02/new-usda-rules-establish-strong-organic-standards-for-pasture-and-livestock.

Cornucopia Institute. "Organic Family Dairies Being Crushed by Rogue Factory Farms." Cornucopia Institute, January 24, 2010. Accessed July 20, 2010. http://www.cornucopia.org/2010/01/organic-family-dairies-being-crushed-by-rogue-factory-farms.

Cowan, Tadlock, and Jody Feder. *The Pigford Case: USDA Settlement of a Discrimination Suit by Black Farmers*. Washington, DC: Congressional Research Service, 2010.

Cox, Craig. *Storefront Revolution: Food Co-Ops and the Counterculture*. New Brunswick, NJ: Rutgers University Press, 1994.

Coyote, Peter. *Sleeping Where I Fall: A Chronicle*. Berkeley, CA: Counterpoint, 1998.

Cronon, William. *Nature's Metropolis: Chicago and the Great West*. New York: Norton, 1991.

Cross, Jennifer. *The Supermarket Trap: The Consumer and the Food Industry*. Bloomington: Indiana University Press, 1976.

Cuellar, Amanda D., and Michael E. Webber. "Wasted Food, Wasted Energy: The Embedded Energy in Food Waste in the United States." *Environmental Science and Technology* 44 (2010): 6464–6469.

Culinary Institute of America. "The Story of the World's Premier Culinary College—A History of Excellence, Professional Advancement, and Innovation," 2010. Accessed January 12, 2011. http://www.ciachef.edu/about/history.asp.

Cummins, Steven, and Sally Macintyre. "'Food Deserts': Evidence and Assumption in Health Policy Making." *British Medical Journal* 325 (2002): 436–438.

Curl, John. *For All the People: Uncovering the Hidden History of Cooperation, Cooperative Movements, and Communalism in America*. Oakland, CA: PM, 2009.

Curl, John. *History of Work Cooperation in America: Worker Cooperatives vs. Wage Slavery*. Berkeley, CA: Homeward, 1980.

Curtin, Philip D. *The Rise and Fall of the Plantation Complex: Essays in Atlantic History*. Cambridge: Cambridge University Press, 1998.

Cutler-Mackenzie, Amy. "Multicultural School Gardens: Creating Engaging Garden Spaces in Learning about Language, Culture, and the Environment." *Canadian Journal of Environmental Education* 14 (2009): 122–135.

Dahlberg, Kenneth A. "Food Policy Councils: The Experience of Five Cities and One County." Paper presented at the Joint Meeting of the Agriculture, Food, and Human Values Society and the Society for the Study of Food and Society, Western Michigan University, Tucson, AZ, June 1994.

Dalton, Rex. "Berkeley's Energy Deal with BP Sparks Unease." *Nature* 445 (2007): 688–689.

Dana, Richard Henry, Jr. *Two Years before the Mast and Other Voyages*. New York: Library of America, 2005.

Danborn, David B. *Resisted Revolution: Urban America and the Industrialization of Agriculture*. Ames: Iowa State University Press, 1979.

Daniel, Cletus E. *Bitter Harvest: A History of California Farmworkers, 1870–1941*. Ithaca, NY: Cornell University Press, 1981.

Davenport, Eugene. "Scientific Farming." *Annals of the American Academy of Political and Social Science* 40 (1912): 45–50.

David, Elizabeth. *French Provincial Cooking*. New York: Harper and Row, 1962.

David, Wier, and Mark Shapiro. *Circle of Poison*. Oakland, CA: Food First Books, 1981.

Davis, Allison. "Le Truc 'Bustaurant' Seats 12, Sells out Fast." *Wired*, November 30, 2010. Accessed January 13, 2011. http://www.wired.com/underwire/2010/11/le-truc/?pid=2135.

Davis, Mike. *Late Victorian Holocausts: El Niño Famines and the Making of the Third World*. London: Verso, 2001.

Dean, Virgil. *An Opportunity Lost: The Truman Administration and the Farm Policy Debate*. Columbia: University of Missouri Press, 2006.

DeBraun, Morgan. "Tomatoes on Campus." *Washington University in St. Louis Student Union Blog* (blog), November 17, 2009. Accessed June 29, 2011. http://su.wustl.edu/blog/no-more-tomatoes.

DeFao, Janine. "Schlepping Off to Market: West Oakland Shoppers Can Hardly Wait for New Store to Open." *San Francisco Chronicle,* July 28, 1999. Accessed September 21, 2011. http://www.sfgate.com/cgi-bin/article.cgi?f=/Chronicle/a/1999/07/28/MN37247.DTL.

Derby, Samara Kalk. "Hmong Presence Growing at Farmer's Market." *Wisconsin State Journal*, July 24, 2010. Accessed July 24, 2010. http://host.madison.com/wsj/news/local/article_9097ee26-9669-11df-86d6-001cc4c002e0.html.

Desmond, Daniel, James Grieshop, and Aarti Subramaniam. *Revisiting Garden Based Learning in Basic Education*. Paris and Rome: International Institute for Educational Planning and Food and Agriculture Organization of the United Nations, 2002.

Deutsch, Tracey. "Untangling Alliances: Social Tensions Surrounding Independent Grocery Stores and the Rise of Mass Retailing." In *Food Nations: Selling Taste in Consumer Societies*, ed. Warren Belasco and Philip Scranton, 156–174. New York: Routledge, 2002.

Deutsch, Tracey. *Building a Housewife's Paradise: Gender, Politics, and American Grocery Stores in the Twentieth Century*. Chapel Hill: University of North Carolina Press, 2010.

Dewey, John. *Enlistment for the Farm*. New York: Division of Intelligence and Publicity of Columbia University, 1917. Accessed May 25, 2011. http://www.foodpolitics.com/wp-content/uploads/John-Dewey-Enlistment-for-the-Farm-1917-Colombia-War-Papers.pdf.

Dicum, Gregory. "Expanding the Frontiers of the Vegetarian Plate." *New York Times*, November 18, 2007. Accessed April 1, 2010. http://travel.nytimes.com/2007/11/18/travel/18Choice.html.

Dicum, Gregory. "At Pop-Ups, Chefs Take Chances with Little Risk." *New York Times*, February 11, 2010. Accessed February 28, 2011. http://www.nytimes.com/2010/02/12/dining/12sfdine.html?_r=2.

Digger Archives. "The Digger Papers," August 1968. Accessed November 4, 2010. http://www.diggers.org/digger_papers.htm.

Dimassa, Cara Mia. "California Endowment Broadens Ambitions and Narrows Scope." *Los Angeles Times*, November 30, 2010. Accessed February 28, 2011. http://articles.latimes.com/2010/nov/30/local/la-me-robert-ross-20101130.

Dirks, Amy, and Kathryn Orvis. "An Evaluation of the Junior Master Gardener Program in Third Grade Classroom." *HortTechnology* 15 (2005): 443–447.

Doherty, Kathleen Elizabeth. "Mediating Critiques of the Alternative Agrifood Movement: Growing Power in Milwaukee." Master's thesis, University of Wisconsin, Milwaukee, 2006.

Dolan, Kerry A. "One Man's Meat." *Forbes* (March 2003): 17.

Donnelly, James S. *The Great Irish Potato Famine*. Stroud, UK: Sutton, 2001.

Donofrio, Gregory Alexander. "Feeding the City." *Gastronomica* 7 (2007): 30–41.

Donohue, Caitlin. "Mandela Food Cooperative Gets the Redford Nod." *San Francisco Bay Guardian,* June 4, 2010. Accessed September 28, 2011. http://www.sfbg.com/pixel_vision/2010/06/04/mandela-food-cooperative-gets-redford-nod.

Douglas, Paul H. "Book Reviews and Notices: Review of James P. Warbasse: Co-Operative Democracy." *Journal of Political Economy* 32 (1924): 383–385.

Dowie, Mark. "Down on the Farm." *Image Magazine, San Francisco Examiner,* July 29, 1990.

Dowie, Mark. "Conservation Refugees." *Orion Magazine*, November–December 2005.

Drew, Jesse. "Call Any Vegetable: The Politics of Food in San Francisco." In *Reclaiming San Francisco: History, Politics, Culture,* ed. James Brook, Chris Carlsson, and Nancy Joyce Peters, 317–331. San Francisco: City Lights Books, 1998.

Dubbink, David. "I'll Have My Town Medium-Rural, Please." *Journal of the American Institute of Planners* 50 (1984): 406–418.

Dubose, René. *The Mirage of Health: Utopia, Progress, and Biological Change*. New York: Harper and Row, 1959.

DuPuis, E. Melanie. "Not in My Body: RBGH and the Rise of Organic Milk." *Agriculture and Human Values* 17 (2000): 285–295.

DuPuis, E. Melanie. *Nature's Perfect Food: How Milk Became America's Drink*. New York: NYU Press, 2002.

DuPuis, E. Melanie, and Daniel Block. "Sustainability and Scale: US Milk-Market Orders as Relocalization Policy." *Environment and Planning A* 40 (2008): 1987–2005.

DuPuis, E. Melanie, and Sean Gillon. "Alternative Modes of Governance: Organic as Civic Engagement." *Agriculture and Human Values* 26 (2009): 43–56.

Dyble, Louise Nelson. "Revolt against Sprawl: Transportation and the Origins of the Marin County Growth Control Regime." *Journal of Urban History* 34 (2007): 38–66.

Dyble, Louise Nelson. "California's Exceptional Farmers' Markets: A Story of Regulation and its Consequences." Paper presented at the Business History Conference, Sacramento CA, April 2008.

Dyble, Louise Nelson. *Paying the Toll: Local Power, Regional Politics, and the Golden Gate Bridge*. Philadelphia: University of Pennsylvania Press, 2009.

Dyl, Joanna. "The War on Rats versus the Right to Keep Chickens: Plague and the Paving of San Francisco, 1907–1908." In *In the Nature of Cities: Culture, Landscape and Urban Space*, ed. Andrew C. Isenberg, 38–61. Rochester, NY: University of Rochester Press, 2006.

Eat at Bill's: Life in the Monterey Market. DVD. Directed by Lisa Brenneis, 2007.

Ecology Center. *A Simple Guide for Electronic Benefits Transfer (EBT) of Food Stamp Benefits at California Farmers' Markets Using a Central Point of Sale (POS) Device and Market Scrip*. Berkeley: Ecology Center, 2008. Accessed January 17, 2011. http://www.ecologycenter.org/ebt/pdf/simpleguide2008.pdf.

Ecology Center Web Site. Accessed March 28, 2012. http://www.ecologycenter.org/ffc.

Econsult Corporation and Penn Institute for Urban Research. *Vacant Land Management in Philadelphia: The Costs of the Current System and the Benefits of Reform*. Philadelphia: Econsult, 2010.

Edgar, Gordon. *Cheesemonger: A Life on the Wedge*. White River Junction, VT: Chelsea Green, 2010.

Ehrlich, Paul. *The Population Bomb*. New York: Ballantine Books, 1968.

Eisenhauer, Elizabeth. "In Poor Health: Supermarket Redlining and Urban Nutrition." *GeoJournal* 53 (2001): 125–133.

Ellick, Adam B. "Boulud Settling Suit Alleging Bias at a French Restaurant." *New York Times*, July 31, 2007. Accessed February 28, 2011. http://www.nytimes.com/2007/07/31/nyregion/31daniel.html.

Engelhardt, Zephryin. *The Missions and Missionaries of California*. San Francisco: James H. Barry, 1908.

Engelhardt, Zephryin. *The Franciscans in California*. Harbor Springs, MI: Holy Childhood Indian School, 1897.

Estabrook, Barry. "Politics of the Plate: The Price of Tomatoes." *Gourmet Magazine*, March 2009. Accessed February 28, 2011. http://www.gourmet.com/magazine/2000s/2009/03/politics-of-the-plate-the-price-of-tomatoes.

Fagan, Kevin. "Bus Ride to the Checkout Lane." *San Francisco Chronicle*, October 15, 1995a. Accessed September 21, 2011. http://www.sfgate.com/cgi-bin/article.cgi?f=/c/a/1995/10/15/SC9400.DTL.

Fagan, Kevin. "Where Markets Aren't So Super." *San Francisco Chronicle*, October 15, 1995b. Accessed September 21, 2011. http://www.sfgate.com/cgi-bin/article.cgi?f=/Chronicle/a/1995/10/15/SC11161.DTL.

Fairfax, Sally K., Lauren Gwin, Mary Ann King, Leigh Raymond, and Laura A. Watt. *Buying Nature: The Limits of Land Acquisition as a Conservation Strategy, 1780–2004*. Cambridge, MA: MIT Press, 2005.

Fallon, Sally. *Nourishing Traditions: The Cookbook That Challenges Politically Correct Nutrition and Diet Dictocrats*. Washington, DC: NewTrends, 1999.

Fanon, Frantz. *Peau Noire, Masques Blancs.* [Black Skin, White Masks] Paris: Editions Du Seuil, 1952.

Fanon, Frantz. *Les Damnés de la Terre.* [The Wretched of the Earth] Paris: François Maspero, 1961.

Feder, Barnaby J. "Nanotech Food Is Potentially Ten Times More Dangerous Than Genetically Engineered." *New York Times*, October 10, 2006. Accessed February 28, 2011. http://www.infowars.com/articles/science/nanotech_food_10x_deadlier_gm_crops.htm.

Federal Trade Commission. *Economic Report on Food Chain Selling Practices in the District of Columbia and San Francisco.* Washington, DC: U.S. Government Printing Office, 1969.

Feenstra, Gail, Sharyl McGrew, and David Campbell. *Entrepreneurial Community Gardens: Growing Food, Skills, Jobs, and Communities.* Oakland, CA: Division of Agriculture and Natural Resources, 1999.

Fernandez, Sarah. *Farmers Markets and Food Co-Ops: Potential Markets for Bay Area Produce.* San Francisco: People for Open Space Farmlands Conservation Project, 1981.

Fine, Janice Ruth. *Worker Centers: Organizing Workers at the Edge of the Dream.* Ithaca, NY: Cornell University Press, 2006.

First National People of Color Environmental Leadership Summit. "The Principles of Environmental Justice (EJ)" October 24–27, 1991. Accessed January 13, 2011. http://www.ejnet.org/ej/principles.pdf.

Fisher, Andy. *Hot Peppers and Parking Lot Peaches: Evaluating Farmers' Markets in Low Income Communities.* San Francisco, Community Food Security Coalition, 1999. Accessed February 11, 2012. http://thrive.preventioninstitute.org/sa/enact/neighborhood/documents/community.farmersmarkets.tools.hotpepperspeaches.pdf.

Fisher, Mary Frances Kennedy. *Serve It Forth.* New York: Harper and Brothers, 1937.

Fisher, Mary Frances Kennedy. *Consider the Oyster.* New York: Duell, Sloan and Pearce, 1941.

Fisher, Mary Frances Kennedy. *How to Cook a Wolf.* New York: Duell, Sloan and Pearce, 1942.

Fisher, Mary Frances Kennedy. *The Gastronomical Me.* New York: Duell, Sloan and Pearce, 1943.

Fisher, Mary Frances Kennedy. *Here Let Us Feast: A Book of Banquets.* New York: Viking Press, 1946.

Fisher, Michael. "Environmental Racism Claims Brought under Title VI of the Civil Rights Act." *Environmental Law (Northwestern School of Law)* 25 (1995): 285–334.

Fitzgerald, Deborah Kay. *Every Farm a Factory: The Industrial Ideal in American Agriculture.* New Haven, CT: Yale University Press, 2003.

Flaccavento, Anthony. "Walmart and the End of the Local Food Movement." *Huffington Post* (blog), October 26, 2010. Accessed October 26 2010. http://www.huffingtonpost.com/anthony-flaccavento/walmart-and-the-end-of-th_b_774350.html.

Flagler, Joel. "The Role of Horticulture in Training Correctional Youth." *HortTechnology* 5 (1995): 185–187.

Flammang, Janet A. *The Taste for Civilization: Food, Politics, and Civil Society*. Champaign: University of Illinois Press, 2009.

Flanagan, Caitlin. "Cultivating Failure: How School Gardens Are Cheating Our Most Vulnerable Students." *Atlantic*, January–February 2010. Accessed February 28, 2011. http://www.theatlantic.com/magazine/archive/2010/01/cultivating-failure/7819.

Fold, Niels, and Bill Pritchard, eds. *Cross-Continental Agro-Food Chains*. London: Routledge, 2005.

Folke, Carl, Steve Carpenter, Thomas Elmqvist, and Lance Gunderson. "C. S. Holling, and Brian Walker. "Resilience and Sustainable Development: Building Adaptive Capacity in a World of Transformation." *AMBIO: A Journal of the Human Environment* 31 (2002): 437–440.

Fong-Torres, Ben. *The Rice Room: Growing up Chinese-American: From Number Two Son to Rock 'N Roll*. New York: Hyperion, 1994.

Food and Agriculture Organization of the United Nations. "Rome Declaration on World Food Security," November 13–17, 1996. Accessed January 13, 2011. http://www.fao.org/Docrep/003/W3613e/W3613e00.htm.

Food and Agriculture Organization of the United Nations. *Trade Reforms and Food Security: Conceptualizing the Linkages*. Rome: FAO, 2003. Accessed February 20, 2012. ftp://ftp.fao.org/docrep/fao/005/y4671e/y4671e00.pdf.

Food and Agriculture Organization of the United Nations. "Supporting Programmes for Food Security." 2010. Accessed March 28, 2012. http://www.fao.org/spfs/en/

Food and Agriculture Organization of the United Nations. *Save and Grow: A Policymaker's Guide to the Sustainable Intensification of Smallholder Crop Production*. Rome: FAO, 2011. Accessed February 20, 2012. http://www.fao.org/docrep/014/i2215e/i2215e.pdf.

Food First. "Global Small-Scale Farmers' Movement Developing New Trade Regimes." *It's Time to Defeat CAFTA, Food First News and Views* 28 (Spring–Summer 2005).

Food, Inc. DVD. Directed by Robert Kenner. Los Angeles and New York: Participant Media and Magnolia Pictures, 2008.

Food Marketing Institute. "Celebrating 75 Years of Supermarkets," 2005. Accessed May 1, 2010. http://www.fmi.org/facts_figs/?fuseaction=75_anniversary.

"Food Movements Unite! Challenges for the Local-Global Transformation of Our Food Systems." Panel discussion with Raj Patel, Nora McKeon, and Nikki Henderson, University of California, Berkeley, Berkeley, Calif., November 2010.

Food Not Bombs Web site. Accessed March 28, 2012. http://www.foodnotbombs.net.

Food Trust. *The Need for More Supermarkets in Philadelphia*. Philadelphia: Food Trust, 2001.

Fowler, Damon Lee. "Dining at Monticello." In *Good Taste and Abundance*, ed. Damon Lee Fowler. Chapel Hill: University of North Carolina Press, 2005.

Fradkin, Philip L. *A River No More: The Colorado River and the West*. Berkeley: University of California Press, 1996.

Francis, Mark. "Children's Garden." *Children's Environments* 12 (1995): 183–191.

Frederick, Donald A. *Co-Ops 101: An Introduction to Cooperatives*. Washington, DC: Rural Business-Cooperative Service, U.S. Department of Agriculture, 1997.

Freidberg, Susanne. *French Beans and Food Scares: Culture and Commerce in an Anxious Age*. New York: Oxford University Press, 2004.

Freidberg, Susanne. *Fresh: A Perishable History*. Cambridge, MA: Harvard University Press, 2009.

French, Michael, and Jim Phillips. *Cheated, Not Poisoned? Food Regulation in the United Kingdom, 1875–1938*. Manchester: Manchester University Press, 2000.

Friedland, William H. "Introduction: Shaping the New Political Economy of Advanced Capitalist Agriculture." In *Towards a New Political Economy of Agriculture*, ed. William H. Friedland, Lawrence Busch, Frederick H. Buttel, and Alan Rudy, 1–34. Boulder, CO: Westview, 1991.

Friedland, William H. "The Global Fresh Fruit and Vegetable System: An Industrial Organization Analysis." In *The Global Restructuring of Agro-Food Systems*, ed. Philip McMichael, 171–172. Ithaca, NY: Cornell University Press, 1994.

Friedmann, Barton, Amy E. Barton, and Robert J. Thomas. *Manufacturing Green Gold: Capital, Labor, and Technology in the Lettuce Industry*. Cambridge: Cambridge University Press, 1981.

Friedmann, Harriet. "The Political Economy of Food: The Rise and Fall of the Postwar International Order." *American Journal of Sociology* 88 (1982): 248–286.

Friedland, Harriet. "Distance and Durability: Shaky Foundations of the World Food Economy." In *The Global Restructuring of Agro-Food Systems*, ed. Philip McMichael, 258–276. Ithaca, NY: Cornell University Press, 1994.

Fromartz, Samuel. *Organic, Inc.: Natural Foods and How They Grew*. Toronto: Harcourt, 2006.

Fukuoka, Masanobu. *The One-Straw Revolution: An Introduction to Natural Farming*. Emmaus, PA: Rodale, 1978.

Fulbright, Leslie. "As PG&E Closes Its Old, Smoky Power Plant, the Neighborhood Breathes a Sigh of Relief." *San Francisco Chronicle*, May 15, 2006. Accessed December 5, 2010. http://www.greenaction.org/hunterspoint/press/sfgate051506.shtml.

Fulda, Carl H. "Food Distribution in the United States: The Struggle between Independents and Chains." *University of Pennsylvania Law Review* 99 (1951): 1051–1161.

Fullerton, Michael, ed. *What Happened to the Berkeley Co-Op? A Collection of Opinions*. Davis, CA: Center for Cooperatives, 1992.

FundingUniverse. "The Vons Companies, Incorporated." Accessed September 7, 2010. http://www.fundinguniverse.com/company-histories/the-vons-companies-incorporated-company-history.html.

Galarza, Ernesto. *Merchants of Labor: The Mexican Bracero Story: An Account of the Managed Migration of Mexican Farm Workers in California 1942–1960*. San Jose: Rosicrucian, 1964.

Galarza, Ernesto. *Tragedy at Chualar*. Santa Barbara, CA: McNally and Loftin, 1977a.

Galarza, Ernesto. *Farmworkers and Agribusiness in California, 1947–1960*. South Bend, IN: University of Notre Dame Press, 1977b.

Gallagher, John. *Reimagining Detroit: Opportunities for Redefining an American City*. Detroit: Wayne State University Press, 2010.

Gardner, Bruce L. *American Agriculture in the Twentieth Century: How It Flourished and What It Cost*. Cambridge, MA: Harvard University Press, 2002.

Garrett, Laurie. *The Coming Plague: Newly Emerging Diseases in a World out of Balance*. New York: Farrar, Straus, and Giroux, 1994.

Garrett, Laurie. *Betrayal of Trust: The Collapse of Global Public Health*. New York: Hyperion, 2000.

Gates, Paul Wallace. *History of Public Land Law Development*. Washington, DC: Zenger, 1968.

Gereffi, Gary, and Miguel Korzeniewicz. *Commodity Chains and Global Capitalism*. Westport, CT: Praeger, 1994.

Gerhart, Matthew. *Accessing Conservation Resources: Land and Politics in Marin and Sonoma Counties*. Berkeley: College of Natural Resources, University of California, 2002.

Getz, Christy, Sandy Brown, and Aimee Shreck. "Class Politics and Agricultural Exceptionalism in California's Organic Agriculture Movement." *Politics and Society* 36 (2008): 478–507.

Getz, Christy, and Aimee Shreck. "What Organic and Fair Trade Labels Do Not Tell Us: Towards a Place-Based Understanding of Certification." *International Journal of Consumer Studies* 30 (2006): 490–501.

Giertz, J. Fred, and Dennis H. Sullivan. "Food Assistance Programs in the Reagan Administration." *Publius* 16 (1986): 133–147.

Gilbert, Jess, and Kevin Wehr. "Increasing Structural Divergence in U.S. Dairying: California and Wisconsin since 1950." *Rural Sociology* 53 (1988): 56–72.

Gillespie, Pat. "Sixth Immokalee Slavery Case Suspect Arrested, Group Accused of Keeping, Beating, Stealing from Immokalee Laborers." *Fort Myers News-Press*, January 18, 2008. Accessed February 28, 2011. http://www.ciw-online.org/slavery_plain_and_simple.html.

Gilliam, Harold. *Island in Time: The Point Reyes Peninsula*. New York: Sierra Club, Scribner, 1974.

Gimenez, Eric H. "The Fight over Food Deserts—Corporate America Smacks Its Way Down." *Huffington Post* (blog), July 14, 2010. Accessed February 28, 2011. http://www.huffingtonpost.com/eric-holt-gimenez/the-fight-over-food-deser_b_646849.html.

Giroux, Henry A. *The University in Chains: Confronting the Military-Industrial-Academic Complex*. Boulder, CO: Paradigm, 2007.

Gladwell, Malcolm. *The Tipping Point*. New York: Little, Brown, 2000.

Glasmeier, Amy K. *Manufacturing Time: Global Competition in the Watch Industry, 1795–2000*. New York: Guilford Press, 2000.

Glassner, Barry. "'The Only Place to Eat in Berkeley': Hank Rubin and the Pot Luck." *Gastronomica* 2 (2002): 26–31.

Glover, Troy D. "Social Capital in the Lived Experiences of Community Gardeners." *Leisure Sciences* 26 (2004): 143–162.

God's Gang. "About Us." Accessed June 29, 2011. http://godsgang1.net/aboutus.aspx.

Golden, Devin R. "Protestors Picket Publix's Saturday Grand Opening over Labor Issues. Publix's Stance: This Is Not the Company's Issue." *BaldwinCountyNow.com,* December 11, 2010. Accessed September 27, 2011. http://m.baldwincountynow.com/articles/2010/12/11/business_news/doc4d0447961c7af468635034.txt.

Goldschmidt, Walter. *As You Sow: Three Studies in the Social Consequences of Agribusiness*. Montclair, NJ: Allanheld, Osmun, and Co, 1947.

Gollan, Jennifer. "Grown in Marin Better Beef, Naturally-Grown in Marin." *Marin Independent Journal,* May 2, 2005.

Goldstein, Jon H., and Joan R. Hartmann. "Agricultural Programs." In *The Impact of Federal Programs on Wetlands*. Vol. 2 of the *Report to Congress by the Secretary of the Interior,* ed. Joan R. Hartmann and Jon H. Goldstein. Washington, DC: U.S. Department of the Interior, 1994. Accessed June 30, 2010. http://www.doi.gov/oepc/wetlands2/v2ch3.html.

Goldstein, Richard. "Earl L. Butz, Secretary Felled by Racial Remark, Is Dead at 98." *New York Times,* February 4, 2008. Accessed June 30, 2011. http://www.nytimes.com/2008/02/04/washington/04butz.html?pagewanted=print.

Goodman, David. "The Quality 'Turn' and Alternative Food Practices: Reflections and Agenda." *Journal of Rural Studies* 19 (2003): 1–7.

Goodman, David, and Michael Watts. "Reconfiguring the Rural or Fording the Divide? Capitalist Restructuring and the Global Agro-Food System." *Journal of Peasant Studies* 22 (1994): 1–49.

Goodman, David, and Michael Watts. *Globalising Food: Agrarian Questions and Global Restructuring*. New York: Routledge, 1997.

Gordon, David M. *An Impressionistic Report about Minority Unemployment in Oakland*. Oakland, CA: Oakland Collection, Ford Foundation Archives, 1966.

Gordon, Kindra. "What's in a Name?" *Beef Magazine,* April 1, 2004. Accessed May 25, 2011. http://beefmagazine.com/mag/beef_whats_name.

Gordon, Robert. "Poisons in the Fields: The United Farm Workers, Pesticides, and Environmental Politics." *Pacific Historical Review* 68 (1999): 51–77.

Gottlieb, Robert. *Forcing the Spring: The Transformation of the American Environmental Movement*. Covelo, CA: Island Press, 1996.

Gottlieb, Robert. *Environmentalism Unbound: Exploring New Pathways for Change*. Cambridge, MA: MIT Press, 2001.

Gottlieb, Robert, and Anupama Joshi. *Food Justice*. Cambridge, MA: MIT Press, 2010.

Grass Farmer Staff Report. *A Rancher's Son Pioneers His Own Grassfed Beef Business on Beautiful Point Reyes Peninsula*. Stockman Grassfarmer, 2009.

Gratz, Roberta Brandes. *The Battle for Gotham: New York in the Shadow of Robert Moses and Jane Jacobs*. New York: Nation Books, 2010.

Grazia, Alfre De, and T. Gurr. *American Welfare*. New York: New York University Press, 1961.

Greenberg, Myron. "Cattle and Taxes under the 1969 Tax Reform Act." *UCLA Law Review* 17 (1970): 1251–1279.

Greene, Maria Louise. *Among School Gardens*. New York: Charities Publication Committee, 1910.

Gregory, James N. *American Exodus: the Dust Bowl Migration and Okie Culture in California*. New York: Oxford University Press, 1991.

Griffin, L. Martin. *Saving the Marin-Sonoma Coast: The Battles for Audubon Canyon Ranch, Point Reyes, and California's Russian River*. Healdsburg, CA: Sweetwater Springs, 1998.

Griffith, David. *Jones's Minimal: Low Wage Labor in the United States*. Albany: State University of New York Press, 1993.

Grimm, Fred. "How About a Side Order of Human Rights?" *Miami Herald,* December 16, 2007. Accessed February 28, 2011. http://ciw-online.org/Fred_Grimm_Side_Order_Rigts.html.

Gross, Harriet, and Nicola Lane. "Landscapes of the Lifespan: Exploring Accounts of Own Gardens and Gardening." *Journal of Environmental Psychology* 27 (2007): 225–241.

Grossman, Margaret Rosso. "Agriculture and the Environment in the United States." *American Journal of Comparative Law* 42 (1994): 291–293.

Grossner, Morris. "Morris Grossner." 2012. Accessed February 15, 2012. http://www.columbia1968.com/morrisgrossner.

Guerrero, Marsha. Edible Schoolyard Tour. January 6, 2011.

Gunderson, Gordon W. "The National School Lunch Program Background and Development." Washington, DC: United States Department of Agriculture, 2012. Accessed January 12, 2010. http://www.fns.usda.gov/cnd/lunch/aboutlunch/programhistory_6.htm.

Guthman, Julie, Amy W. Morris, and Patricia Allen. "Squaring Farm Security and Food Security in Two Types of Alternative Food Institutions." *Rural Sociology* 71 (2006): 662–685.

Gutfreund, Owen D. *Twentieth Century Sprawl: Highways and the Reshaping of the American Landscape*. New York: Oxford University Press, 2004.

Guthey, Greig Tor, Lauren Gwin, and Sally K. Fairfax. "Creative Preservation in California's Dairy Industry." *Geographical Review* 93 (2003): 171–192.

Guthey, Greig Tor. "Terroir and the Politics of Agro-Industry in California's North Coast Wine District." PhD diss., University of California, Berkeley, 2004.

Guthman, Julie. *Agrarian Dreams: The Paradox of Organic Farming in California*. Berkeley: University of California Press, 2004.

Gutierrez, Ramon, and Richard Orsi, eds. *Contested Eden: California Before the Gold Rush*. Berkeley: University of California Press, 1998.

Gwin, Lauren. "New Pastures, New Food: Building Viable Alternatives to Conventional Beef." Ph.D. diss., University of California, Berkeley. 2006.

Gwin, Lauren. "Scaling-up Sustainable Livestock Production: Innovation and Challenges for Grass-Fed Beef in the U.S." *Journal of Sustainable Agriculture* 33 (2009): 189–209.

Gwynne, Robert N. "UK Retail Concentration, Chilean Wine Producers and Value Chains." *Geographical Journal* 174 (2008): 97–108.

Hackel, Steven. *Children of Coyote, Missionaries of St. Francis: Indian-Spanish Relations in Colonial California, 1769–1850*. Chapel Hill: University of North Carolina Press, 2005.

Haight, Anna. "Bread and Butter: Now Everyone Can Dig in at the Fork at Point Reyes." *Marin Independent Journal*, July 13, 2010. Accessed February 28, 2011. http://www.marinij.com/lifestyles/ci_15510679.

Hall, Kevin D., Juen Guo, Michael Dore, and Carson C. Chow. "The Progressive Increase of Food Waste in America and Its Environmental Impact." *PLoS ONE* 4 (2009). Accessed August 17, 2011. http://doi:10.1371/journal.pone.0007940.

Halweil, Brian. *Eat Here: Reclaiming Homegrown Pleasures in a Global Supermarket*. Washington, DC: WorldWatch Institute, 2004.

Hamilton, Lisa. *Deeply Rooted: Unconventional Farmers in the Age of Agribusiness*. Berkeley, CA: Counterpoint, 2009.

Hamilton, Shane. "Cold Capitalism: The Political Ecology of Frozen Concentrated Orange Juice." *Agricultural History* 77 (2003): 557–581.

Hamilton, Shane. "The Economies and Conveniences of Modern-Day Living: Frozen Foods and Mass Marketing, 1945–1965." *Business History Review* 77 (2003): 33–60.

Hamilton, Shane. *Trucking Country: The Road to America's Walmart Economy*. Princeton, NJ: Princeton University Press, 2008.

Hansen, John Mark. *Gaining Access: Congress and the Farm Lobby: 1919–1981*. Chicago: University of Chicago Press, 1991.

Harper, Althea, Annie Shattuck, Eric Holt-Gimenez, Alison Alkon, and Frances Lambrick. *Food Policy Councils: Lessons Learned*. Portland, OR: Community Food Security Coalition, 2009.

Harrell, Ashley. "Straus Dairy Feels the Pinch from Organic Milk Shortage." *Point Reyes Light,* April 4, 2006. Accessed July 17, 2010. http://groups.ucanr.org/GIM/Straus_dairy_feels_the_pinch_from_organic_milk_shortage.

Harrington, Michael. *The Other America*. New York: Macmillan, 1962.

Harris, Gardiner, and William Neuman. "Senate Passes Sweeping Law on Food Safety." *New York Times*, November 30, 2010. Accessed February 28, 2011. http://www.nytimes.com/2010/12/01/health/policy/01food.html.

Hart, John. *Farming on the Edge: Saving Family Farms in Marin County, California*. Berkeley: University of California Press, 1992.

Harvest of Shame. Television. Directed by Fred W. Friendly. New York: CBS Television, 1960.

Harvey, David. *The Condition of Postmodernity: An Enquiry into the Origins of Cultural Change*. London: Blackwell, 1989.

Harvey, David. *The Enigma of Capital: And the Crises of Capitalism*. London: Oxford University Press, 2010.

Harvey, David. *The Limits to Capital*. Oxford: Basil Blackwell, 1982.

Hassanein, Neva. *Changing the Way America Farms: Knowledge and Community in the Sustainable Agriculture Movement*. Lincoln: University of Nebraska Press, 1999.

Hassanein, Neva. "Practicing Food Democracy: A Pragmatic Politics of Transformation." *Journal of Rural Studies* 19 (2003): 77–86.

Hawley, Ellis W. *The New Deal and the Problem of Monopoly*. Princeton, NJ: Princeton University Press, 1966.

Hayden-Smith, Rose. *A Brief History of School Gardens*. Kitchen Gardeners: A Global Community Cultivating Change, 2011. Accessed February 15, 2011. http://kgi.org/blogs/rose-hayden-smith/brief-history-school-gardens.

Hays, Samuel. *Conservation and the Gospel of Efficiency: The Progressive Conservation Movement, 1890–1920*. Cambridge, MA: Harvard University Press, 1959.

Heffernan, Maureen. "The Children's Garden Project at River Farm." *Children's Environments* 11 no. 13 (1994): 221–231.

Heilbron, J. L., and W. Robert Seidel. *Lawrence and His Laboratory: A History of the Lawrence Berkeley Laboratory*. Berkeley: University of California Press, 1989.

Henderson, Anne T., and Karen L. Mapp. *A New Wave of Evidence: The Impact of School, Family, and Community*. Austin, TX: National Center for Family and Community Connections with Schools, 2002.

Henderson, Elizabeth. "Rebuilding Local Food Systems from the Grass-Roots Up." In *Hungry for Profit: The Agribusiness Threat to Farmers, Food, and the Environment*, ed. Fred Magdoff, John Bellamy Foster, and Frederick Buttel, 175–188. New York: Monthly Review, 2000.

Hendrickson, John. *Energy Use in the U.S. Food System: A Summary of Existing Research and Analysis*. Madison, WI: Center for Integrated Agricultural Systems, 1996.

Hendrickson, Mary, William Heffernan, Philip Howard, and Judith Heffernan. *Consolidation in Food Retailing and Dairy: Implications for Farmers and Consumers in a Global Food System. Report to the National Farmers Union*. Columbia: University of Missouri-Columbia, 2001.

Henke, Christopher R. *Cultivating Science, Harvesting Power: Science and Industrial Agriculture in California*. Cambridge, MA: MIT Press, 2008.

Hennessey-Lavery, Susana. "SEFA: Southeast Access Working Group." PowerPoint presentation, 2004. Accessed June 28, 2011. http://webcache.googleusercontent.com/search?q=cache:wPmy-OFyR3oJ:cphan.org/AM08Library/Susanna Hennesy Lavery CPHAN 031308.ppt+Lavery+LEJ&cd=2&hl=en&ct=clnk&gl=us&client=safari.

Henry, Sarah. "People's Community Market Closer to Finding Funding with White House Announcement." *Bay Area Bites* (blog), July 20, 2011. Accessed October 3, 2011. http://blogs.kqed.org/bayareabites/2011/07/20/peoples-community-market-closer-to-finding-funding-with-white-house-announcement.

Hermann, Janice R., Stephany P. Parker, Barbara J. Brown, Youmasu J. Siewe, Barbara A. Denney, and Sarah J. Walker. "After-School Gardening Improves Children's Reported Vegetable Intake and Physical Activity." *Journal of Nutrition Education and Behavior* 38 (2006): 201–202.

Hess, Charlotte, and Elinor Ostrom, eds. *Understanding Knowledge as Commons: From Theory to Practice*. Cambridge, MA: MIT Press, 2007.

Hesterman, Oran B. *Fair Food: Growing a Healthy, Sustainable Food System for All*. New York: Public Affairs, 2011.

Hewitt, Ben. *The Town that Food Saved: How One Community Found Vitality in Local Food*. Emmaus, PA: Rodale Press, 2010.

Hightower, Jim. *Eat Your Heart Out: Food Profiteering in America*. New York: Crown, 1975.

Hightower, Jim. *Hard Tomatoes, Hard Times: A Report of the Agribusiness Accountability Project on the Failure of America's Land Grant College Complex*. Cambridge, MA: Schenkman, 1973.

Hilgard, Eugene Woldemar. *Reports of Experiments on Methods of Fermentation and Related Subjects during the Years 1886–1887*. Berkeley: California Agricultural Experiment Station, 1888.

Hilgard, Eugene Woldemar. *Report on the Arid Regions of the Pacific Coast.* Berkeley: California Agricultural Experiment Station, 1887.

Hilgard, Eugene Woldemar, and Winthrop John Van Leuven Osterhout. *Agriculture for Schools of the Pacific Slope.* New York: Macmillan, 1910.

Hill, Angele, and Josh Richman. "Bakery Enterprise Spirals out of Control: Esteemed Operation's Reputation Tarnished in Recent Years." *Oakland Tribune*, September 18, 2007. Accessed February 28, 2011. http://www.insidebayarea.com/oaklandtribune/localnews/ci_6544356.

Himmelgreen, David A., and Nancy Romero-Daza. "Eliminating 'Hunger' in the U.S.: Changes in Policy Regarding the Measurement of Food Security." *Food and Foodways* 18 (2010): 96–113.

Hirsch, Jessie. "The Bay Citizen: The Food Truck Revolution Revs Up, with a Little Help." *New York Times,* October 2, 2011. Accessed October 3, 2011. http://www.nytimes.com/2011/10/02/us/the-food-truck-revolution-revs-up-with-a-little-help.html.

Hiscock, John. "Meryl Streep Interview for Julie and Julia." *Telegraph*, August 28, 2009. Accessed January 12, 2011. http://www.telegraph.co.uk/culture/film/starsandstories/6100589/Meryl-Streep-interview-for-Julie-and-Julia.html.

Hnuhtsen, Kevin. "With Congress and the Public Concerned: Dairymen for Now Can't Use Milk Hormone." *Point Reyes Light,* November 18, 1993.

Holt-Gimenez, Eric, and Raj Patel. *Food Rebellions! Crisis and the Hunger for Justice.* Oakland, CA: Food First Books, 2009.

Hopkins, Raymond F., and Donald J. Puchala, eds. *The Global Political Economy of Food.* Madison: University of Wisconsin Press, 1978.

Hopkins, Raymond F., and Donald J. Puchala. "Perspectives on the International Relations of Food." *International Organization* 32 (1978): 581–616.

Hoppe, Robert A., and David E. Banker. *Structure and Finances of U.S. Farms: 2005 Family Farm Report.* Washington, DC: Economic Research Service, U.S. Department of Agriculture, 2005.

Hopper, Elly. "Backyard Bounty: Waste Not." *Terrain Magazine,* Fall–Winter 2008. Accessed September 25, 2011. http://ecologycenter.org/terrain/issues/fall-winter-2008/backyard-bounty-waste-not.

Horizon. *Annual Report: Horizon Organic Holding Corporation.* Broomfield, CO: Horizon, 2000.

Howard, Sir Albert. *An Agricultural Testament.* New York: Oxford University Press, 1943.

Howard, Sir Albert, and Yeshwant D. Wad. *The Waste Products of Agriculture.* New York: Oxford University Press, 1931.

Human Rights Center. *Hidden Slaves: Forced Labor in the United States.* Berkeley and Washington, DC: Human Rights Center, Free the Slaves, 2004.

Humphrey, Kim. *Shelf Life: Supermarkets and the Changing Cultures of Consumption.* Cambridge, MA: Cambridge University Press, 1998.

Hundley, Norris. *The Great Thirst: Californians and Water: A History*. Berkeley: University of California Press, 2001.

Hung, Yvonne. "East New York Farms: Youth Participation in Community Development and Urban Agriculture." *Children, Youth and Environments* 14 (2004): 20–31.

Hunter, Robert. *Poverty*. New York: Macmillan, 1904.

Hunters Point Community Web site. Accessed March 28, 2012. http://www.hunterspointcommunity.com.

Hunter Point Naval Shipyard Web site. Accessed March 28, 2012. http://www.hunterspointnavalshipyard.com.

Hurt, R. Douglas. *American Agriculture: A Brief History*. West Lafayette, IN: Purdue University Press, 2002.

Hurt, R. Douglas. *Problems of Plenty: The American Farmer in the Twentieth Century*. Chicago: Ivan R. Dee, 2002.

Hutchins, Wells A. "Pueblo Water Rights in the West." *Texas Law Review* 38 (1959–1960), 748–762.

Hutchins, Wells A. *The California Law of Water Rights*. Sacramento: Production Economics Research Branch, Agricultural Research Service, U.S. Department of Agriculture, 1956.

Igler, David. *Industrial Cowboys: Miller and Lux and the Transformation of the Far West, 1850–1920*. Berkeley: University of California Press, 2001.

Ilbery, Brian, and Moya Kneafsey. "Niche Markets and Regional Specialty Food Products in Europe: Towards a Research Agenda." *Environment and Planning A* 12 (1999): 2207–2222.

Ilbery, Brian, and Moya Kneafsey. "Producer Constructions of Quality in Regional Specialty Food Production: A Case Study from South West England." *Journal of Rural Studies* 16 (2000a): 217–230.

Ilbery, Brian, and Moya Kneafsey. "Registering Regional Specialty Food and Drink Products in the United Kingdom: The Case of PDOs and PGIs." *Area* 32 (2000b): 317–325.

Ilbery, Brian, Carol Morris, Henry Buller, Damian Maye, and Moya Kneafsey. "Product, Process and Place: An Examination of Food Marketing and Labelling Schemes in Europe and North America." *European Urban and Regional Studies* 12 (2005): 116–132.

"Indian Slavery." *Daily Alta California,* August 14, 1862.

Imhoff, Daniel, ed. *The CAFO Reader: The Tragedy of Industrial Animal Factories*. San Rafael, CA: Earth Aware Editions, 2010.

Imhoff, Daniel. *Food Fight: The Citizens' Guide to the Next Food and Farm Bill.* Berkeley: University of California Press, 2007.

Ingalsbe, Gene, and Frank Groves. "Historical Development." In *Cooperatives in Agriculture*, ed. David W. Cobia, 106–210. Englewood Cliffs, NJ: Prentice Hall, 1989.

Institute for Agriculture and Trade Policy. *EBT at Farmers Markets: Initial Insights from National Research and Local Dialogue*. Minneapolis: Institute for Agriculture and Trade Policy, 2010.

Institute for Agriculture and Trade Policy and Minnesota School Nutrition Association. *Farm to School in Minnesota: A Survey of School Foodservice Leaders*. Minneapolis: Institute for Agriculture and Trade Policy, 2010.

Intergovernmental Panel on Climate Change. *Climate Change 2007: Impacts, Adaptation, and Vulnerability. Contribution of Working Group II to the Fourth Assessment Report of the Intergovernmental Panel on Climate Change*. Cambridge: Cambridge University Press, 2007.

International Assessment of Agricultural Knowledge, Science and Technology for Development. *Agriculture at a Crossroads: Synthesis Report: A Synthesis of the Global and Sub-Global IAASTD Reports*. Washington, DC: Island Press, 2009.

Jackson, Robert H., and Edward Castillo. *Indians, Franciscans, and Spanish Colonization: The Impact of the Mission System on California Indians*. Albuquerque: University of New Mexico Press, 1995.

Jacobs, Jane. *The Death and Life of Great American Cities*. New York: Random House, 1961.

Jacobs, Meg. *Pocketbook Politics: Economic Citizenship in Twentieth-Century America*. Princeton, NJ: Princeton University Press, 2005.

Jaffe, Dan. "Beyond the Organic Tomato: Farm Workers and Labor Conditions on Four Northern California Organic Farms." Unpublished paper on file with authors, 2000.

James, Marquis, and Bessie R. James. *The Story of Bank of America: Biography of a Bank*. Washington, DC: Beard Books, 1954.

Jamie Oliver's Food Revolution. *The Unhealthiest City in America*. Television. New York: American Broadcasting Company, 2010.

Jansen, Marius B. *The Making of Modern Japan*. Cambridge, MA: Belknap Press, 2002.

Jarosz, Lucy A. "Understanding Agri-Food Networks as Social Relations." *Agriculture and Human Values* 17 (2000): 279–283.

Jarvis, David, Philip Dunham, and Brian Ilbery. "Rural Industrialization, 'Quality' and Service: Some Findings from South Warwickshire and North Devon." *Area* 34 (2002): 56–69.

Jayaraman, Saru. "Letters: Making Sure Restaurant Workers Get Their Due." *New York Times,* December 24, 2010. Accessed May 26, 2011. http://www.nytimes.com/2010/12/25/opinion/lweb25restaurant.html.

Jeanjean, Catherine A. "The Chefs Collaborative." *Journal of Agriculturaland Food Information* 8 (2008): 3–10.

Jervell, Anne M., and D. A. Jolly. "Beyond Food: Towards a Multifunctional Agriculture." Paper presented at the annual meeting of the Association for the

Study of Agriculture, Food, and Human Values, University of Minnesota, Minneapolis, June 2001.

Johnson, Josee. "Counterhegemony or Bourgeois Piggery? Food Politics and the Case of Foodshare." In *The Fight over Food: Producers, Consumers, and Activists Challenge the Global Food System*, ed. Wynne Wright and Gerad Middendorf, 99–121. University Park: Pennsylvania State University Press, 2008.

Johnson, Wendy. *Gardening at the Dragon's Gate: At Work in the Wild and Cultivated World*. New York: Bantam Books, 2008.

Johnston, Hank, and Bert Klandermans, eds. *Social Movements and Culture*. Minneapolis: University of Minnesota Press, 1995.

Jones, Lewis W. "The South's Negro Farm Agent." *Journal of Negro Education* 22 (1953): 38–45.

Joseph, Cynthia. "Two Big Firsts at Oakland Community Farmers Market on Saturday." *Oakland Local*, March 2, 2010. Accessed February 28, 2011. http://oaklandlocal.com/article/2-big-firsts-oakland-community-farmer%e2%80%99s-market-saturday.

Julie and Julia. DVD. Directed by Nora Ephron. Culver City: Columbia Pictures, 2009.

Jung, Carolyn. "Chicken Lovers Go for the Big Chill." *San Francisco Chronicle*, March 8, 2009. Accessed February 28, 2011. http://articles.sfgate.com/2009-03-08/food/17211540_1_chilled-chicken-lovers-air-chilling.

Kahrl, William L., ed. *The California Water Atlas*. Sacramento: Governor's Office of Planning and Research, 1979.

Kallen, Horace M. *The Decline and Rise of the Consumer: A Philosophy of Consumer Cooperation*. New York: D. Appleton Century, 1936.

Kallet, Arthur, and Frederick John Schlink. *100,000,000 Guinea Pigs: Dangers in Everyday Foods, Drugs, and Cosmetics*. New York: Vanguard, 1933.

Kambara, Kenneth M., and Crispin L. Shelley. *The California Agricultural Direct Marketing Study*. Davis: California Institute of Rural Studies and U.S. Department of Agriculture Agricultural Marketing Service, 2002.

Kamp, David. *The United States of Arugula: How We Became a Gourmet Nation*. New York: Broadway Books, 2006.

Kane, Joe. "The Supermarket Shuffle." *Mother Jones Magazine,* July 1984, 7.

Kann, Kenneth. *Comrades and Chicken Ranchers: The Story of a California Jewish Community*. Ithaca, NY: Cornell University Press, 1993.

Kapoor, Sybil. "Stilton-Crazy after All These Years." *Independent*, September 23, 2000. Accessed February 28, 2011. http://www.independent.co.uk/life-style/food-and-drink/features/stiltoncrazy-after-all-these-years-701774.html.

Kaufman, Jerry, and Martin Bailkey. "Farming inside Cities: Entrepreneurial Urban Agriculture in the United States." Lincoln Institute Working Paper no. 13 (2001).

Kaufman, Phillip R. "Consolidation in Food Retailing: Prospects for Consumers and Grocery Suppliers." *Agricultural Outlook*, August 2000.

Kaufman, Phillip R., James M. Macdonald, Steve M. Lutz, and David M. Smallwood. *Do the Poor Pay More for Food? Item Selection and Price Differences Affect Low-Income Household Food Costs.* Washington, DC: Food and Rural Economics Division, Economic Research Service, U.S. Department of Agriculture, 1997.

Kaynak, Erdener, and S. Tamer Cavusgil. "The Evolution of Food Retailing Systems: Contrasting the Experience of Developed and Developing Countries." *Journal of the Academy of Marketing Science* 10 (1982): 249–268.

Kazak, Don. "From Humble Beginnings in Troubled Times: Financial Problems Dogged Co-Op." *Palo Alto Online*, March 14, 2001. Accessed May 25, 2011. http://www.paloaltoonline.com/news_features/coop_market/beginnings.php.

Kerner Commission (U.S. Riot Commission). *Report of the National Advisory Commission on Civil Disorders*. Washington, DC: U.S. Government Printing Office, 1968.

Kessler, David A. *The End of Overeating: Taking Control of the Insatiable American Appetite*. New York: Rodale Books, 2009.

King, Corn. DVD. Directed by Aaron Woolf. Amherst: Balcony Releasing, 2007.

King, Franklin Hiram. *Farmers of Forty Centuries or Permanent Agriculture in China, Korea, and Japan*. Madison, WI: Carrie Baker King, 1911.

Kisseloff, Jeff. *Generation of Fire: Voices of Protest from the 1960s: An Oral History*. Lexington: University Press of Kentucky, 2006.

Klebaner, Benjamin J. *Poor Relief and Public Works during the Depression of 1857*. New York: Arno Press, 1976.

Klein, Herbert S. *The Atlantic Slave Trade*. Cambridge: Cambridge University Press, 1999.

Klemmer, Cynthia D., Tina M. Waliczek, and Jayne M. Zajicek. "Growing Minds: The Effect of a School Gardening Program on the Science Achievement of Elementary Students." *HortTechnology* 15 (2005): 448–452.

Kloppenburg, Jack, Jr., Sharon Lezberg, Kathryn De Master, George W. Stevenson, and John Hendrickson. "Tasting Food, Tasting Sustainability: Defining the Attributes of an Alternative Food System with Competent, Ordinary People." *Human Organization* 59 (2000): 177–186.

Kneafsey, Moya, Brian Ilbery, and Tim Jenkins. "Exploring the Dimensions of Culture Economies in Rural West Wales." *Sociologia Ruralis* 41 (2001): 296–310.

Knickel, Karlheinz. "The Marketing of Rhongold Milk: An Example of the Reconfiguration of National Relations with Agricultural Production and Consumption." *Journal of Environmental Policy and Planning* 3 (2001): 123–136.

Knuhtsen, Kevin. "With Congress and the Public Concerned: Dairymen for Now Can't Use Milk Hormone." *Point Reyes Light,* November 18, 1993.

Koc, Mustafa, and Kenneth A. Dahlberg. "The Restructuring of Food Systems: Trends, Research, and Policy Issues." *Agriculture and Human Values* 16 (1999): 109–116.

Koch, Shari, Tina M. Waliczek, and Jayne M. Zajicek. "The Effect of a Summer Garden Program on the Nutritional Knowledge, Attitudes, and Behaviors of Children." *HortTechnology* 16 (2006): 620–625.

Konefal, Jason, Carmen Bain, Michael Mascarenhas, and Lawrence Busch. "Supermarkets and Supply Chains in North America." In *Supermarkets and Agri-Food Supply Chains*, ed. David Burch and Geoffrey Lawrence, 270–290. Cheltenham, UK: Edward Elgar, 2007.

Kosseloff, Jeff. *Generation on Fire: Voices of Protest from the 1960s: An Oral History*. Lexington: University Press of Kentucky, 2007.

Kraus, Sibella. *The Farm-Restaurant Project: Regional Self-Reliance and High Quality Produce*. Berkeley, CA: Sustainable Agriculture Education (SAGE), 1983.

Kroc, Ray, and Robert Anderson. *Grinding It Out: The Making of McDonald's*. New York: St. Martin's Press, 1977.

Kuczmarski, Robert J., Katherine M. Flegal, Stephen M. Campbell, and Clifford L. Johnson. "Increasing Prevalence of Overweight among US Adults: The National Health and Nutrition Examination Surveys, 1960 to 1991." *Journal of the American Medical Association* 272 (1994): 205–211.

Kuh, Patric. *The Last Few Days of Haute Cuisine: America's Culinary Revolution*. New York: Viking Penguin, 2001.

Kummer, Corby. "The Great Grocery Smackdown." *Atlantic,* March 2010. Accessed February 28, 2011. http://www.theatlantic.com/magazine/archive/2010/03/the-great-grocery-smackdown/7904.

Kurlansky, Mark. *Cod: The Biography of a Fish That Changed the World*. New York: Penguin, 1997.

Kurlansky, Mark. *Salt: A World History*. New York: Penguin Books, 2003.

Kurlansky, Mark. *1968: The Year That Rocked the World*. New York: Ballantine Books, 2005.

Kurlansky, Mark. *The Big Oyster: History on the Half Shell*. New York: Ballantine Books, 2006.

Kurlansky, Mark. *The Food of a Younger Land: A Portrait of American Food—before the National Highway System, before Chain Restaurants, and before Frozen Food, When the Nation's Food Was Seasonal*. New York: Riverhead Hardcover, 2009.

Kuruvilla, Matthai. "Oakland Urban Farming Prompts Plan to Redo Rules." *San Francisco Chronicle,* May 9, 2011. Accessed September 28, 2011. http://www.sfgate.com/cgi-bin/article.cgi?f=/c/a/2011/05/08/BA7O1J74O5.DTL#ixzz1ZJFrkbxu.

Kuruvila, Matthai, Kevin Fagan, and Jaxon Van Derbeken. "Muslim Bakery Head Wielded Political Clout." *San Francisco Chronicle*, January 27, 2008.

Accessed February 28, 2011. http://articles.sfgate.com/2008-01-27/news/17150782_1_john-bey-bakery-founder-yusuf-bey.

Labao, Linda M. *Locality and Inequality: Farm and Industry Structure and Socioeconomic Conditions*. Albany: State University of New York Press, 1990.

Lage, Ann, and William Duddleson. *Saving Point Reyes National Seashore: An Oral History of Citizen Action in Conservation*. Berkeley, CA: Bancroft Library, 1993.

Landers, Patti S. "The Food Stamp Program: History, Nutritional Education, Impact." *Journal of the American Dietetic Association* 107 (2007): 1945–1951.

Lang, Tim, and David Barling. "The Environmental Impact of Supermarkets: Mapping the Terrain and the Policy Problems in the UK." In *Supermarkets and Agri-Food Supply Chains*, ed. David Burch and Geoffrey Lawrence, 192–215. Cheltenham, UK: Edward Elgar, 2007.

Lang, Tim, and Michael Heasman. *Food Wars: The Global Battle for Mouths, Minds, and Markets*. London: Earthscan, 2003.

Lange, Dorothea, and Paul S. Taylor. *An American Exodus: A Record of Human Erosion*. New York: Reynal and Hitchcock, 1939.

Lappé, Anna. *Diet for a Hot Planet: The Climate Crisis at the End of Your Fork and What You Can Do about It*. New York: Bloomsbury USA, 2010.

Lappé, Anna, and Bryant Terry. *Grub: Ideas for an Urban Organic Kitchen*. London: Penguin, 2006.

Lappé, Francis Moore. *Diet for a Small Planet*. New York: Ballantine Books, 1971.

Lappé, Francis Moore, and Joseph Collins. *Food First: Beyond the Myth of Scarcity*. Boston: Houghton Mifflin, 1977.

Lappé, Francis Moore, Joseph Collins, and David Kinley. *Aid as Obstacle: Twenty Questions about Our Foreign Aid and the Hungry*. San Francisco: Institute for Food and Development Policy, 1981.

Laskawy, Tom. "36 Million Pounds of Proof That Our Food Safety System Is Broken." *Grist*, August 10, 2011a. Accessed August 12, 2011. http://www.grist.org/food-safety/2011-08-10-salmonella-tainted-turkey-food-safety-system-broken.

Laskawy, Tom. "The New Agtivist: Shakirah Simley Wants to Preserve Justice." *Grist*, June 24, 2011b. Accessed February 28, 2011. http://www.grist.org/article/food-the-new-agtivist-shakirah-simley-wants-to-preserve-justice.

Latour, Bruno. *Science in Action: How to Follow Scientists and Engineers through Society*. Cambridge, MA: Harvard University Press, 1987.

Latour, Bruno. "On Recalling ANT." In *Actor Network Theory and After*, ed. John Law and John Hassard, 15–25. Oxford and Keele: Blackwell and the Sociological Review, 1999a.

Latour, Bruno. *Pandora's Hope: Essays on the Reality of Science Studies*. Cambridge, MA: Harvard University Press, 1999b.

Lauck, Jon. *American Agriculture and the Problem of Monopoly: The Political Economy of Grain Belt Farming, 1953–1980*. Lincoln: University of Nebraska Press, 2000.

Lawrence, Felicity. *Not on the Label: What Really Goes into the Food on Your Plate*. London: Penguin, 2004.

Lawrence, Geoffrey, and David Burch. "Understanding Supermarkets and Agri-food Supply Chains." In *Supermarkets and Agri-Food Supply Chains*, ed. David Burch and Geoffrey Lawrence, 1–28. Cheltenham, UK: Edward Elgar, 2007.

Lawson, Laura J. *City Bountiful: A Century of Community Gardening in America*. Berkeley: University of California Press, 2005.

Leavitt, Judith W. *The Healthiest City: Milwaukee and the Politics of Health Reform*. Princeton, NJ: Princeton University Press, 1982.

Lee, Charles. *Toxic Wastes and Race in the United States*. New York: United Church of Christ Commission for Racial Justice, 1987.

Lehman, Yael. Interview by Louise Nelson Dyble. Philadelphia, PA, August 2, 2010.

Le Jaouen, Jean-Claude. *The Fabrication of Farmstead Goat Cheese*. Ashfield, MA: Cheesemakers' Journal, 1987.

Lekies, Kristi S., Marcia Eames-Sheavly, Kimberly J. Wong, and Anne Ceccarini. "Children's Garden Consultants." *HortTechnology* 16 (2006): 139–142.

Leonard, Annie. *The Story of Stuff: How Our Obsession with Stuff Is Trashing the Planet, Our Communities, and Our Health—and a Vision for Change*. New York: Free Press, 2010.

Lerza, Catherine, and Michael Jacobson. *Food for People, Not for Profit: A Sourcebook on the Food Crisis*. New York: Ballantine, 1975.

Levenstein, Harvey. *Revolution at the Table: The Transformation of the American Diet*. Berkeley: University of California Press, 1988.

Levenstein, Harvey. *Paradox of Plenty: A Social History of Eating in Modern America*. New York: Oxford University Press, 1993.

Levine, Susan. *School Lunch Politics: The Surprising History of America's Favorite Welfare Program*. Princeton, NJ: Princeton University Press, 2008.

Levinson, Marc. *The Box: How the Shipping Container Made the World Smaller and the World Economy Bigger*. Princeton, NJ: Princeton University Press, 2006.

Lewis, Leo. "Scientists Warn of Lack of Vital Phosphorus as Biofuels Raise Demand." *Sunday Times,* June 23, 2008. Accessed October 11, 2011. http://knowledge.cta.int/en/Dossiers/S-T-Issues-in-Perspective/Phosphorus-depletion/Documents/Scientists-warn-of-lack-of-vital-phosphorus-as-biofuels-raise-demand.

Libman, Kimberly. "Growing Youth Growing Food." *Applied Environmental Education and Communication* 6 (2007): 87–95.

Liebman, Ellen. *California Farmland: A History of Large Agricultural Holdings*. Totowa, NJ: Rowman and Allenheld, 1983.

Lightfoot, Kent G. *Indians, Missionaries, and Merchants: The Legacy of Colonial Encounters on the California Frontiers*. Berkeley: University of California Press, 2006.

Linder, Mark. "I Gave My Employer a Chicken That Had No Bone: Joint Firm-State Responsibility for Line-Speed-Related Occupational Injuries." *Case Western Reserve Law Review* 46 (1995): 33–95.

Lineberger, Sarah E., and Jayne M. Zajicek. "Can a Hands-On Teaching Tool Affect Students' Attitudes and Behavior Regarding Fruit and Vegetables?" *Hort-Technology* 10 (2000): 593–597.

Linsenmeyer, Helen Walker. *From Fingers to Finger Bowls: A Sprightly History of California Cooking*. San Diego: Union-Tribune, 1976.

Lippman, Morris. "The Palo Alto Co-Op." In *What Happened to the Berkeley Co-Op? A Collection of Opinions*, ed. Michael Fullerton, 83–86. Davis, CA: Center for Cooperatives, 1992.

Literacy for Environmental Justice Web Site. Accessed March 28, 2012. http://www.lejyouth.org/.

Livingston, Dewey. *Ranching on the Point Reyes Peninsula: A History of the Dairy and Beef Ranches within Point Reyes National Seashore, 1834–1992*. Point Reyes, CA: Point Reyes National Seashore, National Park Service, 1993.

Livingston, Dewey. *A Good Life*. San Francisco: National Park Service, U.S. Department of the Interior, 1995.

Lobao, Linda. *Locality and Inequality: Farm Structure, Industry Structure, and Socioeconomic Conditions*. Albany, NY: The State University of New York Press, 1990.

Loftis, Anne. *Witnesses to the Struggle: Imagining the 1930s California Labor Movement*. Reno: University of Nevada Press, 1998.

Lohr, Virginia I., and Caroline H. Pearson-Mims. "Children's Active and Passive Interactions with Plants Influence Their Attitudes and Actions toward Trees and Gardening as Adults." *HortTechnology* 15 (2005): 472–476.

Longstreth, Richard. *The Drive-In, the Supermarket, and the Transformation of Commercial Space in Los Angeles, 1914–1941*. Cambridge, MA: MIT Press, 1999.

Looney, J. W. "The Changing Focus of Government Regulation of Agriculture in the United States." *Mercer Law Review* 44 (1993): 763–771.

Lovell, Margaretta. "Food Photography and Inverted Narratives of Desire." *Exposure* 33 (2001): 21–26.

Lowe, Marcy, and Gary Gereffi. *A Value Chain Analysis of the U.S. Beef and Dairy Industry*. Durham, NC: Center on Globalization, Governance, and Competitiveness, Duke University, 2009.

Lucchesi, Paolo. "Oakland Dilemma: Longtime Butcher Enzo's OUT." *San Francisco Chronicle*, June 1, 2010. Accessed February 28, 2011. http://insidescoopsf.sfgate.com/blog/2010/06/01/oakland-dilemma-longtime-butcher-enzos-out-more-local-marin-sun-farms-in/.

Luck, J. Murray. *The War on Malnutrition and Poverty*. New York: Harpers, 1946.

Luttrell, Clifton B. *The Russian Wheat Deal—Hindsight vs. Foresight*. St. Louis: Federal Reserve Bank of St. Louis, 1973.

Lyman, Howard. *Mad Cowboy: Plain Truth from the Cattle Rancher Who Won't Eat Meat*. New York: Scribner, 2001.

Lynch, Kermit. *Inspiring Thirst: Vintage Selections from the Kermit Lynch Wine Brochure*. Berkeley, CA: Ten Speed Press, 2004.

Lyson, Thomas A. *Civic Agriculture: Reconnecting Farm, Food, and Community*. University Park: Pennsylvania State University Press, 2004.

Lyson, Thomas A., and Charles C. Geisler. "Toward a Second Agricultural Divide: The Restructuring of American Agriculture." *Sociologia Ruralis* 32 (1992): 248–263.

Lyson, Thomas A., G. W. Stevenson, and Rick Welsh, eds. *Food and the Mid-Level Farm: Renewing an Agriculture of the Middle*. Cambridge, MA: MIT Press, 2008.

Mack, Adam. "Good Things to Eat in Suburbia." Ph.D. diss., University of South Carolina, 2006.

Mackintosh, Barry. "Harold L. Ickes and the National Park Service." *Journal of Forest History* 29 (1985): 78–84.

MacCannell, Dean. *Technology, Public Policy, and the Changing Structure of American Agriculture*. Washington, DC: U.S. Government Printing Office, 1986.

MacLeod, Gordon. "'Institutional Thickness' and Industrial Governance in Lowland Scotland." *Area* 29 (1997): 299–311.

Maher, Neil M. *Nature's New Deal: The Civilian Conservation Corps and the Roots of the American Environmental Movement*. New York: Oxford University Press, 2008.

Makintosh, Barry. *The National Parks: Shaping the System*. Washington, DC: U.S. Department of the Interior, 1984.

Maney, Ardith. *Still Hungry after All These Years: Food Assistance Policy From Kennedy to Reagan*. Westport, CT: Greenwood Press, 1989.

Marable, Manning. *Malcolm X: A Life of Reinvention*. New York: Viking Press, 2011.

Marcello, Patricia Cronon. *Ralph Nader: A Biography*. Westport, CT: Greenwood Press, 2004.

Marin Agricultural Land Trust. *Land Preservation Report*. Point Reyes Station, CA: Marin Agricultural Land Trust, 2000.

Marin Agricultural Land Trust. "East Meets West (Marin): Teaching Cheesemaking, Promoting Conservation." *MALT News* 25 (2009).

Marin County Planning Department. *Can the Last Place Last? Preserving the Environmental Quality of Marin*. San Rafael, CA: Marin County Planning Department, 1971.

Marin County Planning Department. *The Viability of Agriculture in Marin*. San Rafael, CA. Marin County Planning Department, 1971.

Marin County Planning Department. *Marin Countywide Plan*. San Rafael, CA: Marin County Planning Department, 1973.

Mark, Jason. "Willow Rosenthal." *Earth Island Journal*, Spring 2008. Accessed February 28, 2011. http://www.earthisland.org/journal/index.php/eij/article/willow_rosenthal.

Markin, Rom J. *The Supermarket: An Analysis of Growth, Development, and Change*. Pullman, WA: Washington State University Press, 1963.

Markowitz, Lisa. "Expanding Access and Alternatives: Building Farmers' Markets in Low-Income Communities." *Food and Foodways* 18 (2010): 66–80.

Marsden, Terry K., and Alberto Arce. "Constructing Quality: Emerging Food Networks in the Rural Transition." *Environment and Planning A* 27 (1995): 1261–1279.

Marsden, Terry K., and Sarah Whatmore. "Finance Capital and Food System Restructuring: National Incorporation of Global Dynamics." In *The Global Restructuring of Agro-Food Systems*, ed. Philip McMichael, 107–128. Ithaca, NY: Cornell University Press, 1994.

Marsh, George Perkins. *Man and Nature: Physical Geography as Modified by Human Action*. New York: Charles Scribner, 1864.

Marshall, Alfred. *Principles of Economics*. 8th ed. London: Macmillan and Co., Ltd., 1920. (orig. pub. 1890). Accessed February 19, 2012. http://www.econlib.org/library/Marshall/marP.html.

Martin, Jiff. Interview by Louise Nelson Dyble. Hartford, CT, July 29, 2010.

Martin, Philip L. *Promises to Keep: Collective Bargaining in California Agriculture*. Ames: Iowa State University Press, 1996.

Martin, Philip L. *Promise Unfulfilled: Unions, Immigration, and the Farm Workers*. Ithaca, NY: Cornell University Press, 2003.

Martin, Philip L. *California Hired Farm Labor 1960–2010: Change and Continuity*. Davis, CA: University of California, Davis, 2011. Accessed February 28, 2012. http://migration.ucdavis.edu/cf/files/2011-may/martin-california-hired-farm-labor.pdf.

Martin, Philip L., and Elizabeth Midgley. "Immigration: Shaping and Reshaping America revised and updated 2nd edition." *Population Bulletin* 61 (2006): 5–46. http://www.prb.org/pdf06/61.4USMigration.pdf. Accessed October 15, 2011.

Martin, Philip L., Michael Fix, and J. Edward Taylor. *The New Rural Poverty*. Washington, DC: Urban Institute, 2006.

Matusow, Allen J. *Farm Policies and Politics in the Truman Years*. Cambridge, MA: Harvard University Press, 1967.

Mayer-Smith, Jolie, Oksana Bartosh, and Linda Peterat. "Teaming Children and Elders to Grow Food and Environmental Consciousness." *Applied Environmental Education and Communication* 6 (2007): 77–85.

McAdam, Doug, and David Snow. "Introduction: Social Movements: Conceptual and Theoretical Issues." In *Social Movements: Readings on Their Emergence, Mobilization, and Dynamics*, ed. Doug McAdam and David Snow, xviii–xxvi. Los Angeles: Roxbury, 1997.

McAleese, Jessica D., and Linda L. Rankin. "Garden-Based Nutrition Education Affects Fruit and Vegetable Consumption in Sixth Grade Adolescents." *Journal of the American Dietetic Association* 107 (2007): 663–665.

McCue, Susan. "Straus Family Creamery." *Small Farm News*, Winter 1999. Accessed March 3, 2010. http://www.sfc.ucdavis.edu/pubs/sfnews/winter99/straus.html.

McClintock, Nathan. "From Industrial Garden to Food Desert: Unearthing the Root Structure of Urban Agriculture in Oakland, California." Berkeley: Institute for the Study of Social Change, University of California, Berkeley, 2008.

McClintock, Nathan, and Jenny Cooper. *Cultivating the Commons: An Assessment of the Potential for Urban Agriculture on Oakland's Public Land*. Berkeley: Department of Geography, University of California, 2010.

McFadden, Steve. "The New Farm: The History of Community Supported Agriculture, Part 1." Kutztown, PA: Rodale Institute, 2003. Accessed February 8, 2011. http://newfarm.rodaleinstitute.org/features/0104/csa-history/part1_print.shtml.

McGinty, Brian. *Strong Wine: The Life and Legend of Agoston Haraszthy*. Stanford, CA: Stanford University Press, 1998.

McHenry, Keith, and C. T. Lawrence Butler. "The 25th Anniversary of Food Not Bombs, Parts 1 and 2." *Z Magazine*, 2005. Accessed January 17, 2011. http://www.foodnotbombs.net/z_25th_anniversary_1.html.

McGeehan, Patrick. "What, No Tip? Service Charge Faces Struggle at Restaurants." *New York Times*, August 15, 2005. Accessed May 25, 2011. http://www.nytimes.com/2005/08/15/nyregion/15tips.html.

McLaughlin, Katy. "The King of the Streets Moves Indoors." *Wall Street Journal*, January 15, 2010. Accessed February 28, 2011. http://online.wsj.com/article/SB10001424052748704842604574642420732091490.html.

McMichael, Philip, ed. *The Global Restructuring of Agro-Food Systems*. Ithaca, NY: Cornell University Press, 1994.

McNamee, Thomas. *Alice Water and Chez Panisse: The Romantic, Impractical, Often Eccentric, Ultimately Brilliant Making of a Food Revolution*. New York: Penguin, 2007.

McPhee, John. *Oranges*. New York: Farrar, Straus, and Giroux, 1975.

McWilliams, Carey. *Factories in the Field: The Story of Migratory Farm Labor in California*. Boston: Little, Brown, 1939.

McWilliams, James E. *A Revolution in Eating: How the Quest for Food Shaped America*. New York: Columbia University Press, 2005.

McWilliams, James E. "Food That Travels Well." *New York Times*, August 6, 2007. Accessed February 28, 2011. http://www.nytimes.com/2007/08/06/opinion/06mcwilliams.html.

McWilliams, James E. *Just Food: Where Locavores Get It Wrong and How We Can Truly Eat Responsibly*. New York: Little, Brown, 2009.

Medvitz, Albert G., Avin D. Sokolow, and Cathy Lemp, eds. *California Farmland and Urban Pressures: Statewide and Regional Perspectives*. Davis: University of California, Davis, Agricultural Issues Center, 1999.

Melosi, Martin V. *The Sanitary City: Urban Infrastructure in America From Colonial Times to the Present*. Baltimore, MD: John Hopkins University Press, 2000.

Merenlender, A. M., Lynn Huntinsinger, Greig Tor Guthey, and Sally K. Fairfax. "Land Trusts and Conservation Easements: Who Is Conserving What for Whom?" *Conservation Biology* 18 (2004): 65–75.

Meyer, Amy, and Randolph Delehanty. *New Guardians for the Golden Gate: How America Got a Great National Park*. Berkeley: University of California Press, 2006.

Meyer, E. "Cultivating Change: A Historical Overview of the School Garden Movement." Unpublished graduate seminar paper, University of California, Davis, 1997.

Michalski, Marie-Caroline. "On the Supposed Influence of Milk Homogenization on the Risk of CVD, Diabetes, and Allergy." *British Journal of Nutrition* 97 (2007): 598–610.

Miller, Louise Klein. *Children's Garden for School and Home: A Manual of Cooperative Gardening*. New York: Appleton, 1904.

Miller, Timothy. *The 60s Communes: Hippies and Beyond*. Syracuse, NY: Syracuse University Press, 1999.

Minister of Information JR. "Village Bottoms Farms: Growing Our Food, Owning Our Future." *San Francisco Bay View*, May 28, 2009. Accessed February 28, 2011. http://sfbayview.com/2009/village-bottoms-farms-growing-our-food-owning-our-future.

Mintz, Sidney. *Sweetness and Power: The Place of Sugar in Modern History*. New York: Viking Penguin, 1985.

Mitchell, D. "Rain Causes Manure Pond to Overflow into Tomales Bay." *Point Reyes Light*, August 23, 1989.

Mitchell, Dave. "Zoning Creates an Olive-Oil Industry." *Point Reyes Light*, December 12, 1996.

Montague-Jones, Guy. "Chicken Survey Finds Two-Thirds Harbour Salmonella, Campylobacter." *Meatprocess.com*, December 1, 2009. Accessed August 8, 2010. http://www.meatprocess.com/safety-legislation/chicken-survey-finds-two-thirds-harbour-salmonella-campylobacter.

Montgomery, David R. *Dirt: The Erosion of Civilizations*. Berkeley: University of California Press, 2008.

Moran, Warren. "The Wine Appellation as Territory in France and California." *Annals of the Association of American Geographer* 83 (1993): 694–717.

Morgan, Kevin, Terry Marsden, and Jonathan Murdoch. *Worlds of Food: Place, Power, and Provenance in the Food Chain*. New York: Oxford University Press, 2006.

Morris, Jennifer L., and Sheri Zidenberg-Cherr. "Garden-Enhanced Nutrition Curriculum Improves Fourth-Grade School Children's Knowledge of Nutrition and Preferences for Some Vegetables." *Journal of the American Dietetic Association* 102 (2002): 91–93.

Morse, Sierra. "Growing the Future: Leaders in Agriculture: Albert Straus Interview." October 14, 2008. Accessed March 10, 2010. http://www.marinhistory.org/audio/albert_straus.pdf.

Moss, Michael. "Safety of Beef Processing Method Is Questioned." *New York Times*, December 30, 2009. Accessed February 28, 2011. http://www.nytimes.com/2009/12/31/us/31meat.html.

Mowbray, A. Q. *The Thumb on the Scale, or, The Supermarket Shell Game*. Philadelphia: J. B. Lippincott, 1967.

Moynihan, Daniel P. *The Politics of a Guaranteed Income: The Nixon Administration and the Family Assistance Plan*. New York: Random House, 1973.

Muhammad, Elijah. *How to Eat to Live*. Chicago: Muhammad Mosque of Islam No. 2, 1967.

Munro, William A., and Rachel Schurman. "Sustaining Outrage: Motivating Sensibilities in the U.S. Anti-GE Movement." In *The Fight over Food: Producers, Consumers, and Activists Challenge the Global Food System*, ed. Wynne Wright and Gerard Middendorf, 145–176. University Park: Pennsylvania State University Press, 2008.

Muraoka, Sharen. *California's Direct Marketing Program: The First Two Years*. Sacramento, CA: California Department of Agriculture, 1978.

Murch, Donna. *Living for the City: Migration, Education and the Rise of the Black Panther Party in Oakland, California*. Chapel Hill: University of North Carolina Press, 2010.

Murdoch, Jonathan, Terry Marsden, and Jo Banks. "Quality, Nature, and Embeddedness: Some Theoretical Considerations in the Context of the Food Sector." *Economic Geography* 26 (2000): 107–125.

Murray, Douglas L. "The Abolition of El Cortito, the Short-Handled Hoe: A Case Study in Social Conflict and State Policy in California." *Social Problems* 30 (1982): 26–39.

Murray, Sarah. *Moveable Feasts: From Ancient Rome to the 21st Century, the Incredible Journeys of the Food We Eat*. New York: St. Martin's Press, 2007.

Nader, Ralph. "Watch That Hamburger." *New Republic,* August 1967a.

Nader, Ralph. "We're Still in the Jungle." *New Republic,* July 1967b.

Nader, Ralph. *Unsafe at Any Speed: the Designed-In Dangers of the American Automobile*. New York: Grossman, 1965.

Nally, David P. *Human Encumbrances: Political Violence and the Great Irish Famine*. Notre Dame: University of Notre Dame Press, 2011.

National Academy of Science Research Council. *Pesticides in the Diets of Infants and Children*. Washington, DC: National Academy, 1993.

National Environmental Policy Act of 1969. 42 U.S.C. 4321.

Neptune, Robert. *California's Uncommon Markets: The Story of the Consumer Cooperatives 1935–1971*. Richmond, CA: Associated Cooperatives, 1971.

Ness, Carol. "Pop-Up General Store in Oakland." *San Francisco Chronicle*, May 27, 2010. Accessed February 28, 2011. http://articles.sfgate.com/2010-05-27/entertainment/21455973_1_soul-food-farm-alice-waters-cook.

Nestle, Marion. "Hunger in America." *Social Research* 66 (1999): 257–282.

Nestle, Marion. "Ethical Dilemmas in Choosing a Healthful Diet: Vote with Your Fork!" *Proceedings of the Nutrition Society* 59 (2000): 619–629.

Nestle, Marion. *Food Politics: How the Food Industry Influences Nutrition and Health*. Berkeley: University of California Press, 2002.

Nestle, Marion. *Safe Food: Bacteria, Biotechnology, and Bioterrorism*. Berkeley: University of California Press, 2003.

Nestle, Marion. *Pet Food Politics: The Chihuahua in the Cole Mine*. Berkeley: University of California Press, 2008.

Niman, Nicolette Hahn. *Righteous Pork Chop: Finding a Life and Good Food beyond Factory Farms*. New York: HarperCollins, 2009.

Nixon, Richard. "Remarks at the White House Conference on Food, Nutrition, and Health, December 2, 1969." Washington, DC, December 1969. Accessed January 12, 2011. http://www.presidency.ucsb.edu/ws/index.php?pid=2349.

Nobel, Justin. "The Birthing of Point Reyes National Seashore." *North Coaster*, Fall 2010.

Nordhaus, Ted, and Michael Shellenberger. *Break Through: From the Death of Environmentalism to the Politics of Possibility*. Boston: Houghton Mifflin, 2007.

Norris, Frank. *The Octopus: A Story of California*. Garden City, NY: Doubleday, Page and Company, 1901.

Nosrat, Samin. "In the Bay Area, Cooks Are Breaking the Rules." *Atlantic* (June 2011): 18. http://www.theatlantic.com/food/archive/2010/06/in-the-bay-area-cooks-are-breaking-the-rules/58374. Accessed February 28, 2011.

Novak, Laura. "Givebacks: Pizza or Go, Shares to Stay." *New York Times*, April 10, 2007. Accessed July 28, 2011. http://www.nytimes.com/2007/04/10/business/retirement/10esop.html.

Nygard, Berit, and Oddveig Storstad. "De-Globalization of Food Markets? Consumer Perceptions of Safe Food: The Case of Norway." *Sociologia Ruralis* 37 (1998): 35–53.

Oberholtzer, Lydia, and Shelly Grow. *Producer-Only Farmers' Markets in the Mid-Atlantic Region: A Survey of Market Managers*. Arlington: Henry A. Wallace Center for Agricultural and Environmental Policy, Winrock International, 2003.

Occidental Arts and Ecology Center Web site. Accessed March 28, 2012. http://www.oaec.org.

O'Connell, Linda, Gregg Langlois, and Dale Hopkins. *Tomales Bay Shellfish Technical Advisory Committee Final Report: Investigation of Nonpoint Pollution Sources Impacting Shellfish Growing Areas in Tomales Bay, 1995–1996*. Sacramento, CA: California State Water Resoruces Control Board, 2000.

Odom, E. Dale. "Associated Milk, Incorporated: Testing the Limits of Capper-Volstaed." *Agricultural History* 59 (1985): 40–55.

Ogden, Cynthia L., Cheryl D. Fryar, Margaret D. Carroll, and Katherine M. Felgal. "Mean Body Weight, Height, and Body Mass Index, United States 1960–2002." *Advance Data* 347 (2004): 1–17.

O'Hara, Jeffrey K. *Market Forces: Creating Jobs through Public Investment in Local and Regional Food Systems*. Cambridge, MA: Union of Concerned Scientists, 2011.

Older, Morris. "The People's Food System." In *The History of Collectivity in the San Francisco Bay Area*, ed. John Curl, Judy Berg, and Allen Cohen, 38–44. San Francisco: Homeward, 1982.

Olmstead, Alan R., and Paul W. Rhode. *Creating Abundance: Biological Innovation and American Agricultural Development*. Cambridge: Cambridge University Press, 2008.

Orman, Larry. Interview with Louise Nelson Dyble. San Francisco, April 11, 2010.

Orsi, Richard J. *Sunset Limited: The Southern Pacific Railroad and the Development of the American West, 1850–1930*. Berkeley: University of California Press, 2005.

Ostrom, Elinor. *Governing the Commons: The Evolution of Institutions for Collective Action*. Cambridge: Cambridge University Press, 1990.

Outdoor Recreation Resources Review Commission. *Outdoor Recreation for America: A Report to the President and the Congress*. Washington, DC: U.S. Government Printing Office, 1962.

Oxfam. *Like Machines in the Fields: Workers without Rights in American Agriculture*. Boston: Oxfam America, 2004. Accessed February 13, 2012. http://www.oxfamamerica.org/files/like-machines-in-the-fields.pdf.

Paddison, Joshua, ed. *Accounts of California before the Gold Rush*. Berkeley, CA: Heyday Books, 1999.

Parrott, Nicholas, Natasha Wilson, and Jonathan Murdoch. "Spatializing Quality: Regional Protection and the Alternative Geography of Food." *European Urban and Regional Studies* 9 (2002): 241–261.

Parson, Sean Michael. "Ungovernable Force: Food Not Bombs, Homeless Activism and Politics in San Francisco." Ph.D. diss., University of Oregon, 2010.

Pastor, Manuel, Jim Sadd, and John Hipp. "Which Came First? Toxic Facilities, Minority Move-in and Environmental Justice." *Journal of Urban Affairs* 23 (2001): 1–21.

Patel, Raj. *Stuffed and Starved: Markets, Power and the Hidden Battle for the World Food System*. London: Portobello Books, 2007.

Patel, Raj. "Five Things You Didn't Know about Supermarkets." *Foreign Policy*, November 2010a. Accessed February 28, 2011. http://www.foreignpolicy.com/articles/2010/10/11/supermarkets?

Patel, Raj. Interview with Steve Inskeep. *NPR Business*. NPR, October 21, 2010b.

Patel, Raj. "What Does Food Sovereignty Look Like?" In *Food Sovereignty: Reconnecting Food, Nature, and Community*, ed. Hannah Wittman, Annette Desmarais, and Nettie Wiebe, 186–195. Oakland, CA: Food First Books, 2010c.

Payne, Tim. *U.S. Farmers Markets—2000: A Study of Emerging Trends*. Washington, DC: U.S. Department of Agriculture, Agricultural Marketing Service, 2002.

Pearson, Elaine. *Human Traffic, Human Rights: Redefining Victim Protection*. London: Anti-Slavery International, 2002.

Peirce, Pam. "San Francisco's Community Gardens: Interview with Pam Peirce." *SF's Community Garden Newsletter*, Spring–Summer 1994. Accessed February 28, 2011. http://www.foundsf.org/index.php?title=San_Francisco's_Community_Gardens.

Pellow, David N., and Robert J. Brulle, eds. *Power, Justice, and the Environment: A Critical Appraisal of the Environmental Justice Movement*. Cambridge, MA: MIT Press, 2005.

Pendergrast, Mark. *Uncommon Grounds: The History of Coffee and How It Transformed Our World*. New York: Basic Books, 2000.

Peninou, Ernest P. *History of the Sonoma Viticultural District: The Grade Growers, the Wine Makers, and the Vineyards*. Santa Rosa, CA: Nomis, 1998.

Pennsylvania Horticultural Society. "About PHS." Accessed January 17, 2011. http://www.pennsylvaniahorticulturalsociety.org/aboutus/index.html.

People for Open Space Farmlands Conservation Project. *Farmland and Farming in Bay Region: A Description*. San Francisco: People for Open Space Farmlands Conservation Project, 1979.

People for Open Space Farmlands Conservation Project. *Bay Area Agricultural Production Issues: Conclusions and Findings*. San Francisco: People for Open Space Farmlands Conservation Project, 1980a.

People for Open Space Farmlands Conservation Project. *Bay Area Farmland Loss: Trends and Case Studies*. San Francisco: People for Open Space Farmlands Conservation Project, 1980b.

People for Open Space Farmlands Conservation Project. *The Functions of Bay Area Farmland*. San Francisco: People for Open Space Farmlands Conservation Project, 1980c.

People for Open Space Farmlands Conservation Project. *The Productivity of Bay Area Rangeland*. San Francisco: People for Open Space Farmlands Conservation Project, 1980d.

People for Open Space Farmlands Conservation Project. *Endangered Harvest: The Future of Bay Area Farmland*. San Francisco: People for Open Space Farmlands Conservation Project, 1980e.

Perelman, Michael. *Farming for Profit in a Hungry World: Capital and the Crisis in Agriculture*. Montclair, NJ: Allanheld, Osmun, 1977.

Perry, Mitch. "Coalition of Immokalee Workers Hails Agreement Group as 'Watershed Moment in the History of Agriculture.'" *Daily Loaf* (blog), November 16, 2010. Accessed February 28, 2011. http://blogs.creativeloafing.com/dailyloaf/2010/11/16/coalition-of-immokalee-workers-hails-agreement-with-trade-group-as-watershed-moment-in-the-history-of-florida-agriculture.

Peryea, Francis J. "Historical Use of Lead Arsenate Insecticides, Resulting Soil Contamination and Implications for Social Remediation." In *Proceedings, 16th World Congress of Social Science*. Montpelier, VT: International Soil Science Society, 1998. CD-ROM.

Petrini, Carlo. *Slow Food Nation: Why Our Food Should Be Good, Clean, and Fair*. New York: Rizzoli Ex Libris, 2007.

Pfeiffer, Dale Allen. *Eating Fossil Fuels: Oil, Food and the Coming Crisis in Agriculture*. Gabriola Island, BC: New Society, 2006.

Phillips, William Emerson. *The Conservation of the California Tule Elk*. Edmonton, AB: University of Alberta Press, 1976.

Philpott, Tom. "The Butz Stops Here: A Reflection on the Lasting Legacy of 1970s USDA Secretary Earl Butz." *Grist*, February 7, 2008. Accessed May 25, 2011. http://www.grist.org/article/the-butz-stops-here.

Pimentel, David and Marcia Pimentel. *Food, Energy and Society*. New York: Wiley-Halsted, 1979.

Pimentel, David, and Marcia Pimentel. *Food, Energy and Society*. 3rd ed. New York: Little, Brown and Company, 2009.

Piore, Michael, and Charles Sabel. *The Second Industrial Divide: Possibilities for Prosperity*. New York: Basic Books, 1984.

Piper, Odessa. Commencement address, University of Wisconsin, Madison, Spring 2006.

Pirog, Rich. *Checking the Food Odometer: Comparing Food Miles for Local Versus Conventional Produce Sales to Iowa Institutions*. Ames: Leopold Center for Sustainable Agriculture, 2003.

Pirog, Rich, Timothy Van Pelt, Kamyar Enshayan, and Ellen Cook. *Food, Fuel and Freeways: An Iowa Perspective on How Far Food Travels, Fuel Usage, and Greenhouse Gas Emissions*. Ames: Leopold Center for Sustainable Agriculture, 2001.

Pisani, Donald J. *From the Family Farm to Agribusiness: The Irrigation Crusade in California and the West, 1850–1931*. Berkeley: University of California Press, 1984.

Pisani, Donald J., and Hal K. Rothman. *Water, Land and Law in the West: The Limits of Public Policy, 1850–1920*. Lawrence: University Press of Kansas, 1996.

Point Reyes Light. "Marshall Rancher Charged in Spill." *Point Reyes Light*, August 29, 1989.

Polakovic, G. "Fumes Force Dairies out to Pasture: Environmentalists Approve Because of Particulate Pollution Downwind from Farms." *Los Angeles Times (Inland Empire Edition)*, May 6, 2003.

Pollan, Michael. *Second Nature: A Gardener's Education*. New York: Atlantic Monthly, 1991.

Pollan, Michael. *The Botany of Desire: A Plant's Eye View of the World*. New York: Random House, 2001.

Pollan, Michael. "Power Steer." *New York Times Magazine*, March 31, 2002. Accessed February 28, 2011. http://www.nytimes.com/2002/03/31/magazine/power-steer.html.

Pollan, Michael. "When a Crop Becomes King." *New York Times*, July 19, 2002b. Accessed February 28, 2011. http://www.nytimes.com/2002/07/19/opinion/when-a-crop-becomes-king.html.

Pollan, Michael. *The Omnivore's Dilemma: A Natural History of Four Meals*. New York: Penguin, 2006.

Pollan, Michael. *In Defense of Food*. New York: Penguin, 2008.

Pollan, Michael. *Food Rules: An Eater's Manual*. New York: Penguin, 2009.

Pons, Elena, and Maud-Alison Long. *Promoting Sustainable Food Systems through Impact Investing*. Corte Madera, CA: Springcreek Foundation, 2011.

Poppendieck, Janet. *Breadlines Knee-Deep in Wheat: Food Assistance in the Great Depression*. New Brunswick, NJ: Rutgers University Press, 1985.

Poppendieck, Janet. *Sweet Charity? Emergency Food and the End of Entitlement*. New York: Penguin, 1998.

Poppendieck, Janet. *Free for All: Fixing School Food in America*. Berkeley: University of California Press, 2010.

Porter, Michael E. *On Compeitition*. Boston: Harvard Business School, 1998.

Pothukuchi, Kameshwari. "Hortaliza: A Youth 'Nutrition Garden' in Southwest Detroit." *Children, Youth and Environments* 14 (2004): 124–155.

Pothukuchi, Kameshwari, and Jerome L. Kaufman. "Placing the Food System on the Urban Agenda: The Role of Municipal Institutions in Food Systems Planning." *Agriculture and Human Values* 16 (1999): 213–224.

Pothukuchi, Kameshwari, and Jerome L. Kaufman. "The Food System—A Stranger to the Planning Field." *APA Journal* 66 (2000): 112–124.

Potter, Clive. *Against the Grain: Agri-Environmental Reform in the United States and European Union*. Wallingford, Oxfordshire: CAB International Wallingford, 1998.

Powell, Bonnie A. "It's about Volume, Not Price: How Straus Family Creamery Is Weathering the Organic Dairy Storm." *Ethicurean* (blog), June 2, 2009. Accessed July 20, 2010. http://www.ethicurean.com/2009/06/02/straus.

Powell, John Wesley. *Report on the Lands of the Arid Regions of the United States, with a More Detailed Account of the Lands of Utah*. Washington DC: U.S. Government Printing Office, 1879.

Powell, Maria, and Greg Lawless. *Select! Sonoma County: A Long-Lived Marketing Program Faces Hard Times*. Madison: University of Wisconsin Center for Cooperatives, 2002.

Poynter, John R. *Society and Pauperism: English Ideas on Poor Relief, 1795–1834*. London: Routledge and K. Paul, 1969.

President's Task Force on Food Assistance. *Report of the President's Assistance*. Washington, DC: U.S. Government Printing Office, 1984.

Press, Eyal, and Jennifer Washburn. "The Kept University." *Atlantic*, March 2000. Accessed June 28, 2011. http://www.theatlantic.com/magazine/archive/2000/03/the-kept-university/6629.

Preston, Hilary. "Drift of Patented Genetically Engineered Crops: Rethinking Liability Theories." *Texas Law Review* 81 (2003): 1153–1175.

Price, Weston Andrew. *Nutrition and Physical Degeneration: A Comparison of Primitive and Modern Diets and Their Effects*. New York: P. B. Hoeber, 1939.

Pridgen, A. "Flap over Learning Source of Bay *E. coli*." *Point Reyes Light*, August 7, 2003.

Rainbow Grocery Web Site. "History." Accessed October 19, 2010. http://www.rainbow.coop/history.

Ranhofer, Charles. *The Epicurean: A Complete Treatise of Analytical and Practical Culinary Art including Table and Wine Service, How to Prepare and Cook Dishes, an Index for Marketing, a Great Variety of Bills of Fare for Breakfasts, Luncheons, Dinners, Buffets, etc., and a Selection of Interesting Bills of Fare of Delmonico's from 1862 to 1894. Making a Franco-American Culinary Encyclopedia*. New York: C. Ranhofer, 1894.

Ratatouille. DVD. Directed by Brad Bird and Jan Pinkava. Emeryville, CA: Pixar Animation Studios, 2007.

Rauzon, Suzanne, May Wang, Natalie Studer, and Pat Crawford. *Changing Students' Knowledge, Attitudes and Behavior in Relation to Food: An Evaluation of the School Lunch Initiative*. Berkeley, CA: Chez Panisse Foundation, 2010.

Ravallion, Martin. "The Two Poverty Enlightenments: Historical Insights from Digitized Books Spanning Three Centuries." *Poverty and Public Policy* 3 (2011). doi:10.2202/1944-2858.1173. Accessed September 16, 2011.

Raven, Hugh, Tim Lang, and Caroline Dumonteil. *Off Our Trolleys? Food Retailing and the Hypermarket Economy*. London: Institute for Public Policy Research, 1995.

Reed, Julia. "Book Review: 'Heat' by Bill Buford: Will Work for Food." *New York Times,* May 28, 2006. Accessed September 27, 2011. http://www.nytimes.com/2006/05/28/books/review/28reed.html.

Reese, Joel. "Johnson's Lays Off All Oyster Workers." *Point Reyes Light,* January 19, 1995.

Regenerative Design Institute. "Core Instructors: James Stark." Accessed July 20, 2010. http://www.regenerativedesign.org/jamesbio.

Register, Cheri. *Packinghouse Daughter: A Memoir*. St. Paul: Minnesota Historical Society, 2000.

Reichl, Ruth, and Gourmet Magazine Editors, eds. *The Best of Gourmet: Sixty-Five Years, Sixty-Five Favorite Recipes*. New York: Random House, 2007.

Reinhardt, Richard. "Parsley and Produce." *San Francisco Chronicle*, October 29, 1954.

Restaurant Opportunities Centers United. *Dining Out, Dining Healthy: The Link Between Public Health and Working Conditions in New York City's Restaurant Industry.* New York: Restaurant Opportunities Centers United, 2006.

Restaurant Opportunities Centers United. *The Great Service Divide: Occupational Segregation and Inequality in the New York City Restaurant Industry.* New York: Restaurant Opportunities Centers United, 2009.

Restaurant Opportunities Centers United. *Serving While Sick: High Risks and Low Benefits for the Nation's Restaurant Workforce, and Their Impact on the Consumer.* New York: Restaurant Opportunities Centers United, 2010.

Restaurant Opportunities Centers United. *Waiting on Equality: The Role and Impact of Gender in the New York City Restaurant Industry*. New York: Restaurant Opportunities Centers United, 2010.

Restaurant Opportunities Centers United. *Behind the Kitchen Door: A Multi-Site Study of the Restaurant Industry*. New York: Restaurant Opportunities Centers United, 2011.

Rhul, J. B. "Farms, Their Environmental Harms, and Environmental Law." *Ecology Law Quarterly* 27 (2000): 263–349.

Richardson, David. "Slavery, Trade and Economic Growth in Eighteenth Century New England." In *Slavery and the Rise of the Atlantic System*, ed. Barbara L. Solow, 237–264. Cambridge: Cambridge University Press, 1991.

Richardson, Jill. "I Met an Activist I've Wanted to Meet for Three Years!" *La Vida Locavore* (blog), May 1, 2009. Accessed October 10, 2010. http://www.lavidalocavore.org/diary/1579/i-met-an-activist-ive-wanted-to-meet-for-three-years.

Richardson, Jill. "A Note about Strauss Dairy." *La Vida Locavore* (blog), January 29, 2010. Accessed March 3, 2010. http://www.lavidalocavore.org/diary/3181/a-note-about-strauss-dairy.

Richter, Robert W. *The Battle over Hetch Hetchy: America's Most Controversial Dam and the Birth of Modern Environmentalism*. New York: Oxford University Press, 2006.

Rilla, Ellie, ed. *Group Memory of the Sonoma County Agriculture Summit*. Paradise Ridge Winery: University of California Cooperative Extension, 1996.

Rilla, Ellie, ed. *Amazing But True: Facts about Marin County Agriculture*. Novato: University of California Cooperative Extension, 2008.

Rilla, Ellie. *Coming of Age: The Status of North Bay Artisan Cheesemaking*. Novato: University of California Cooperative Extension, 2011.

Rilla, Ellie, and Lisa Bush. *The Changing Role of Agriculture in Point Reyes National Seashore*. Novato: University of California Cooperative Extension, 2009.

Rilla, Ellie, and Stephanie Larson. *Final Report of the Marin Coastal Watershed Improvement Project*. Novato: University of California Cooperative Extension, 1995.

Rinaldi, Alberto. The Emilian Model Revisited: Twenty Years After. In *Materiali di discussione del Dipartimento di Economia politica, no. 417*. Modena, Italy: University of Modena and Reggio Emilia, 2002.

Roach, David. "Healthful, Safe, Sustainable? How to Provide Good Food for Everyone." Presentation at the Commonwealth Club, San Francisco, December 2006.

Robbins, Jim. "Balancing Cattle, Land and Ledgers." *New York Times*, October 8, 2003. Accessed February 28, 2011. http://www.nytimes.com/2003/10/08/dining/balancing-cattle-land-and-ledgers.html?pagewanted=all&src=pm.

Robbins, William. *Colony and Empire: The Capitalist Transformation of the American West*. Lawrence: University Press of Kansas, 1994.

Robert Wood Johnson Foundation. "Robert Wood Johnson Foundation Announces Winners of the 2010 Community Health Leaders Award." Accessed January 17, 2011. http://www.communityhealthleaders.org/news_features/pr/62569.

Roberts, Paul. *The End of Food: The Coming Crisis in the World Food Industry*. Boston: Houghton Mifflin Harcourt, 2008.

Robertson, Laurel, Carol Flinders, and Bronwen Godfrey. *Laurel's Kitchen: A Handbook for Vegetarian Cookery and Nutrition*. Berkeley, CA: Nilgiri, 1976.

Robinson, Carolyn W., and Jayne M. Zajicek. "Growing Minds: The Effects of a One-Year School Garden Program on Six Constructs of Life Skills of Elementary School Children." *HortTechnology* 15 (2005): 453–457.

Robinson-O'Brien, Ramona, Mary Story, and Stephanie Heim. "Impact of Garden-Based Youth Nutrition Intervention Programs." *Journal of the American Dietetic Association* 109 (2009): 273–280.

Rodale Institute Web Site. "The Rodale Story." Accessed September 7, 2010. http://www.rodale.com/rodale-story.

Rodale Institute Web Site. "Rodale Directory of Student Farms." Accessed September 7, 2010. http://newfarm.rodaleinstitute.org/features/0104/studentfarms/directory.shtml.

Rodale, Jeremy I. *Pay Dirt: Farming and Gardening with Composts*. New York: Devin-Adair, 1945.

Rogers, Rob. "Cheese Producers Hailed as Model for West Marin Development." *Marin Independent Journal*, January 17, 2007. Accessed February 28, 2011. http://www.malt.org/news/press.php?item=press_01_17_07_marinij.

Rogers, Rob. "College of Marin's Organic Farming Students Get Hands-On Experience." *Marin Independent Journal*, March 20, 2009. Accessed February 28, 2011. http://www.marinij.com/marinnews/ci_11961588?iadid=search-www.marinij.com-www.marinij.com.

Rogers, Rob. "Mature at Last, Marin County's Cheeses Stand Alone." *Marin Independent Journal*, August 9, 2010. Accessed February 28, 2011. http://www.marinij.com/ci_15725747.

Rohrs, Sarah. "It Was No Summer of Love for West Marin Hippies." *Point Reyes Light*, May 28, 1987.

Rolland, David. "Giacomini Shaped Future of West Marin." *Point Reyes Light*, January 2, 1997.

Rome, Adam W. *The Bulldozer in the Countryside: Suburban Sprawl and the Rise of American Environmentalism*. Cambridge: Cambridge University Press, 2001.

Ronald Grant Archive Web Site. Accessed March 28, 2012. http://www.ronaldgrantarchive.com/.

Rorabaugh, William J. *The Alcoholic Republic: An American Tradition*. New York: Oxford University Press, 1981.

Rose, Joseph B., and Gary N. Chaison. "Unionism in Canada and the United States in the 21st Century: The Prospects for Revival." *Relations Industrielles* 56 (2001): 34–65.

Rose, Mark. *Interstate: Express Highway Politics, 1939–1989*. Knoxville: University of Tennessee Press, 1990.

Rose, Mark, Bruce E. Seely, and Paul Barrett. *The Best Transportation System in the World*. Columbus: Ohio State University Press, 2007.

Rosen, Laurel, and Sally McGrane. "The Revolution Will Not Be Catered: How Bay Area Food Collectives of the '60s Set the Stage for Today's Sophisticated Tastes." *San Francisco Chronicle*, March 8, 2000. Accessed February 15, 2012. http://www.sfgate.com/cgi-bin/article.

Rossett, Peter. "Global Small-Scale Farmers' Movement Developing New Trade Regimes." *Food First News and Views* 28 (2005): 2.

Royte, Elizabeth. 2009. "Street Farmer." *New York Times*, July 1, 2009. Accessed February 28, 2011. http://www.nytimes.com/2009/07/05/magazine/05allen-t.html?_r=2&em.

Ruffolo, Jennifer. *Tmdls: The Revolution in Water Quality Regulation*. Sacramento: California State Library, California Research Bureau, 1999.

Ruhl, J. B. "Farms, Their Environmental Harms, and Environmental Law." *Ecology Law Quarterly* 27 (2000): 263–349.

Runge, C. Ford. "Environmental Protection from Farm to Market. In *Thinking Ecologically: The Next Generation of Environmental Policy*, ed. Marian R.

Chertow and Daniel C. Esty, 200–216. New Haven, CT: Yale University Press, 1997.

Russell, Hope. "Ranch Workers Win Concessions." *Point Reyes Light,* July 30, 1992.

Russo, Mike. *Apples to Twinkies: Comparing Federal Subsidies of Fresh Produce and Junk Food.* Boston: U.S. PIRG Education Fund, 2011. Accessed October 1, 2011. http://www.uspirg.org/uploads/a0/6c/a06c37a077f3152e839e4b3fbbfc8a0a/Apples-to-Twinkies-web-vUS.pdf.

Sackman, Douglas C. *Orange Empire: California and the Fruits of Eden.* Berkeley: University of California Press, 2007.

Sadin, Paul. *Managing a Land in Motion: An Administrative History of Point Reyes National Seashore.* Seattle: Historical Research Associates, 2007.

Saekel, Karola. "Grand Old Men of Food: Bay Area Pioneers Who Changed the Way America Eats." *San Francisco Chronicle*, August 8, 2001. Accessed June 29, 2011. http://articles.sfgate.com/2001-08-08/food/17611346_1_alfred-peet-great-chefs-food-san-francisco-chronicle.

Said, Carolyn. "S.F. Food Bank Keeping Up with the Times." *San Francisco Chronicle*, December 1, 2010. Accessed February 28, 2011. http://www.sfgate.com/cgi-bin/article.cgi?f=/c/a/2010/11/30/BAD61GASI6.DTL.

Salatin, Joel. *Pastured Poultry Profits.* Swoope, VA: Polyface, 1996.

Sandler, Ronald, and Phaedra C. Pezzullo, eds. *Environmental Justice and Environmentalism: The Social Justice Challenge to the Environmental Movement.* Cambridge, MA: MIT Press, 2007.

Saxenian, Annalee. *Regional Advantage: Culture and Competition in Silicon Valley and Route 128.* Cambridge, MA: Harvard University Press, 1994.

Schanbacher, William D. *The Politics of Food: The Global Conflict between Food Security and Food Sovereignty.* Westport, CT: Praeger, 2010.

Schell, Orville. *Modern Meat: Antibiotics, Hormones, and the Pharmaceutical Farm.* New York: Random House, 1984.

Schlosser, Eric. *Fast Food Nation: The Dark Side of the All-American Meal.* New York: Harper Perennial, 2001.

Schmid, Ron. *The Untold Story of Milk: Green Pastures, Contented Cows and Raw Dairy Foods.* Winona Lake, IN: NewTrends, 2003.

Scott, Mel. *The San Francisco Bay Area: A Metropolis in Perspective.* Berkeley: University of California Press, 1985.

Seale, Bobby. *Seize the Time: The Story of the Black Panther Party and Huey P. Newton.* New York: Random House, 1970.

Sealy, M. R. "A Garden for Children at Family Road Care Center." Master's thesis, Louisiana State University and Agricultural Mechanical College, 2001.

Segrave, Kerry. *Tipping: An American Social History of Gratuities.* Jefferson, MO: Mcfarland, 1998.

Self, Robert O. *American Babylon: Race and the Struggle for Postwar Oakland.* Princeton, NJ: Princeton University Press, 2003.

Self, Robert O. "The Black Panther Party and the Long Civil Rights Movement." In *In Search of the Black Panther Party*, ed. Jama Lozero and Wohuru Williams, 15–55. Durham, NC: Duke University Press, 2006.

Selg, Peter. *The Agriculture Course Koberwitz, Whitsun 1924: Rudolph Steiner and the Beginnings of Biodynamics*. Forest Row, UK: Temple Lodge, 2010.

Sen, Amartya. *Poverty and Famines: An Essay on Entitlement and Deprivation*. Oxford: Clarendon Press, 1981.

Sen, Rinku, and Fekkak Mamdouh. *The Accidental America: Immigration and Citizenship in the Age of Globalization*. San Francisco: Berrett-Koehler, 2008.

Seth, Andrew, and Geoffrey Randall. *The Grocers: The Rise and Rise of the Supermarket Chains*. Dover, UK: Kogan Page, 2001.

Seth, Andrew, and Geoffrey Randall. *Supermarket Wars: Global Strategies for Food Retailers*. Houndsmills, UK: Palgrave Macmillan, 2005.

Severson, Kim. "The Rise and Fall of a Star: How the King of California Cuisine Lost an Empire." *San Francisco Chronicle*, September 29, 1999.

Severson, Kim. "With Goat, a Rancher Breaks Away from the Herd." *New York Times*, October 15, 2008. Accessed July 17, 2010. http://www.nytimes.com/2008/10/15/world/americas/15iht-15goat.16964683.html.

Severson, Kim, and Adam B. Ellick. "A Top Chef's Kitchen Is Far Too Hot, Some Workers Say." *New York Times*, January 17, 2007. Accessed February 28, 2011. http://www.nytimes.com/2007/01/17/dining/17prom.html?_r=2&pagewanted=1.

Sewell, Bryan H., and Robin M. Whyatt. *Intolerable Risk: Pesticides in Our Children's Food*. New York: Natural Resources Defense Council, 1989.

Seydel, Robin. "Creating Regional Capital to Grow Regional Food Systems." Presentation at the EcoFarm Conference, Pacific Grove, CA., January 2011.

Shapiro, Laura. *Perfection Salad: Women and Cooking at the Turn of the Century*. New York: Farrar, Straus, and Giroux, 1986.

Shaw, D. John. *World Food Security: A History since 1945*. New York: Palgrave Macmillan, 2007.

Sherriff, Carol. *The Artificial River: The Erie Canal and the Paradox of Progress, 1817–1862*. New York: Hill and Wang, 1996.

Shieh, Y., Stephan S. Monroe, R. L. Fankhauser, Gregg W. Langlois, William Burkhardt III, and Ralph S. Baric. "Detection of Norwalk-Like Virus in Shellfish Implicated in Illness." *Journal of Infectious Diseases* 181 (2000): Supplement 2, S360–366.

Schrader, Wayne L. "Organic Agriculture." University of California, Cooperative Extension, Vegetable Research and Information Center (n.d.). Accessed February 19, 2012. http://vric.ucdavis.edu/pdf/Organic/organic_agriculture.pdf

Silverglade, Bruce, and Irene Ringel Heller. *Food Labeling Chaos: The Case for Reform*. Washington, DC: Center for Science in the Public Interest, 2010.

Simon, Michele. *Appetite for Profit: How the Food Industry Undermines Our Health and How to Fight Back*. New York: Nation Books, 2006.

Sims, Judith. "Rolling in Radicchio: California Farmers Cultivate the Palate of the Future with Vegetables of a Different Color—and Edible Flowers." *Los Angeles Times Magazine*, July 24, 1988.

Sinclair, Upton. *The Jungle*. New York: Doubleday, Page 1906.

Sinton, Peter. "Granting Financial Wishes under One Roof." *San Francisco Chronicle*, August 18, 1999. Accessed February 28, 2011. http://articles.sfgate.com/1999-08-18/business/17696560_1_niman-ranch-west-oakland-east-oakland-sba.

Skaggs, Jimmy M. *Prime Cut*. College Station: Texas A&M University Press, 1986.

Skelly, Sonya M., and Jayne M. Zajicek. "The Effect of an Interdisciplinary Garden Program on the Environmental Attitudes of Elementary School Students." *HortTechnology* 8 (1998): 579–583.

Smart, Maya Payne. "Seeds of Change: Buying Black Farm Products Enriches the Health and Wealth of Our Community." *Black Enterprise*, January 1, 2007. Accessed February 28, 2011. http://www.thefreelibrary.com/seeds+of+change:+buying+black+farm+products+enriches+the+health+and...-a0157033225.

Smelser, Neil J., and Richard Swedberg, eds. *The Handbook of Economic Sociology*. Princeton, NJ: Princeton University Press, 1994.

Smith, Andrew F. *The Tomato in America: Early History, Culture, and Cookery*. Urbana: University of Illinois Press, 2001.

Smith, Leanna L., and Carl Motsenbocke. "Impact of Hands-On Science through School Gardening in Louisiana Public Elementary Schools." *HortTechnology* 15 (2005): 439–443.

Smoothe, V. "Where Is East Oakland?" *A Better Oakland* (blog), December 12, 2008. Accessed January 12, 2011. http://www.abetteroakland.com/where-is-east-oakland/2008-12-12.

Snyder, Lynne Page. "The 'Death-Dealing Smog' over Donora, Pennsylvania: Industrial Air Pollution, Public Heath Policy and the Politics of Expertise, 1948–1949." *Environmental History Review* 18 (1994): 117–139.

Soil Association Web Site. Accessed March 28, 2012. http://www.soilassociation.org/.

Solari, Elaine-Maryse. "The Making of an Archaeological Site and the Unmaking of a Community in West Oakland, California." In *The Archaeology of Urban Landscapes: Explorations in Slumland*, ed. Alan Mayne and Tim Murray, 22–38. Cambridge: Cambridge University Press, 2001.

Solow, Barbara L., ed. *Slavery and the Rise of the Atlantic System*. Cambridge: Cambridge University Press, 1993.

Sonoma County Agricultural Preservation and Open Space District. *A Decade of Preservation*. Santa Rosa, CA: Sonoma County Agricultural Preservation and Open Space District, 2003. Accessed September 25, 2003. http://www.sonomaopenspace.org/docManager/1000001687/410quarterlyreport.pdf.

Southern Sustainable Agriculture Working Group. *Food Security Begins at Home: Creating Community Food Coalitions in the South*. Fayetteville, AR: Southern Sustainable Agriculture Working Group.

Spargo, John. *The Bitter Cry of the Children*. New York: Macmillan, 1906.

Spector, Robert. *Category Killers: The Retail Revolution and Its Impact on Consumer Culture*. Boston: Harvard Business School Press, 2005.

Spelter, Siggy. "Will PRSVA Be Opened?" *Point Reyes Light*, February 18, 2010.

Starchefs.Com. "Woman and Food, Laura Chenel—Founder and Owner of Laura Chenel's Chevre, Inc." n.d. Accessed October 1, 2010. http://www.starchefs.com/features/women/html/bio_chenel.shtml.

Starr, Kevin. *Endangered Dreams: The Great Depression in California*. New York: Oxford University Press, 1997.

Starrs, Paul F., and Peter Goin. *Field Guide to California Agriculture*. Berkeley: University of California Press, 2010.

Stassart, Pierre, and Sarah J. Whatmore. "Metabolising Risk: Food Scares and the Un/Re-Making of Belgian Beef." *Environment and Planning A* 35 (2003): 449–462.

Stegner, Wallace Earle. *Beyond the Hundredth Meridian: John Wesley Powell and the Second Opening of the West*. New York: Penguin, 1954.

Stein, Walter J. *California and the Dust Bowl Migration*. Westport, CT: Greenwood Press, 1976.

Steinhart, John, and Carol Steinhart. "Energy Use in the U.S. Food System." *Science* 184 (1974): 307–316.

Sterngold, James. "Urban Sprawl Benefits Dairies in California." *New York Times,* October 22, 1999.

Stevenson, George W., Kathryn Ruhf, Sharon Lezberg, and Kate Clancy. "Warrior, Builder, and Weaver Work: Strategies for Changing the Food System." In *Remaking the North American Food System,* ed. C. Clare Hinrichs and Thomas A. Lyson, 33–62. Lincoln: University of Nebraska Press, 2008.

Stumbos, Paul, ed. *The Small Farm Program: The First Fifteen Years*. Davis: University of California Small Farm Center, 1990.

Stock, Catherine McNicol, and Robert D. Johnston. *The Countryside in the Age of the Modern State: Political Histories of Rural America*.Ithaca, NY: Cornell University Press, 2001.

Stoll, Steven. *The Fruits of Natural Advantage: Making the Industrial Countryside in California*. Berkeley: University of California Press, 1998.

Storper, Michael. *The Regional World: Territorial Development in a Global Economy*. New York: Guilford Press, 1997.

Strand, Oliver, and Joe DiStefano. "Chefs Look for Wild Ingredients Nobody Else Has." *New York Times*, November 23, 2010. Accessed February 28, 2011. http://www.nytimes.com/2010/11/24/dining/24forage.html?_r=1.

Straus Family Creamery. "Straus Organic Dairy Keeps Working to Save the Family Farms." Press release, November 30, 2006. Accessed February 28, 2011. http://www.strausfamilycreamery.com/?id=68&mdid=24.

Straus Family Creamery. "Tales from the Family Farm." n.d. Accessed July 20, 2010. http://www.strausfamilycreamery.com/?section=farm%20tales.

Straus, Albert. *Dairies and Feedlots: What's the Good News?*. San Francisco: Commonwealth Club of California, August 2008.

Stinton, Peter. "Granting Financial Wishes under One Roof." *San Francisco Chronicle,* August 18, 1999. Accessed April 15, 2010. http://articles.sfgate.com/1999-08-18/business/17696560_1_niman-ranch-west-oakland-east-oakland-sba.

Striffler, Steve. *Chicken: The Dangerous Transformation of America's Favorite Food*. New Haven, CT: Yale University Press, 2005.

Strochlic, Ron, and Kari Hammerschlag. *Best Labor Practices on Twelve California Farms: Toward a More Sustainable Food System*. Davis, CA. California Institute for Rural Studies, 2005.

Stukenberg, D., D. Blayney, and J. Miller. "Major Advances in Milk Marketing: Government and Industry Consolidation." *Journal of Dairy Science* 89 (2006): 1195–1206.

Suback, Susan. "Global Environmental Costs of Beef Production." *Ecological Economics* 30 (1999): 79–91.

Sullivan, Dan. "Organics in the News: Harvey V. Veneman's Spectre of Unintended Consequences Roils Organic Waters." Rodale Institute, March 31, 2005. Accessed July 20, 2010. http://newfarm.rodaleinstitute.org/columns/org_news/2005/0405/harvey.shtml.

Sumner, Daniel A., Henrich Brunke, and Jose E. Bervejillo. "The 2002 Census of Agriculture: A Wealth of Useful Data." *Agricultural Issues Center Issues Brief* 26 (2004): 1–4. Accessed February 19, 2012. http://aic.ucdavis.edu/pub/briefs/brief26.pdf.

Super Size Me. DVD. Directed by Morgan Spurlock. New York and Los Angeles: Samuel Goldwyn Films, Roadside Attractions, 2004.

Tabor, George M. *Judgment of Paris: California vs. France and the Historic 1976 Paris Tasting That Revolutionized Wine*. New York: Scribner, 2005.

Takaki, Ronald T. *Strangers from a Different Shore: A History of Asian Americans*. New York: Little, Brown, 1998.

Tanner, Champ B., and Roy W. Simonson. "Franklin Hiram King—Pioneer Scientist." *Soil Science Society of America Journal* 57 (1993): 286–292.

Tasch, Woody. *Inquiries into the Nature of Slow Money*. White River Junction, VT: Chelsea Green, 2008.

Taubes, Gary. *Good Calories, Bad Calories: Challenging the Conventional Wisdom on Weight, Diet and Disease*. New York: Knopf, 2007.

Taubes, Gary. "Is Sugar Toxic?" *New York Times,* April 13, 2011. Accessed April 30, 2011. http://www.nytimes.com/2011/04/17/magazine/mag-17Sugar-t.html?pagewanted=all.

Taylor, Benjamin. "Years in the Making, Mandela Foods Cooperative Still a Secret." *East Bay Express*, June 16, 2010. Accessed February 28, 2011. http://www.eastbayexpress.com/ebx/years-in-the-making-mandela-foods-cooperative-still-a-secret/content?oid=1834167.

Taylor, Dorceta E. "The Rise of the Environmental Justice Paradigm: Injustice Framing and the Social Construction of Environmental Discourse." *American Behavioral Scientist* 43 (2000): 508–580.

Tedlow, Richard. *New and Improved: The Story of Mass Marketing in America.* New York: Basic Books, 1990.

Teicholz, Nina. "The World According to Sam." *Gourmet*, June 2005. Accessed February 28, 2011. http://www.gourmet.com/magazine/2000s/2005/06/walmart.

Terry, Bryant. *Vegan Soul Kitchen: Fresh, Healthy, and Creative African-American Cuisine*. Cambridge, MA: Da Capo Press, 2009.

Thiboumery, Arion, and Mike Lorentz. "Marketing Beef for Small-Scale Producers." Accessed February 28, 2011. http://www.extension.org/mediawiki/files/0/00/Marketing_Beef_for_Small-Scale_Producers.pdf.

Thomas, Sarah L. "The Politics of Growth: Private Rights, Public Amenities, and Land Use Debates in Seasonal Cities, 1955–1985." Ph.D. diss., University of California, Berkeley, 2009.

Thomas, Sarah L. "When Equity Almost Mattered: Outdoor Recreation, Land Acquisition, and Mid-Twentieth-Century Conservation Politics." *Natural Resources Journal* 50 (2010): 501–516.

Thompson, David. "What's Next for California's Consumer Co-ops?" In *What Happened to the Berkeley Co-Op? A Collection of Opinions*, ed. Michael Fullerton, 87–92. Davis, CA: Center for Cooperatives, 1992.

Thompson, David. "Co-Op Principles Then and Now (Part 2)." *Cooperative Grocer for Retailers and Cooperator,* July–August 1994. Accessed January 4, 2011. http://www.cooperativegrocer.coop/articles/2004-01-09/co-op-principles-then-and-now-part-2.

Thompson, Edward, Jr., Alethea Marie Harper, and Sibella Kraus. *Think Globally—Eat Locally: San Francisco Foodshed Assessment.* Davis, CA: American Farmland Trust, 2008.

Thrupp, Martin. "The School Mix Effect: The History of an Enduring Problem in Education Research, Policy and Practice." *British Journal of Sociology of Education* 16 (1995): 183–203.

Time Magazine. "A Mess However It's Sliced." *Time Magazine*, January 4, 1982. Accessed February 28, 2011. http://www.time.com/time/magazine/article/0,9171,953288,00.html.

Tolbert, Lisa C. "The Aristocracy of the Market Basket: Self-Service Food Shopping in the New South." In *Food Chains: From Farmyard to Shopping Cart*, ed. Warren Belasco, 179–195. Philadelphia: University of Pennsylvania Press, 2009.

Tomales Bay Association. *Proceedings of the Fourth Conference on the State Tomales Bay*. Point Reyes Station, CA: Inverness Yacht Club, October 2000.

Tomales Bay Watershed Council. *The Tomales Bay Watershed Stewardship Plan: A Framework for Action*. Point Reyes Station, CA: Tomales Bay Watershed Council, 2003.

Tower, Jeremiah. *California Dish: What I Saw (and Cooked) at the American Culinary Revolution*. New York: Free Press, 2003.

Trubek, Amy B. *Haute Cuisine: How the French Invented the Culinary Profession*. Philadelphia: University of Pennsylvania Press, 2000.

Trubek, Amy B. *The Taste of Place: A Cultural Journey into Terroir*. Berkeley: University of California Press, 2008.

Trueman, Kerry. "Smart Cities Are (Un)Paving the Way for Urban Farmers." *Grist*, August 30, 2010. Accessed February 28, 2011. http://www.grist.org/article/food-smart-cities-are-unpaving-the-way-for-urban-farmers-and-locavores.

Tryphonopoulos, Nickolas Sypros. "An Investigation of the Economic Structure of a Small Area: Napa County, California." Ph.D. diss., University of California, Berkeley, 1967.

Tsai, Luke. "Reading, Writing and Replanting." *East Bay Express,* April 28, 2010. Accessed January 12, 2011. http://www.eastbayexpress.com/ebx/reading-writing-and-replanting/Content?oid=1711990.

Ulrich, Roger S. "Effects of Gardens on Health Outcomes." In *Healing Gardens: Therapeutic Benefits and Design Recommendations*, ed. Clare Cooper Marcus and Marni Barnes, 27–86. New York: Wiley, 1999.

Unger, Serena, and Heather Wooten. *Oakland Food Systems Assessment*. Berkeley and Oakland: University of California, Berkeley, and Oakland Mayor's Office, 2006.

United Farm Workers. "UFW Chronology." 2006. Accessed January 12, 2011. http://www.ufw.org/_page.php?menu=research&inc=_page.php?menu=research&inc=history/01.html.

United Nations World Commission on Environment and Development. *Our Common Future*. New York: Oxford University Press, 1987.

U.S. Congress, House Committee on Government Operations. *Consumer Problems of the Poor: Supermarket Operations in Low-Income Areas and the Federal Response*. Washington, DC: U.S. Government Printing Office, 1968.

U.S. Department of Agriculture. *Comparison of Prices Paid for Selected Foods in Chain Stores in High and Low Income Areas of Six Cities*. Washington, DC: U.S. Department of Agriculture, 1968.

U.S. Department of Health and Human Services, National Institutes of Health, and National Cancer Institute. *Reducing Environmental Cancer Risk: What We Can Do Now.* President's Cancer Panel, 2008–2009 Annual Report, 2010. Accessed May 25, 2011. http://deainfo.nci.nih.gov/advisory/pcp/annualReports/pcp08-09rpt/PCP_Report_08-09_508.pdf.

U.S. Department of Transportation, Federal Highway Commission. "Environmental Justice Cypress Freeway Replacement Project: California Department of Transportation." Accessed January 12, 2011. http://www.fhwa.dot.gov/environment/environmental_justice/case_studies/case5.cfm.

U.S. Federal Trade Commission. *Economic Report on Food Chain Selling Practices in the District of Columbia and San Francisco: A Staff Report*. Washington, DC: U.S. Government Printing Office, 1969.

U.S. General Accounting Office. *Dairy Termination Program: A Perspective on Its Participants and Milk Production*. Washington, DC: GAO/RCED, 1988.

U.S. Office of the Federal Register. *The National Organic Program Regulation and Policies*. Washington, DC: U.S. Government Printing Office, 2002.

U.S. Office of the Federal Register. "Proposed Rules." *Federal Register* 75 (2010): 35338–35354. Accessed December 13, 2010. http://archive.gipsa.usda.gov/rulemaking/fr10/06-22-10.pdf.

University of California Cooperative Extension. *Proceedings of the Marin County Agriculture Summit*. Novato: University of California Cooperative Extension, 2010.

Van Cleef, Lisa. "Gardening Conquers All: How to Cut Your Jail Recidivism Rates by Half." *San Francisco Chronicle*, December 18, 2002. Accessed February 28, 2011. http://articles.sfgate.com/2002-12-18/home-and-garden/17573369_1_garden-project-jail-counselor-recidivism-rate.

Van Den Bosch, Robert. *The Pesticide Conspiracy*. Berkeley: University of California Press, 1978.

Vanishing of the Bees. DVD. Directed by George Langworth and Maryam Henein. Los Angeles: Hive Mentality Films, 2009.

Vaught, David. *Cultivating California: Growers, Specialty Crops, and Labor, 1875–1920*. Baltimore, MD: John Hopkins University Press, 1999.

Ver Ploeg, Michele, Vince Breneman, Tracey Farrigan, Karen Hamrick, David Hopkins, Phil Kaufman, Biing-Hwan Lin, et al. *Access to Affordable and Nutritious Food—Measuring and Understanding Food Deserts and Their Consequences*. Washington, DC: U.S. Department of Agriculture, 2009.

Vina, Stephen R. *Harvey V. Veneman and the National Organic Program: A Legal Analysis*. Washington, DC: Congressional Research Service, 2006.

Vishal, Anoothi. "The Organic Option." *Telegraph (Calcutta)*, April 4, 2010. Accessed September 10, 2011. http://www.telegraphindia.com/1100404/jsp/graphiti/story_12295898.jsp.

Vollman, William T. *Imperial*. New York: Viking Press, 2008.

Vorasarun, Chaniga. "Ten Top-Earning Celebrity Chefs." *Forbes Magazine*, August 8, 2008. Accessed November 1, 2010. http://www.forbes.com/2008/08/08/celebrity-chef-earners-forbeslife-cx_cv_0808food.html.

Wade, Louise C. *Chicago's Pride: The Stockyards, Packingtown and Environs in the Nineteenth Century*. Champaign: University of Illinois Press, 1987.

Walden, GraceAnn. "Restoring an S.F. Treasure: Culinary Plans for the Ferry Building Are Coming Along." *San Francisco Chronicle*, October 2, 2002a. Accessed February 28, 2011. http://articles.sfgate.com/2002-10-02/food/17564970_1_san-francisco-s-ferry-building-morning-market-ferry-plaza-farmers-market.

Walden, GraceAnn. "Farmers Come to the Bay." *San Francisco Bay Crossings*, October 9, 2002b. Accessed February 28, 2011. http://www.baycrossings.org/archives/2002/09_october/farmers_come_to_the_bay__.htm.

Walker, Richard A. *The Conquest of Bread: 150 Years of Agribusiness in California*. New York: New Press, 2004a.

Walker, Richard A. "Industry Builds Out the City: The Suburbanization of Manufacturing in the San Francisco Bay Area, 1850-1940." In *The Manufactured Metropolis*, ed. Robert Lewis, 92–123. Philadelphia: Temple University Press, 2004b.

Walker, Richard A. *The Country in the City: The Greening of the San Francisco Bay Area*. Seattle: University of Washington Press, 2008.

Walker, Rob. "Shared Tastes." *New York Times*, October 8, 2010. Accessed February 28, 2011. http://www.nytimes.com/2010/10/10/magazine/10FOB-Consumed-t.html.

Walsh, Bryan. "Why Global Warming Portends a Food Crisis." *Time Magazine*, January 13, 2009. Accessed September 10, 2010. http://www.time.com/time/health/article/0,8599,1870766,00.html.

Walsh, John P. *Supermarkets Transformed: Understanding Organizational and Technological Innovations*. New Brunswick, NJ: Rutgers University Press, 1993.

Walsh, William I. *The Rise and Decline of the Great Atlantic and Pacific Tea Company*. Secaucus, NJ: Lyle Stuart, 1986.

Warbasse, James Peter. *Cooperative Democracy*. New York: Macmillan, 1923.

Wargo, John P. *Our Children's Toxic Legacy: How Science and Law Fail to Protect Us from Pesticides*. New Haven, CT: Yale University Press, 1996.

Warshall, Peter. "Can You Imagine a Better Food System? It's Easy If You Try." Presentation at the EcoFarm Conference, Pacific Grove, CA, January 2011.

"Waterfront Food Mart Proposed." *San Francisco Chronicle*, May 14, 1992.

Waters, Alice. "The Farm Restaurant Connection." In *Our Sustainable Table*, ed. Robert Clark, 113–222. San Francisco: North Point, 1990.

Waters, Alice. *Edible Schoolyard: A Universal Idea*. San Francisco: Chronicle Books, 2008.

Waters, Alice, and Katrina Heron. "Op-Ed: No Lunch Left Behind." *New York Times*, February 19, 2009. Accessed September 9, 2011. http://www.nytimes.com/2009/02/20/opinion/20waters.html

Waters, Alice, and friends. *Forty Years of Chez Panisse: The Power of Gathering*. New York: Clarkson Potter, 2011.

Watt, Laura A. "Managing Cultural Landscapes: Reconciling Local Preservation and Institutional Ideology in the National Park Service." Ph.D. diss., University of California, Berkeley, 2001.

Weber, Warren. Interview by Lauren Gwin. Bolinas, CA, March 23, 2004.

Weinburg, Carl. "Big Dixie Chicken Goes Global: Exports and the Rise of the North Georgia Poultry Industry." *Business and Economic History Online* 1 (2003). Accessed September 7, 2010. http://www.h-net.org/~business/bhcweb/publications/BEHonline/2003/Weinberg.pdf.

Wells, Evelyn. *Champagne Days of San Francisco*. New York: D. Appleton Century, 1939.

Welte, Jim. "West Marin Dairyman Straus Gets New Poop-Powered Car." *Marin Independent Journal*, August 4, 2009. Accessed July 20, 2010. http://www.marinij.com/ci_12993042?source=most_emailed.

Wendell, Berry. *What Are People For?* New York: North Point, 1990.

Whatmore, Sarah. "Agro-Food Complexes and the Refashioning of Rural Europe." In *Globalization, Institutions, and Regional Development in Europe*, ed. Ash Amin and Nigel Thrift, 46–67. New York: Oxford University Press, 1994.

Whatmore, Sarah, Pierre Stassart, and Henk Renting. "Guest Editorial: What's Alternative about Alternative Food Networks?" *Environment and Planning A* 35 (2003): 389–391.

Whiren, Alice Phipps. "Planning a Garden From a Child's Perspective." *Children's Environments* 12 (1995): 250–255.

White House. *World Food Problem: A Report on the President's Science Advisory Committee*. Vol. 2 of the *Report on the Panel of the World Food Supply*. Washington, DC: White House, 1967.

Whyte, William H. *City: Rediscovering the Center*. New York: Doubleday, 1988.

Whyte, William H. *The Social Life of Small Urban Spaces*. Washington, DC: Conservation Foundation, 1980.

Wiebe, Robert. *The Search for Order, 1877–1920*. New York: Hill and Wang, 1967.

Wilkerson, Isabel. *The Warmth of Other Suns: The Epic Story of America's Great Migration*. New York: Random House, 2010.

William, Rice. "The Heat Is on the Great Chefs of Tomorrow." *Washington Post*, April 29, 1979.

Williams, Susan. "The Ferry Building Marketplace: A Food Lovers Mecca and Mastermind of the Marketplace Mix." *San Francisco Bay Crossings*, December 11, 2003. Accessed February 28, 2011. http://www.baycrossings.com/archives/2003/11_december/the_ferry_building_marketplace_a_food_lovers_mecca.htm.

Wilson, James Q. *Bureaucracy: What Government Agencies Do and Why They Do It*. New York: Basic Books, 1989.

Winders, Bill. *The Politics of Food Supply: U.S. Agricultural Policy in the World Economy*. New Haven, CT: Yale University Press, 2009.

Winne, Mark. *Closing the Food Gap: Resetting the Table in the Land of Plenty*. Boston: Beacon, 2008.

Winne, Mark. *Food Rebels, Guerrilla Gardeners, and Smart—Cookin' Mamas: Fighting Back in an Age of Industrial Agriculture*. Boston: Beacon Press, 2010a.

Winne, Mark. "Food System Planning: Setting the Community's Table." *Planner's Network*, January 9, 2010b. Accessed February 28, 2011. http://www.plannersnetwork.org/publications/2004_winter/winne.htm.

Winter, Michael. "Embeddedness, the New Food Economy and Defensive Localism." *Journal of Rural Studies* 19 (2003): 23–32.

Wittman, Hannah, Annette Desmarais, and Nettie Wiebe. "The Origins and Potential of Food Sovereignty." In *Food Sovereignty: Reconnecting Food, Nature, and Community*, ed. Hannah Wittman, Annette Desmarais, and Nettie Wiebe, 1–12. Oakland, CA: Food First Books, 2010.

Witzel, Michael Karl. *The American Drive-In Restaurant*. St. Paul, MN: MBI, 2002.

Woelfel, Charles J., ed. *Encyclopedia of Banking and Finance*. 9th ed. Chicago: Probus Professional, 1991.

Woeste, Victoria Saker. *The Farmer's Benevolent Trust: Law and Agricultural Cooperation in Industrial America*. Chapel Hill: University of North Carolina Press, 1998.

Wollman, Cynthia. "Big Cheese of Chevre Owes It All to Her Kids: Laura Chenel Says Goats Keep Her Grounded." *San Francisco Chronicle*, June 14, 2002. Accessed January 12, 2011. http://articles.sfgate.com/2002-06-14/news/17549400_1_laura-chenel-chevre-goat-milk-milk-cheese.

Wong, Carina. *Lunch Matters: How to Feed Our Children Better: The Story of the Berkeley School Lunch Initiative*. Berkeley, CA: Chez Panisse Foundation, 2008.

Worster, Daniel. *Rivers of Empire: Water, Aridity, and the Growth of the American West*. New York: Pantheon, 1985.

Wright, Wynne, and Gerard Middendorf, eds. *The Fight over Food*. University Park: Pennsylvania State University Press, 2008.

Wrigley, Neil. "The Consolidation Wave in U.S. Food Retailing: An European Perspective." *Agribusiness* 17 (2001): 489–513.

Wyman, Carolyn. *Better Than Home Made: Amazing Foods That Have Changed the Way We Eat*. Philadelphia: Quirk Books, 2004.

Yale University Program in Agrarian Studies. "The Chicken: Its Biological, Social, Cultural, and Industrial History from Neolithic Middens to McNuggets" Conference, Yale University, New Haven, CT, May 2002.

Yamamoto, B. T. "But Who's Going to Water? Complexity and Thick Explanation on a Critical Ethnographic Study of Two School Garden Projects." Master's thesis, University of California, Davis, 2000.

Yasukochi, George. "The Berkeley Co-Op—Anatomy of a Nobel Experiment." In *What Happened to the Berkeley Co-Op? A Collection of Opinions,* ed. Michael Fullerton, 23–46. Davis, CA: Center for Cooperatives, 1992.

Yost, Bambi, and Louise Chawla. *Fact Sheet #3: Benefits of Gardening for Children.* Denver: University of Colorado at Denver and Health Sciences Center, 2009.

Young Workers United. *Five Year Report, 2002–2007.* San Francisco: Young Workers United, 2009.

Young Workers United. Dining with Justice Guide 2011: A Guide to Guilt Free Eating in San Francisco. San Francisco: Young Workers United, 2011. Accessed February 2012. Available at http://youngworkersunited.org/section.php?id=76

Zagat, Tim, and Nina Zagat. "Adding Fairness to the Tip." *New York Times,* December 13, 2010. Accessed May 26, 2011. http://www.nytimes.com/2010/12/14/opinion/14zagat.html.

Zahler, Helen Sara. *Eastern Workingmen and National Land Policy, 1829–1862.* New York: Columbia University Press, 1941.

Zigler, Edward F., Katherine Marsland, and Heather Lord. *The Tragedy of Child Care in America.* New Haven: Yale University Press, 2009.

Zimmerman, Mark. *The Supermarket: A Revolution in Distribution.* New York: McGraw-Hill, 1955.

Zuckerman, Larry. *The Potato: How the Humble Spud Rescued the Western* World. Essex, UK: Faber and Faber, 1998.

Index